3rd International Conference on Quenching and Control of Distortion Engineering (QDE 2025)

Vancouver, Canada
6-7 May 2025

ISBN: 979-8-3313-2174-1

Printed from e-media with permission by:

Curran Associates, Inc.
57 Morehouse Lane
Red Hook, NY 12571

Some format issues inherent in the e-media version may also appear in this print version.

ASM International
9639 Kinsman Road
Materials Park, Ohio 44073-0002
USA

Phone: +1 440.338.5151

memberservicecenter@asminternational.org

Additional copies of this publication are available from:

Curran Associates, Inc.
57 Morehouse Lane
Red Hook, NY 12571 USA
Phone: 845-758-0400
Fax: 845-758-2633
Email: curran@proceedings.com
Web: www.proceedings.com

TABLE OF CONTENTS

QDE 2025: Proceedings of the 3rd International Conference on
Quenching and Distortion Engineering
May 6–7, 2025, Vancouver, Canada
https://doi.org/10.31399/asm.cp.qde2025fm01

2025 Organizing Committee

Conference Chair

Rainer Fechte-Heinen, IWT

Lesley Frame, University of Connecticut

Scott MacKenzie, FASM, Retired, Quaker Houghton International Inc.

COMMITTEE MEMBERS:
Rajan Bhambroo, Tenneco
Thomas Christiansen, WPI
Lesley Frame, University of Connecticut
Robert Goldstein, FASM, Fluxtrol Inc.
Scott MacKenzie, FASM, Quaker Houghton
Mohammed Maniruzzaman, Caterpillar
Michael Nappi, Bodycote Thermal Processing, Inc.
Greg Petrus, Scientific Forming Technologies Corp.
Larissa Vilela Costa, Nucor Steel Brandenburg

INTERNATIONAL ADVISORY COMMITTEE (IFHTSE EXECUTIVE COMMITTEE):
Rafael Colás, FASM, Universidad Autónoma de Nueva Leon
Rainer Fechte-Heinen, Leibniz-IWT
Imre Felde, Obuda University
Jianfeng Gu, Shanghai Jiao Tong University
Stefan Hock, IFHTSE
Bernard Kuntzmann, Listemann AG
Masahiro Okumiya, Toyota Technological Institute
Massimo Pellizzari, University of Trento
Reinhold Schneider, University of Applied Sciences
Marcel Somers, Technical University of Denmark
Christophe Stocky, R&D Centre of ABS Acciaierie Bertoli Safau
Eva Troell, RISE IVF Research Institutes of Sweden
Professor Mufu Yan, Harbin Institute of Technology

QDE 2025: Proceedings of the 3rd International Conference on
Quenching and Distortion Engineering
May 6–7, 2025, Vancouver, Canada
https://doi.org/10.31399/asm.cp.qde2025p0001

Distortion Minimization using DCGQ Process for Reduced Cost and Improved Quality

Zhichao (Charlie) Li, and Jason Meyer
DANTE Solutions, Inc., Cleveland, Ohio, USA
Charlie.Li@Dante-Solutions.com

Jason Fetty
U.S. Army Combat Capabilities Development Command Aviation & Missile Center, Fort Eustis, VA, USA

Abstract

Quenching is a thermal process that hardens steels, and increases component mechanical properties, by way of phase transformation. A high quenching rate, where a part is rapidly cooled, is typically desired for optimal mechanical properties. For high alloy aerospace steels (e.g. Ferrium C64) with high hardenability and high thermal resistance properties, though, the effect of cooling rate during martensitic transformation on the obtained hardness and mechanical properties is expected to be insignificant. One challenge with quenching for all steels is shape and size distortion. Distortion control for complex aerospace parts with complex geometric features, such as rotorcraft gears with thin-wall sections, can be very challenging and costly. A patented DANTE Controlled Gas Quench (DCGQ) process was developed for high precision distortion control during quench hardening. The DCGQ process follows a recipe of quenching time and the ambient gas temperature designed by computer modeling, with a specified maximum allowed temperature difference in the part section during the martensitic transformation. A prototype DCGQ chamber was fabricated and demonstrated under contracts funded by the U.S. Army Combat Capabilities Development Command Aviation & Missile Center (DEVCOM AvMC) PR2025012. The experimental results from coupon samples have validated its effectiveness on obtaining hardness, mechanical properties, and fatigue properties equivalent to those from a standard high pressure gas quench (HPGQ) process for various high alloy aerospace steels. For aerospace gears with challenging thin-wall sections, the DCGQ process can quench the part fast enough in the high temperature zone above the martensite start temperature to avoid potential diffusive phase and carbide formation issues. The DCGQ process also makes it possible to successfully heat treat aerospace gears with reduced distortion compared to traditional press quenching or high-pressure gas quenching (HPGQ).

Introduction

Quench hardening of rotorcraft steel components such as gears is used to increase the hardness and mechanical properties for improved performance. This process includes heating and cooling steps to induce phase transformations of the steel gears in order to obtain the desired properties, with lower bainite and tempered martensite as the preferred phases for high hardness and mechanical properties. During heating, the initial phases transform to austenite, followed by dissolution of carbides in austenite. During quenching, the austenite transforms to the high hardness phases of martensite or lower bainite. Gradients of temperature and phase transformations in the part during both heating and quenching steps generate plastic strain, which leads to distortion in the processed part. This distortion can be separated into shape distortion and size distortion [1]. Size distortion is from the density or volume difference of the part before and after heat treatment process, as lower bainite and tempered martensite has a lower density than the initial phases of the steel. This size distortion can be compensated for during the design process of a gear, as the amount of size distortion can be predicted by existing modeling techniques. Shape distortion is caused by thermal stress and phase transformation stress during the process. The consistency and control of shape distortion is challenging because of the many factors that can affect it [2]. Thermal stress and phase transformation stress are two of the leading factors that contribute to shape distortion during quenching processes [3].

Thermal stress during quenching is generated by the thermal gradient in the steel part, and there is no plastic strain (distortion) generated if the thermal stress is below the yield strength of the material at the given temperature. When phase transformations happen during the heat treatment process, the plastic strain is inevitable due to the transformation induced plasticity effect (TRIP) and the transformation strain [4]. Distortion control during quench hardening can be achieved by applying mechanical constraints, such as a press quench or plug quench process [5]. Another approach for distortion control is to minimize the temperature gradient in the part while martensitic transformation is happening. Using both computer modeling and experimental approaches, a patented DANTE Controlled Gas Quench (DCGQ) process was developed for hardening steel components with minimized distortion [6]. The DCGQ process was developed to precisely control the temperature difference in a part during quenching, in the martensitic phase transformation temperature range, by controlling the ambient gas temperature and holding time. The gas temperature of the quenching chamber was designed as a time-temperature curve and implemented by a program-controlled chamber inlet gas temperature. A DCGQ prototype unit was built and has demonstrated a 50% reduction in shape distortion. Further modeling investigation demonstrated that the main portion of the distortion was from the straight heating and austenitizing process [6]. Plans for future efforts are being developed to demonstrate a reduction in shape distortion for rotorcraft gears, as well as integrating the DCGQ process with a vacuum furnace.

Fabrication of DCGQ Prototype

A DCGQ prototype was designed and fabricated by Atmosphere Engineering in order to test the effectiveness of the invented quench hardening process using a program-controlled gas temperature in the quenching chamber. The fabricated prototype located at Akron Steel Treating in Akron, OH is shown in Figure 1, and it contains the following major components:

- Quenching chamber: 36" x 36" x 36" for quenching parts
- Chamber gas agitator: for more turbulent gas flow and more uniform temperature in the chamber
- Gas heater: for heating gas to a specified temperature (used to adjust the inlet gas temperature)
- Gas blower: for blowing gas into the quenching chamber
- Heat exchanger: for reducing gas temperature exiting the quenching chamber
- Gas mixture: for mixing hot and cold gas with programed ratio to obtain the specified gas temperature
- Human Machine Interface (HMI) control panel: for inputting the designed DCGQ recipe

Figure 1: DCGQ prototype (a) HMI control panel, and (b) front view

Using this prototype, steel parts are austenitized in an external furnace, then transferred to this prototype for hardening. The quenching chamber is heated to a temperature slightly above the martensite start temperature (Ms) while the parts are austenitized externally. During quenching, the chamber gas temperature is controlled by the inlet gas temperature, following a recipe designed by computer modeling of a specific part geometry and steel grade. This computer modeling was all performed using DANTE, which is a software package that ties into either Ansys or Abaqus solvers. The DCGQ recipe is a calculated time-temperature curve of the inlet gas, as demonstrated by the black curve in Figure 2. The dotted red curve in Figure 2 is a collected inlet gas temperature from a DCGQ experimental run, which follows the recipe curve closely. The fabricated prototype and the control system are effective to control the quenching gas temperature following the design quenching recipe. The DCGQ recipe allows both "step" and "ramp" features for controlling the inlet temperature flexibly. The "step" recipe feature is shown in Figure 2, and an "ramp" recipe feature is shown in the case study of this paper.

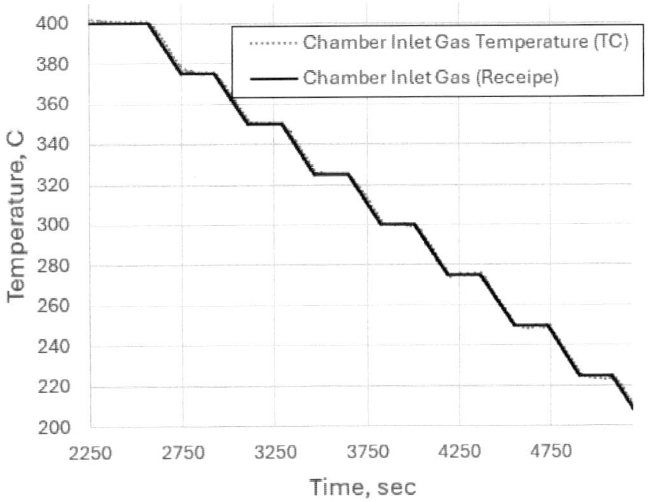

Figure 2: Demonstration effectiveness of chamber temperature control and the two types of DCGQ recipe design

During liquid (e.g. oil) quench or High Pressure Gas Quench (HPGQ) processes, the extraction of heat from the part surface is controlled by the heat transfer coefficient and the quenchant temperature. The targeting quenchant temperature is a constant below the Ms. Therefore, the thin section of a complex part geometry cools faster than the thick section, which leads to nonuniform martensitic transformation and distortion. During a DCGQ process, the gas temperature in the quenching chamber is controlled in the martensitic transformation temperature range by a programmed recipe. The temperature difference between the thin section and the thick section is controlled by the chamber gas temperature and holding time at that temperature. Unlike HPGQ process, where high quenching rate increases the temperature gradient in the part, a higher convection coefficient in DCGQ process doesn't decrease the temperature uniformity, because the temperature uniformity in the part is controlled by the chamber gas temperature instead of the convection coefficient. For the same reason, a directional gas flow doesn't affect the quenching distortion during the DCGQ process. The temperature uniformity in the part during DCGQ quench is controlled by the uniformity of the chamber gas temperature and the applied recipe. This is an important reason why DCGQ process can minimize distortion and obtain consistent quenching results.

Mechanical Property Experiments of DCGQ Processed Steels

With a more uniform temperature in the part cross section during quenching, the distortion is reduced. Distortion coupons made of Ferrium C64 were processed to demonstrate the effectiveness of distortion control [7]. This demonstrated the ability to reduce shape distortion by 50%. Since the DCGQ process has a slower cooling rate than the HPGQ process, it is also necessary to investigate the effect of DCGQ process on the mechanical properties, including Charpy, tensile, and hardness. Charpy and tensile samples made of Ferrium C64 were processed using the DCGQ prototype. The results are compared with the 2 bar HPGQ process, as shown in Figure 3. There is no noticeable difference between HPGQ and DCGQ processed samples for impact strength, yield strength, elongation and reduction area.

Longitudinal Charpy Impact Test				
	HPGQ		DCGQ	
	SAMPLE ID	CVN ENERGY (ft. lbs.)	SAMPLE ID (tested at 70°F)	CVN ENERGY (ft. lbs.)
Set 1 Comparison	11L	17.0	1-1	18.0
	12L	20.0	1-2	21.0
	13L	17.0	1-3	20.0
	AVG	18.0	AVG	19.7
Set 2 Comparison	21L	19.0	2-1	19.0
	22L	15.0	2-2	18.0
	23L	18.0	2-3	16.0
	AVG	17.3	AVG	17.7
TOTAL AVG	HPGQ	17.7	DCGQ	18.7

LONGITUDINAL TENSILE TEST									
SAMPLE ID		TENSILE (psi)		YIELD OS .2% (psi)		ELONGATION (%)		RA (%)	
HPGQ	DCGQ	HPGQ	DCGQ	HPGQ	DCGQ	HPGQ	DCGQ	HPGQ	DCGQ
01L	1-1	236000	237000	203000	202000	17.00	18.00	71.00	73.00
02L	2-1	236000	236000	203000	200000	17.00	18.00	71.00	70.00
03L	3-1	235000	237000	203000	199000	17.00	18.00	71.00	70.00
04L	4-1	236000	234000	204000	214000	16.00	17.00	71.00	72.00
AVG		235750	236000	203250	203750	16.75	17.75	71.00	71.25

Figure 3: Mechanical property comparison between HPGQ and DCGQ processes for Ferrium C64

Carburized bending coupons were processed using DCGQ to compare the residual stresses and hardness with HPGQ processed coupons. The coupon size was about 50 mm x 10 mm x 15 mm with a tooth designed to represent the geometric effect of actual gear teeth. The residual stresses were measured by X-ray diffraction at two locations: 1) circumferential direction of the tooth root, and 2) longitudinal direction of the top flat surface, as shown in Figure 4(a). The tooth root has a higher compressive stress than the flat surface, and there is no noticeable difference between DCGQ and HPGQ processed parts. The hardness comparison is shown in Figure 4(b), and there is no noticeable difference between DCGQ and HPGQ processed coupons.

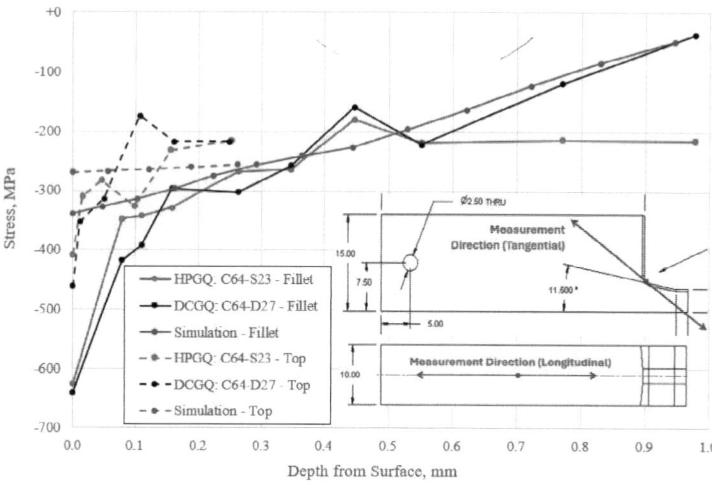

(a)

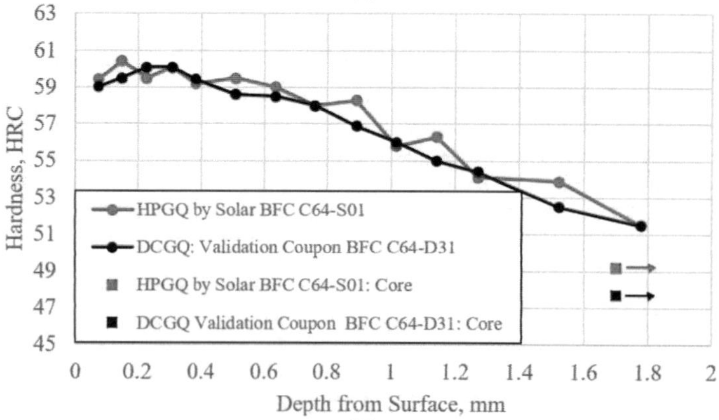

(b)

Figure 4: Measured (a) residual stress, and (b) hardness comparison between HPGQ and DCGQ processed coupons with carburization

The DCGQ process is suitable for high hardenability and high temperature resistant steel grades. Besides Ferrium C64, three more steel grades, Pyrowear 53, M50NiL, and Pyrowear 675 were studied for their responses to DCGQ process. Figure 5(a) compares the hardness between DCGQ and HPGQ processed coupons with carburization. The coupons used are carburized cylinders with 10 mm diameter, and all coupons are low pressure carburized, followed by reheating and DCGQ or HPGQ processes. There is no noticeable hardness difference between DCGQ and HPGQ for M50NiL and Pyrowear 675 steels. For Pyrowear 53 steel, both the case hardness and the core hardness from DCGQ process are equivalent to those from HPGQ process. However, the hardness at the case core interaction region is slightly lower from the DCGQ process, which is caused by the self-tempering of martensite during quenching. The self-tempering effect can be controlled or eliminated by controlling the quenching rate. M50NiL and Pyrowear 675 steels don't show self-tempering because their tempering temperatures are higher than their Ms temperatures. Figure 5(b) compares the tensile properties between DCGQ and HPGQ for uncarburized Pyrowear 53, M50NiL and Pyrowear 675 steels. There is no noticeable difference in the yield strength, elongation and reduction area between DCGQ and HPGQ for all the three steel candidates. Figure 5(c) compares the impact strength between DCGQ and HPGQ for the three steel candidates. The experimental data for M50NiL steel samples shows no noticeable difference between DCGQ and HPGQ processes. For Pyrowear 675 steel, the data shows that DCGQ samples have a higher impact strength than HPGQ, and a DCGQ process with a slower quenching rate has a higher impact strength. For Pyrowear 53 steel, the data shows that DCGQ samples have a lower impact strength than HPGQ, and a faster DCGQ process leads to a higher impact strength.

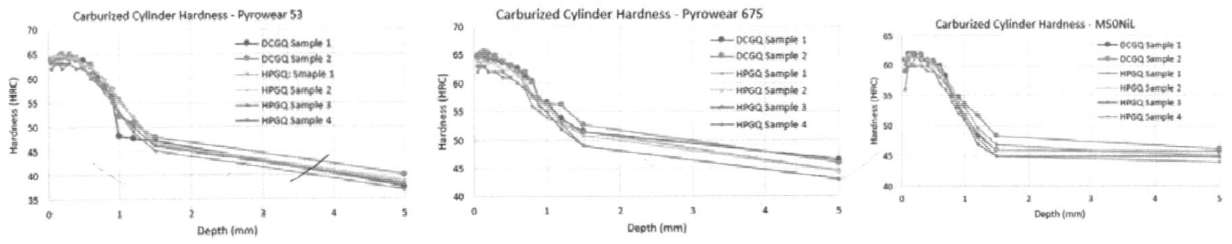

(a)

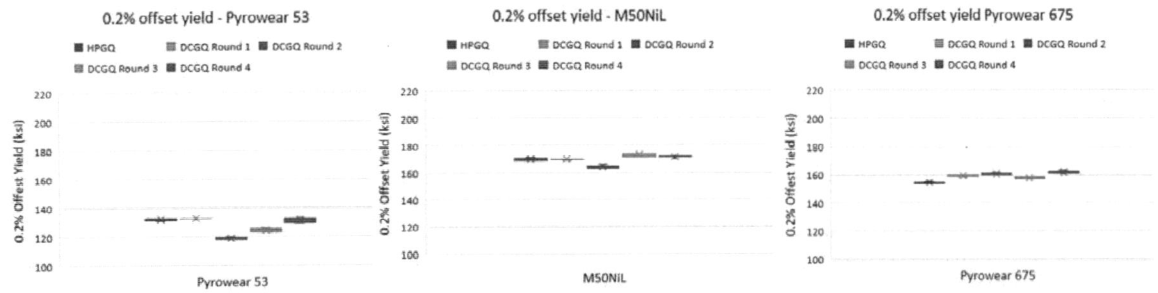

(b)

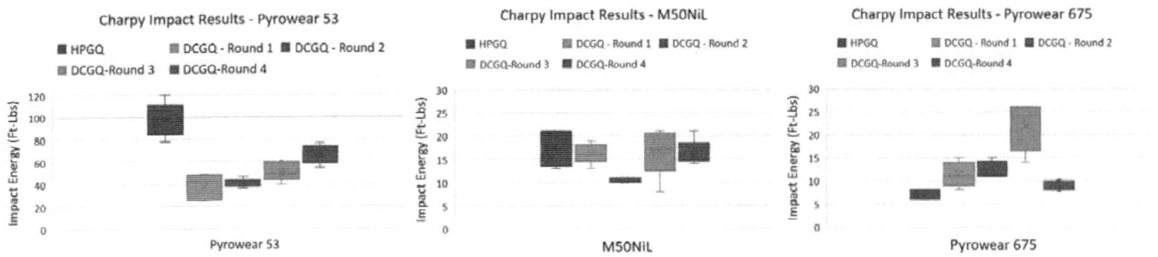

(c)

Figure 5: Comparison of (a) hardness, (b) tensile properties, and (c) impact properties for Pyrowear 53, M50NiL and Pyrowear 675 between DCGQ and HPGQ processes

Modeling Case Study

A pinion gear shown in Figure 6(a) was selected to demonstrate the effectiveness of the DCGQ process on distortion control. The pinion gear is made of carburized M50NiL steel. Its diameter is about 60 mm, and its height is about 160 mm. Figure 6(b) shows the CAD model of the gear with several teeth removed to view the inner geometry. This gear has a significant difference on the wall thickness, with section A being about 3 mm, and section B being about 20 mm.

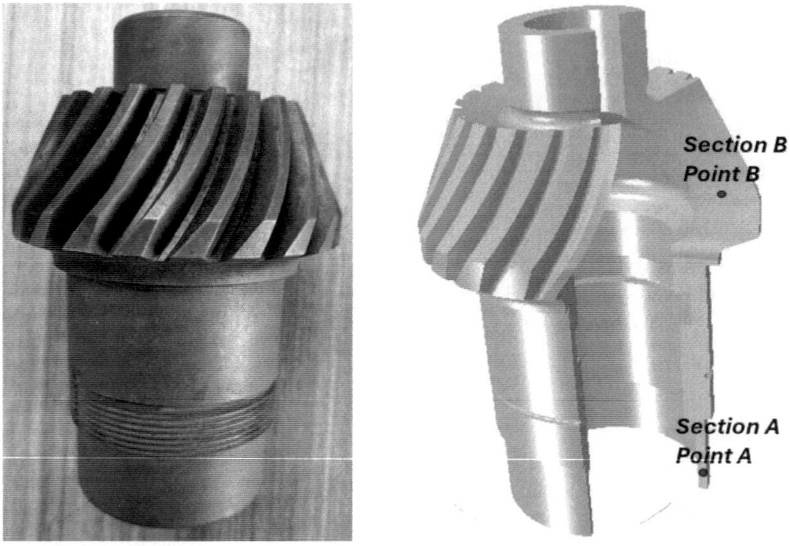

(a) *(b)*

Figure 6: (a) Photo of pinion gear, and (b) CAD model with portion removed to view internal geometric features

The nominal composition of M50NiL steel is listed in Table 1. M50NiL has a high hardenability due to its high alloy content. The high thermal resistant property is caused by the secondary hardening during tempering forming alloy carbides. The recommended tempering temperature is above Ms.

Table 1: Nominal composition of M50NiL

C	*Mn*	*Si*	*Ni*	*Cr*	*Mo*	*V*	*Fe*
0.13	0.20	0.20	3.31	4.10	4.18	1.25	balance

The gear was selectively carburized by copper plating the uncarburized surfaces. The simulated carbon distribution contour using an LPC process is shown in Figure 7(a), with red color representing carburized surfaces. Figure 7(b) shows the simulated carbon distribution in terms of depth and the effective case depth is about 1.25 mm.

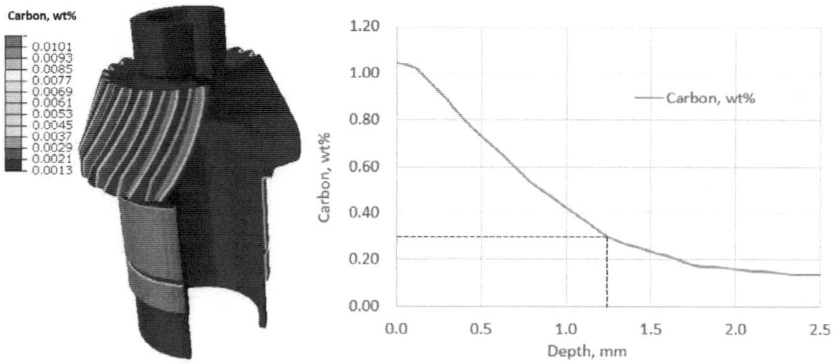

(a) (b)

Figure 7: Simulated carbon distribution profile from an LPC process. (a) Carbon distribution contour showing selective carburizing regions (red), and (b) carbon distribution in terms of depth showing an ECD about 1.25 mm

Figure 8 shows a dilatometry strain curve calculated from DANTE material database for M50NiL steel. The X-axis is temperature, and the Y-axis is the linear strain (roughly 1/3 of the volume strain). The material is heated up from the room temperature of 20 °C, and the volume increases till reaching ~750 °C when austenitizing transformation starts. With the transformation of initial phases to austenite, the volume decreases. The material is heated up further for carbide dissolution before quenching. During quenching (cooling), the material volume in austenite phase decreases. The martensitic phase transformation starts at a temperature slightly below 400 °C, and the material volume expands.

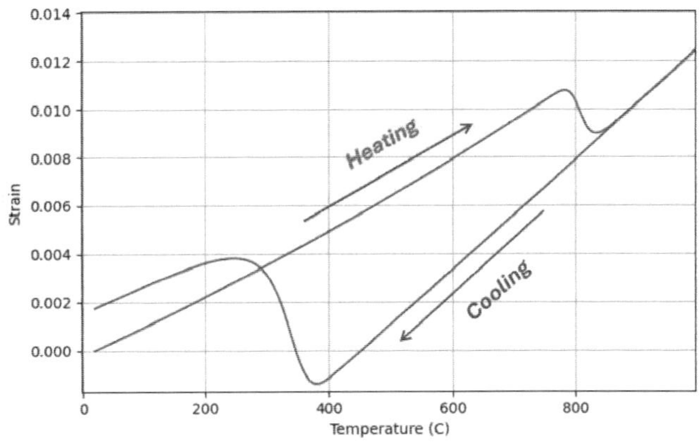

Figure 8: Continuous cooling temperature and strain data with phase transformations calculated from DANTE material database for M50NiL

During heating, the part surface was heated up earlier than the core, and a thinner section was heated up earlier than a thicker section. With a controlled heating rate, the thermal stress during heating can be lower than the yield strength without causing plastic strain. However, when austenitizing phase transformation occurs, with the surface or thinner section transforming to austenite earlier, plastic strain occurs due to the nonuniform material volume shrinkage and the TRIP effect. To effectively reduce the distortion caused by heating, the temperature uniformity, especially while austenitizing phase transformation is happening, should be controlled.

During quenching, the material volume decreases due to its thermal shrinkage, as shown by the "cooling" curve in Figure 8. The material volume increases while martensitic transformation occurs. The nonuniform phase transformation to martensite and TRIP effect lead to distortion in the part. The temperature uniformity in the part, especially while martensitic transformation is happening should be controlled for minimized distortion [6]. The austenitizing process can contribute to distortion for parts with complex geometries, and a stepped heating process can be used to effectively control the distortion caused by heating. In this paper, two heat treatment processes, HPGQ and DCGQ as listed in Table 2, are modeled to demonstrate the relation of temperature, phase transformation and distortion during both heating and quenching process steps.

Table 2: Description of HPGQ and DCGQ processes

Process Steps	HPGQ	DCGQ
Heating / Austenitizing	***Straight heating*** Step 1: 1080 °C for 1800 seconds	***Step heating*** Step 1: 700 °C for 1200 seconds Step 2: 800 °C for 900 seconds Step 3: 850 °C for 900 seconds Step 4: 1080 °C for 900 seconds
Quenching	***One quenching step*** Step 1: 6 bar nitrogen quench for 600 seconds; nitrogen gas is cooled by heat exchanger	***Two quenching steps*** Step 1: Quench and hold at 400 °C for 900 seconds Step 2: DCGQ quench following the recipe: Ambient gas temperature drops linearly from 400 °C to 250 °C in 900 seconds, Step 3: Ambient gas drops from 250 °C to 20 °C in 900 seconds
Cool to room	Take part out and cool in air	Take part out and cool in air

Figure 9 compares the simulated temperature and austenite fraction at point A and point B between the straight heating and stepped heating processes described in Table 2. With the straight heating, the temperature difference between point A and point B in the austenitizing temperature range (750 °C to 800 °C) is over 200 °C, as shown in Figure 9(a). The transformation to austenite is completed already at point A before transformation at point B begins. With the stepped heating process, the temperature difference between point A and point B is greatly reduced by holding at 700 °C before further heating to the austenitizing temperature range. The stepped heating can effectively control the distortion caused by the austenitizing process for complex gears.

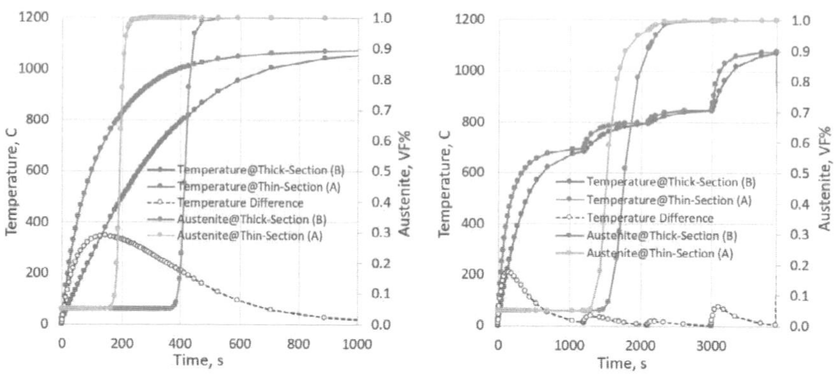

Figure 9: Comparison of simulated temperature and phase transformation difference between (a) straight heating process, and (b) stepped heating process.

During an HPGQ process, the gear is quenched as fast as possible with the ambient gas cooled by an external heat exchanger. The temperature of the quenching gas is uncontrolled and is affected by the load, heat exchanger, and the quenching chamber. The simulated temperature and martensitic transformation using the HPGQ process described in Table 2 are shown in Figure 10(a). The maximum temperature difference between point A and point B exceeds 500 °C during HPGQ. The timing of martensitic transformation between the two points has a significant difference, which is one of the main factors of quenching distortion. During a DCGQ process, the part is quenched to a temperature above the Ms, held for a specific time, then quenched following the DCGQ recipe. The DCGQ recipe is designed based on the maximum allowed differences of temperature and martensite fraction between the thinnest section (section A) and the thickest section (section B). Figure 10(b) shows the simulated temperature and martensite fraction at points A and B for the DCGQ process. The maximum temperature difference during quench to the Ms temperature is about 300 °C. During this step, the in-process stress is from the thermal gradient only because there are no phase transformations. The part and cooling rate should be designed to avoid excessive stress exceeding material's yield strength for minimum distortion. During cooling below the Ms, distortion is inevitable due to martensitic transformation. However, by minimizing the temperature difference in the part section, the distortion can be effectively controlled. In this case study, the DCGQ recipe is designed to control the temperature difference between point A and point B within 15.0 °C, as shown by the blue curve with hollow marks in Figure 10(b).

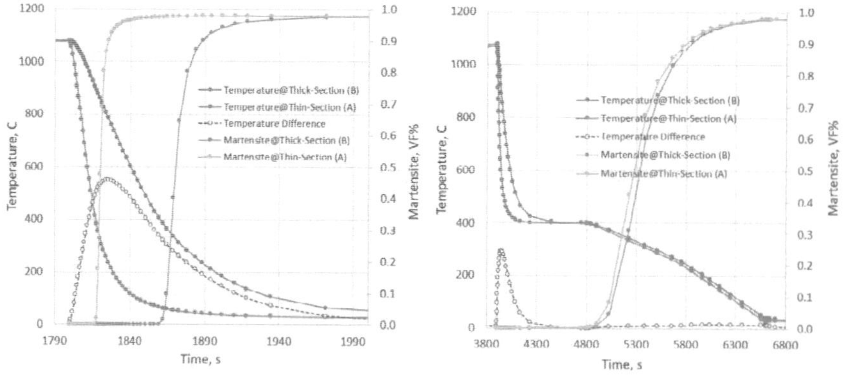

Figure 10: Comparison of simulated temperature and phase transformation histories between (a) HPGQ process, and (b) DCGQ process.

The simulated radial distortion between the HPGQ and DCGQ processes are compared in Figure 11. For both processes, uniform thermal boundary conditions are applied on the entire part surface. The simulated radial distortion for the HPGQ process is approximately 70 μm, as shown in Figure 11(a), compared to the radial distortion less than 10 μm for the DCGQ process in Figure 11(b). It is worth mentioning that nonuniform convection coefficients on the part surface will greatly affect consistency and temperature in the part during an HPGQ process. However, the effect of nonuniform convection coefficient on the temperature uniformity in the part during DCGQ process can be negligible because the temperature uniformity is controlled by the ambient gas temperature. In a gas quench process, the directional gas flow is inevitable, and it is not possible to eliminate the nonuniform convection coefficient along the part. By using the ambient gas temperature to dominate the quenching, the temperature uniformity in the part during a DCGQ process can be greatly improved to reduce distortion.

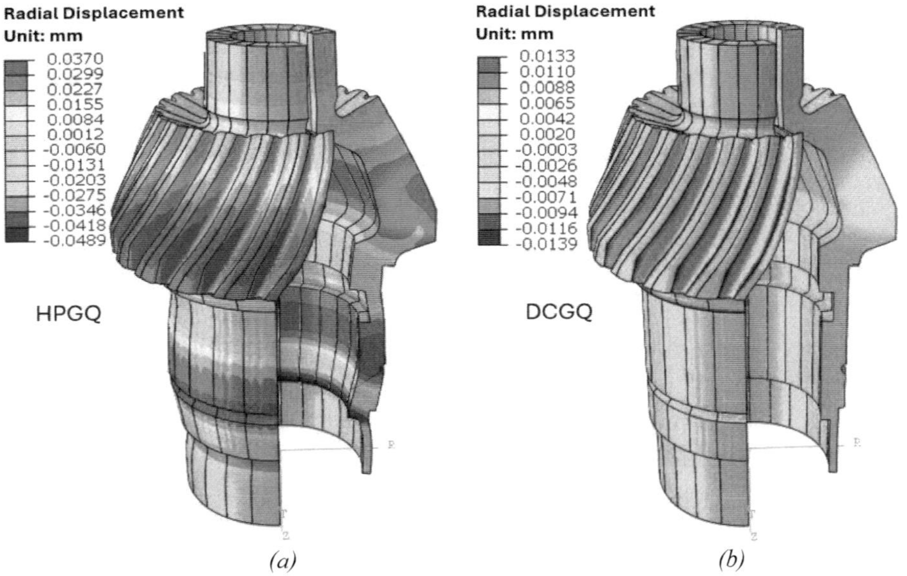

Figure 11: Simulated radial distortion for (a) HPGQ, and (b) DCGQ processes

Conclusions

The DANTE Controlled Gas Quench (DCGQ) process is an innovative hardening process for minimizing distortion suitable for high hardenability and high temperature resistant steels. The primary objective of DCGQ is to control the temperature uniformity in the part during martensitic transformation temperature range by applying a program controlled ambient gas temperature recipe. The recipe is designed as time vs. gas temperature curve using computer modeling targeting allowed differences of temperature and martensite transformation in the part cross section. A DCGQ prototype unit was fabricated successfully to demonstrate the effectiveness of quenching using controlled ambient gas temperature, and multiple steels grades have been processed for testing mechanical properties. Experimental results have demonstrated a 50% reduction in shape distortion for Ferrium C64 samples, including the contribution to distortion from a straight austenitizing process in a hot furnace. Experimental results from tensile, Charpy and hardness testing has demonstrated that DCGQ properties for Ferrium C64, M50NiL, Pyrowear 675, and Pyrowear 53 steels are equivalent to those with the HPGQ process. The DCGQ recipe was also designed for an aerospace pinion gear made of carburized M50NiL steel. The distortion response of this gear is compared between 6 bar HPGQ and DCGQ processes. Plans for future efforts are being developed to demonstrate a reduction in shape distortion for rotorcraft gears, as well as integrating the DCGQ process with a vacuum furnace. With the well-controlled distortion from DCGQ process integrated into a vacuum furnace, it is possible to eliminate the need of post heat treat machining for reduced cost and improved mechanical properties.

Acknowledgements

This research was partially funded by the Government under Contract No. W911W6-16-D-0004 and W911W6-23-D-0003. The views and conclusions contained in this document are those of the authors and should not be interpreted as representing the official policies, either expressed or implied, of the U.S. Army Combat Capabilities Development Command Aviation & Missile Center or the U.S. Government.

References

[1] O. Kessler, C. Prinz, T. Sackmann, L. Nowag, H. Surm, F. Frerichs, Th. Lubben, H.-W. Zoch, "*Experimental study of distortion phenomena in manufacturing chains*", Mat.-wiss. u. Werkstofftech. 2006, 37, No. 1

[2] Z. Li, B. L. Ferguson, E. Lee, S. Habean, and J. Meyer, "*Sources of Heat Treatment Distortion and Approaches for Distortion Reduction during Quench Hardening Process*", Proceedings of the 29th International Federation for Heat Treatment and Surface Engineering World Congress, September 30–October 3, 2024, Cleveland, Ohio, USA

[3] H. Surm, F. Frerichs, T. Lubben, F. Hoffmann, H.-W. Zoch, "*Distortion of rings due to inhomogeneous temperature distribution*", Mat.-wiss. u.Werkstofftech. 2012, 43, No. 1–2

[4] T. Antretter, D. Zhang, and E. Parteder, "*Modelling Transformation Induced Plasticity – an Application to Heavy Steel Plates*", steel research int. 81 (2010) No. 8

[5] M. Vidoni, S. Sarkar, H. Farivar, J. Sims, M. Jordan, D. Münter, "*Simulation of Fixture Hardening for Tool and Process Design Optimization*", HTM J. Heat Treatm. Mat. 76 (2021) 5

[6] J. Sims, B.L. Ferguson, and Z. Li, "*Gas quenching system and method for minimizing distortion of heat treated parts*", United States Patent 11,306,371

[7] J. Sims, Z. Li, and B.L. Ferguson, "*Process to Minimize Distortion during High Pressure Gas Quenching Processes*", Proceedings of the 31st ASM Heat Treating Society Conference, September 14–16, 2021

QDE 2025: Proceedings of the 3rd International Conference on Quenching and Distortion Engineering
May 6–7, 2025, Vancouver, Canada
https://doi.org/10.31399/asm.cp.qde2025p0011

Heat treatment simulation and quenching test considering the influence of heterogeneous composition

***Keisuke NISHIMOTO*[1], *Kohei KAIGAISHI*[1], *Eisuke KUROSAWA*[2]**
[1] *Kobe Steel, Ltd., 1-1-5, Takatsukadai, Nishi-ku, Kobe, Hyogo, Japan*
[2] *Kobe Steel, Ltd., 5-9-12, Kitashinagawa, Shinagawa-ku, Tokyo, Japan*

Abstract

It is generally known from various studies that quenching crack generation is greatly influenced by the chemical composition of quenched material. Due to generation of segregation in the inner parts during casting process, materials made from especially large steel ingots unavoidably have compositional heterogeneity. Numerical simulation is effective for investigating the relationship between compositional heterogeneity and quenching crack. However, in previous research, the influence of compositional heterogeneity was not focused on in the simulation. Therefore, in this study, we developed the heat treatment simulation that can consider the difference in phase transformation behavior and strength characteristics in response to heterogeneous composition by calculating the TTT diagram and stress-strain curve corresponding to the composition at each position of the simulation model. Using the developed heat treatment simulation, the calculation of the quenching of cylindrical test specimens was conducted for both the homogeneous composition material and the heterogeneous composition material from segregation zone. As a result, it was confirmed that extremely high stress occurs in the heterogeneous composition material. In addition, quenching test using cylindrical test specimens, which were taken from the segregated part and the less-segregated part in a large steel ingot, was conducted to verify the validity of the calculation result. As a result, no cracks occurred at all specimens from the less-segregated part, whereas cracks occurred more frequently in the segregated part. Those results reveal that compositional heterogeneity has a significant influence on the quenching crack.

1. Introduction

Forged products are generally subjected to heat treatment to ensure required mechanical properties. Problems such as quenching crack can occur during cooling, and various cracking factors have been examined in past studies[1)]. One factor identified is the chemical composition of quenched material; Arikawa et al. [2)] focused on segregation occurring in large steel ingots, investigating the relationship between segregation zones and quenching crack generation. Segregation inevitably occurs within large steel ingots, generally with V-segregation and negative segregation at the center, and reverse V-segregation at the 1/2R position. The above previous studies[2)] have shown that reverse V-segregation increases the risk of quenching crack through various experiments. However, many details of the crack occurrence mechanism remain unclear. To investigate the mechanism, in addition to experimental approaches, heat treatment simulation[3)] by coupling temperature, phase transformation, and stress-strain is effective. However, it was limited to evaluations of homogeneous composition materials. Therefore, this study aims to develop the heat treatment simulation method that takes into account compositional heterogeneity, enabling consideration of differences in phase transformation behavior and strength properties due to component distribution within the same material. Cylinder specimen quenching experiments were conducted, and quenching stresses with and without component heterogeneity were analyzed using the developed simulation method.

2. Methods

2.1 Quenching test

In order to evaluate the influence of inverse V-segregation on the occurrence of quenching crack, quenching experiments were conducted. Inverse V-segregation occurs when the solidification rate is slow, so test specimens were taken from a large 110-ton steel ingot. The representative chemical composition is shown in Table 1. The test specimens were cylindrical with dimensions of $\phi 12\times 36$, and five specimens each were taken from the 1/2R position and the surface layer of the large steel ingot. Specimens were taken from these positions because it is generally known that inverse V-segregation occurs at the 1/2R position of large steel ingots. For comparison, specimens were also taken

Table 1 : Chemical composition of steel used (mass%).

C	Si	Mn	Cr	Mo
0.38	0.3	1.23	0.23	0.02

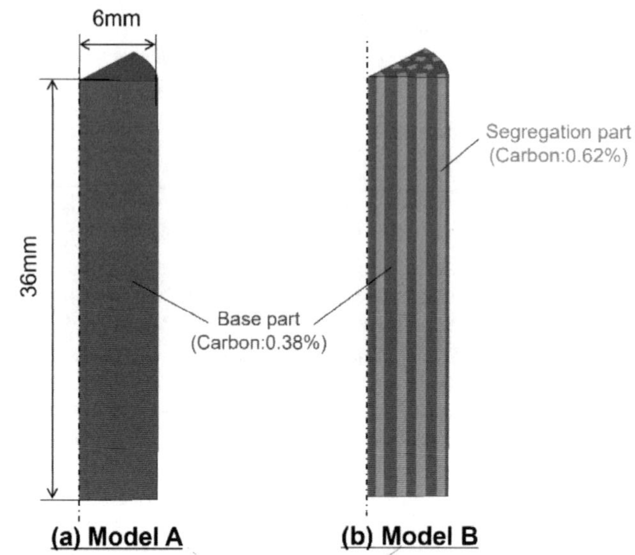

Fig.1 : Computational model of cylindrical specimen (3D, 1/8).

from the surface layer, where segregation is less likely to occur. In the experiment, the specimens were heated up to 900°C, held for 30 minutes, and then water quenched. The occurrence or absence of quench cracking was determined by visual inspection of the specimen surfaces after water quenching.

2.2 Heat treatment simulation

By directly modeling inverse V-segregation, the influence of segregation on quenching stress is evaluated. The heat treatment simulation consists of two steps: heat conduction analysis to calculate the temperature history and phase fraction changes during cooling, and thermo-elastoplastic analysis to calculate thermal strain, transformation strain, and transformation plastic strain from the temperature history and phase fraction. These are performed using the finite element method (FEM). In this study, to consider the difference in phase transformation behavior and strength characteristics in response to heterogeneous composition, each element of the computational model is assigned chemical composition. In the heat conduction analysis, the TTT diagram corresponding to the composition concentration of each element is calculated, and the change of diffusion-type phase transformations is computed to obtain the phase fraction considering the composition. Additionally, since martensitic transformation is a diffusionless-type transformation, the relationship between temperature and phase fraction remains constant regardless of the cooling rate. Using the Koistinen-Marburger rule, it is expressed by:

$$\xi = \xi_\gamma[1 - \exp\{-A_M(M_s - T)\}] \quad (1)$$

where ξ and ξ_γ are the volume fractions of martensite and retained austenite, respectively. A_M is a material parameter, M_s is the martensitic transformation start temperature (Ms point), and T shows the temperature. The Ms point is expressed using a composition-dependent prediction formula, which takes into account the compositional influence on the martensitic transformation behavior for each element. On the other hand, in the thermo-elastoplastic analysis, the stress-strain curves for each phase are calculated based on the composition concentration of each element.

The model used for the analysis is shown in Fig. 1. The model shape is a 1/8 model of a cylindrical test specimen with dimensions ϕ12×36, using symmetric boundary conditions, and hexahedral elements were used. To evaluate the quenching stress due to the occurrence or absence of compositional heterogeneity, two types of models were constructed. Model A is a homogeneous composition model, while Model B simulates inverse V-segregation by aligning linear carbon-enriched regions of several millimeters in size based on the base part. The carbon concentration given here is based on the results measured by an electron probe microanalyzer (EPMA) on samples taken from the inverse V-segregation regions. The stress-strain curves, linear expansion coefficients, and transformation expansion amounts used in the analysis were calculated using the material property software JMatPro, based on the target composition values. Additionally, Fig. 2 shows the strain-temperature relationship during martensitic transformation. These curves were calculated using equation (1), the composition-dependent prediction formula of Ms point[4], and the

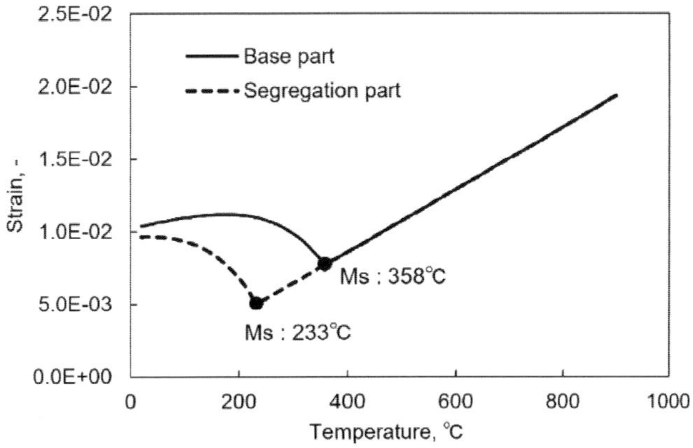

Fig.2 : Relationship between strain and temperature during martensite transformation.

Table 2 : Quenching test results.(○:No crack, ●:Crack)

	results
specimens from the less-segregated part (from surface part in large ingot)	○○○○○
specimens from the segregated part (from 1/2R part in large ingot)	○○●●●

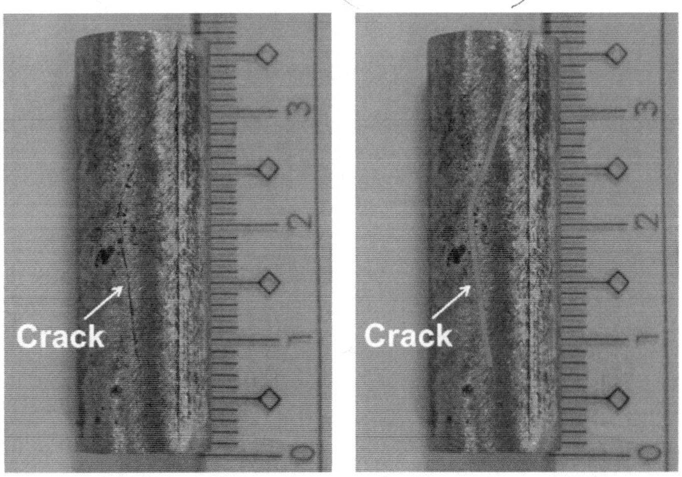

Fig.3 : Photo of quenching crack after water quenching.

linear expansion coefficients of each phase. The different curves were assigned due to the composition concentration of each element. The transformation plastic coefficients were set to 6.3×10^{-5} MPa^{-1} for bainitic transformation and 8.0×10^{-5} MPa^{-1} for martensitic transformation, referencing literature values[5)]. The quenching start temperature is 900°C and the heat transfer coefficient during water quenching is used as the boundary condition. This was determined by inverse identification from the results of temperature measurement experiments conducted on cylindrical test specimens of the same shape as the analysis model.

3. Results

3.1 Quenching test

The quenching test results for cylindrical specimens taken from the 1/2R position and the surface layer are shown in Table 2. While no cracking occurred in any of the specimens taken from the surface layer, cracking was observed in the specimens taken from the 1/2R position. A photograph of the quenching crack in a cylindrical test specimen taken from the 1/2R position is shown in Fig. 3. As shown in this photograph, the direction of crack occurrence is axial, and

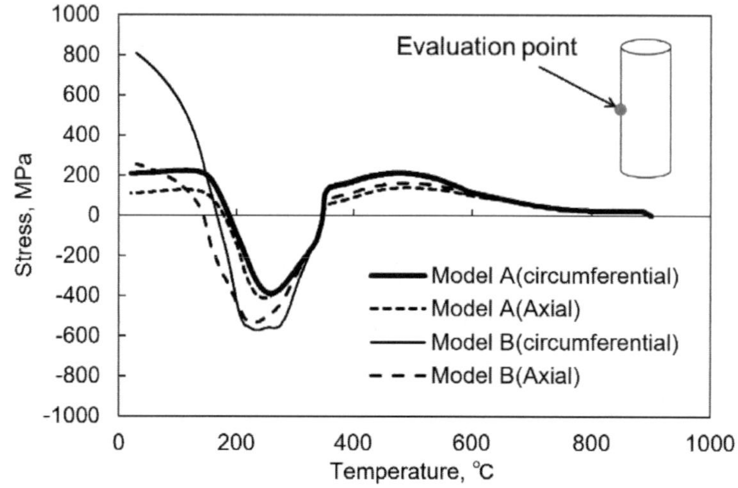

Fig.4: Relationship between temperature and quenching stress at the center of the surface for Model A and Model B.

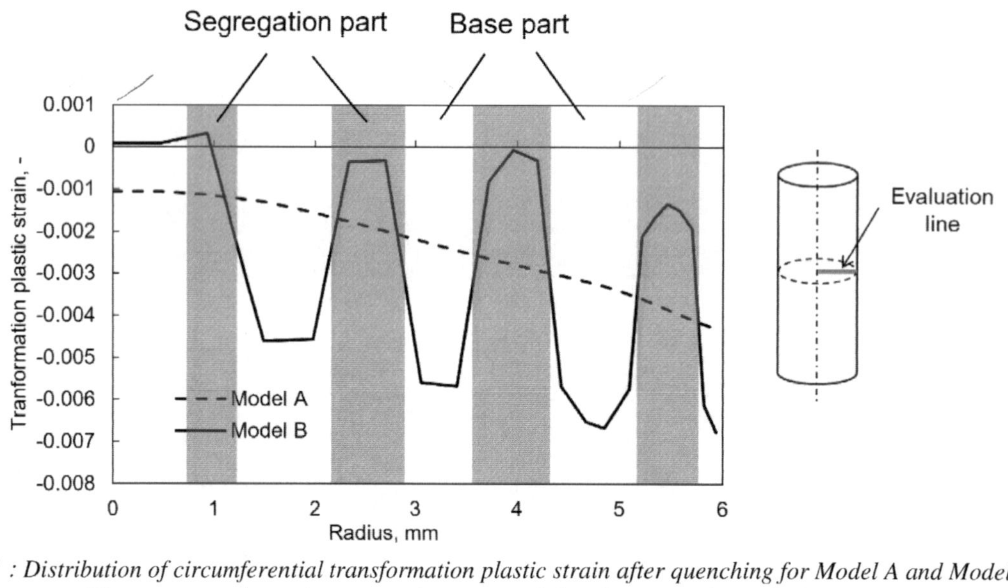

Fig.5 : Distribution of circumferential transformation plastic strain after quenching for Model A and Model B. The gray area shows the segregation part, and the white area shows the base part.

cracks occurred in the same direction in all specimens. These results revealed that inverse V-segregation increases the risk of quench cracking. Furthermore, it can be inferred that the circumferential stress caused the cracks, based on the direction in which the cracks occurred.

3.2 Heat treatment simulation

The relationship between temperature and quenching stress at the center of the surface for Model A and Model B are shown in Fig. 4. Here, the solid line represents circumferential stress, and the dashed line represents axial stress. Regardless of the direction of stress, tensile stress due to thermal contraction is generated immediately after the start of quenching. As martensitic transformation progresses at the surface, the generated stress shifts to compression due to transformation expansion. Finally, tensile stress occurs due to the influence of plastic strain caused by thermal stress and transformation stress, as well as transformation strain and transformation plastic strain. At the completion of quenching, both models show the highest stress, with Model A having a stress magnitude of several hundred MPa in both the circumferential and axial directions, whereas Model B shows an increase in circumferential stress up to approximately 800 MPa. As a result of investigating the cause of this difference, it was found that transformation plastic strain had a significant influence among various types of strains. The distribution of circumferential transformation plastic strain after quenching for both models are shown in Fig. 5. While Model A shows a smooth distribution, Model B exhibits localized changes in the behavior of transformation plastic strain according to compositional heterogeneity. In the case of uniform composition such as Model A, the Ms point is the same throughout the entire region, and the only factor causing differences in the timing of martensitic transformation is the cooling rate. On the other hand, in Model B, as shown in Fig. 2, there is a difference of 125°C in the Ms point due to compositional concentration differences, so the influence of heterogeneous composition is significant in addition to the cooling rate. Due to this influence, the magnitude of the stress during phase transformation differs considerably between segregated and non-segregated regions. As a result, local differences in transformation plastic strain occur, leading to increased generated stress.

The above analysis results are consistent with the quenching test results, which indicate that the risk of quench cracking increases in materials with heterogeneous composition taken from regions with inverse V-segregation. Furthermore, all quench cracks occurred in the axial direction, which aligns with the analysis showing that the circumferential stress is larger than the axial stress. This is caused by the difference in transformation plastic strain occurring in the circumferential direction, making it clear that compositional heterogeneity significantly affects the occurrence of quenching crack.

Using the heat treatment analysis method, it is expected that the threshold for crack occurrence can be quantitatively evaluated. However, when applying this analysis method to actual product scales, there may be issues related to computational cost. To address the issue, we plan to develop heat treatment simulation method that combines macro-scale analysis with local-scale analysis of segregation regions.

4. Conclusions

In this study, with the aim of developing a heat treatment analysis method that considers compositional heterogeneity, we developed an analysis method capable of considering phase transformation behavior and strength characteristics according to the compositional distribution within the same component. By analyzing quenching stress with and without compositional heterogeneity, it was found that high circumferential stress occurs in models that consider compositional heterogeneity, such as inverse V-segregation. The quenching tests also showed that specimens taken from regions with inverse V-segregation had a higher risk of quench cracking, verifying the validity of the analysis method. Additionally, it was clarified that the increased risk of crack occurrence was due to differences in the timing of martensitic transformation caused by compositional heterogeneity, which led to larger gaps in transformation plastic strain.

References

[1] Chapman R. D. et.al., "A Test for Measuring Quench Crack Sensitivity of Engineering Alloy Steels", *Metal Progress* (1953), p 67-72.

[2] Arikawa, T., et.al., "Relation between ingot segregation part of large forged product and quenching crack", *Transactions of the Japan Society of Mechanical Engineers*, Vol.83, No.845 (2017), p 16-00442(in Japanese). doi:10.1299/transjsme.16-00442

[3] Inoue, T. and Arimoto, K., "Development and implementation of CAE system "HEARTS" for heat treatment simulation based on metallo-thearmo-mechanics", *Journal of the Society of Materials Science*, Vol.44, No.496(1995), p 103-109(in Japanese).

[4] Kunitake, T., "Prediction of Ac1, Ac3 and Ms Temperatures of Steel by Empirical Formulas", *Journal of The Japan Society for Heat Treatment*, Vol.41, No.1 (2001), p 164-169(in Japanese).

[5] Okamura, K., "Material properties and database", *Journal of the society of material science*, Vol.55, No.5 (2006), p 529-535(in Japanese).

QDE 2025: Proceedings of the 3rd International Conference on
Quenching and Distortion Engineering
May 6–7, 2025, Vancouver, Canada
https://doi.org/10.31399/asm.cp.qde2025p0016

Quenching Distortion Simulation Consider with Residual Stress due to Machining Process

Tsuyoshi Sugimoto and Kouki Hata
National Institute of Technology, Asahikawa College, Asahikawa, Hokkaido, Japan
t_sugi@asahikawa-nct.ac.jp, +81-166-55-8005

Abstract

As the trend toward CASE in automobiles and the power fields of automobiles, ships and aircraft progress in response to life cycle CO2 (LCCO2), power transmission parts are required to have ultra-long-life materials, high precision shapes, and rapid response to new design shapes. On the other hand, the number of materials and heat treatment engineers is decreasing due to the declining birthrate, and the retirement of experienced engineers is making it difficult to manufacture products that rely on intuition and skill. To address this issue, research and development of model-based development (MBD) and process integration (PI) is being promoted every day in the materials and process fields, and the same is true for heat treatment processes, which are the basis of the strength of steel parts. In MBD, heat treatment conditions are determined by heat treatment simulations, so simulations with a level of quality that can replace experiments are required.

As a result of investigating the prediction quality of heat treatment simulations, it has been found that cooling conditions, materials, and effects from previous processing processes are important factors. This study derived the mechanism by which previous processing processes affect heat treatment quality, especially "variation in deformation," and clarified a method for predicting it using heat treatment simulations. When machining is performed in the upstream process, residual stress distribution and anisotropic metal structure are generated on the surface of the part, which generates heat treatment deformation when heated and affects the final heat-treated shape. We investigated the metal structure and residual stress at the completion of machining for a round bar and performed a heat treatment deformation simulation with this in mind and were able to reproduce the heat treatment deformation.

Introduction

Background

In the heat treatment process, the part supplied from the previous machining process is treated under specific heat treatment conditions in the process itself, realize the heat treatment hardness, metal structure, residual stress, etc., and causing heat treatment deformation. Hardness, metal structure, residual stress, etc. govern the durability of the finished part, so they are the objective variables when determining the heat treatment conditions. On the other hand, heat treatment deformation changes the accuracy of the finished product shape, so it is necessary to predict the heat treatment deformation as shown in Figure 1 and form the part in a shape that subtracts the heat treatment deformation from the finished product shape, and the preparation of the heat treatment process involves a lot of labor and CO_2 emissions. Predicting the amount of heat treatment deformation using heat treatment simulation and determining processing conditions without prototyping and identifying conditions with small heat treatment deformation will play a major role in reducing $LCCO_2$ related to heat treatment production preparation and mass production of 1.35 billion tons/year [1].

What is important here is that heat treatment deformation has a large repetitive variation depending on the part shape and processing conditions. In Figure 2, it shows an example of the variation of heat treatment deformation [2]. Gear tooth profile deformation has a large degree of repetitive variation compared to the actual deformation amount, and performing a single analysis with a heat treatment simulation is meaningless in identifying the conditions. Based on the above "background," the purpose of this research is to clarify the factors that cause variation in heat treatment deformation and incorporate them into heat treatment simulation.

The applicant has conducted the research described below and proposed various parameters as factors that cause variation in heat treatment deformation, as shown in Figure 3. Furthermore, with support from Grant-in-Aid for Japanese Scientific Research KAKENHI JSPS21K14061 , the applicant has conducted research into (i) material composition variation and (ii) cooling variation that occur within the process, and has established measures to address the factors of variation and a simulation of that variation. This application aims to clarify the mechanism of the influence of (iii) the previous process and improve the quality of heat treatment deformation variation simulation by incorporating a constitutive equation.

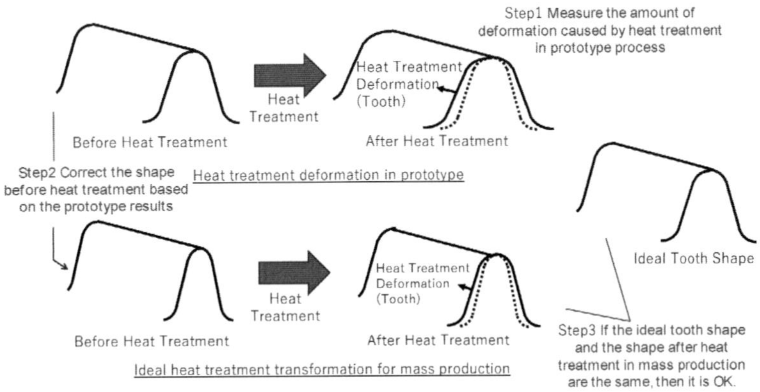

Figure 1: Concept of shape correction before heat treatment

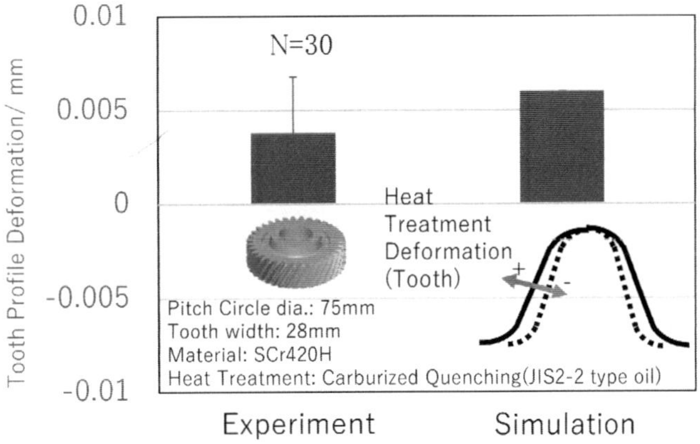

Figure 2: Variation in heat treatment deformation

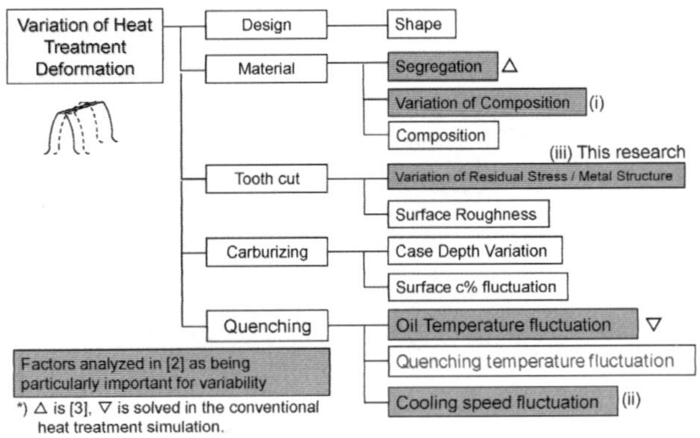

Figure 3: Factors affecting variation in heat treatment deformation

Preliminary investigation

Figure 4 shows examples of the influence of the pre-processing on heat treatment deformation, based on the applicant's preliminary investigation. For example, if the workpiece is clamped with a three-jaw chuck on a lathe during the peripheral processing in the pre-processing, a large deformation occurs in the chuck part when the workpiece is carburized and hardened. Since it is difficult to match the clamping position of the three-jaw chuck with the phase of the finished product, when the finished product is measured, it is observed as a variation in heat treatment deformation.

In addition, if the pre-processing is performed by lathe processing, a pre-processing structure as shown in Figure 5 is formed. This structure is predicted to undergo some kind of deformation during the austenite transformation that occurs during heating, and is a cause of heat treatment deformation. As shown in Figure 3, it is also necessary to verify the influence of post-processing residual stress. Figure 4. Heat treatment deformation caused by the pre-processing Figure 5. Post-processing structure generated in the processed part This research formulates as a "core question" how the metal structure affected by the residual stress and processing given in such pre-processing affects the heat treatment quality, especially the heat treatment deformation, and incorporates this into a heat treatment simulation that is already in practical use, thereby realizing a heat treatment simulation that incorporates the influence of the pre-processing

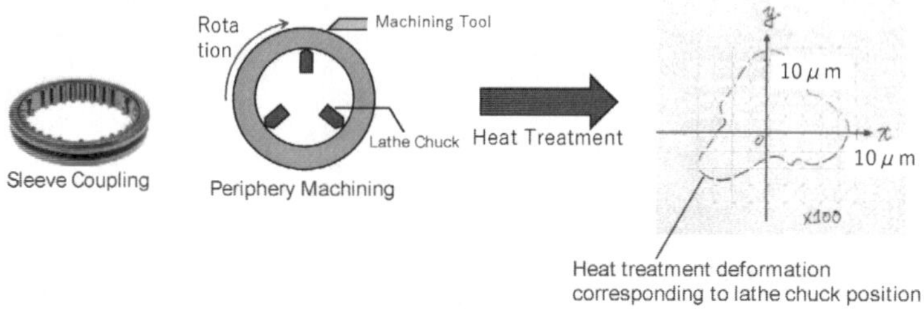

Figure 4: Example of heat treatment deformation caused by previous process

Figure 5: Post-processing structure occurring in processed areas

Figures 6 and 7 show an example of how the work-affected layer remains after processing changes with heat treatment. A 5mm diameter round bar (JIS S45C) was turned on a lathe with 1020 rpm and 0.60mm/r feed rate condition, heated to 930°C, and then cooled in a furnace to obtain a slowly cooled metal structure. Between processes, the round bar was cut into large diameter cross sections near the processed surface, one side was used to observe the structure as is, and the other side was heated and then slowly cooled. The structure at the completion of processing is shown in Figure 6, and the structure after heat treatment is shown in Figure 7. When a structure with a work-affected layer on the surface was heat treated, the structure after heat treatment. Therefore, it was shown that machining affects the condition during heat treatment.

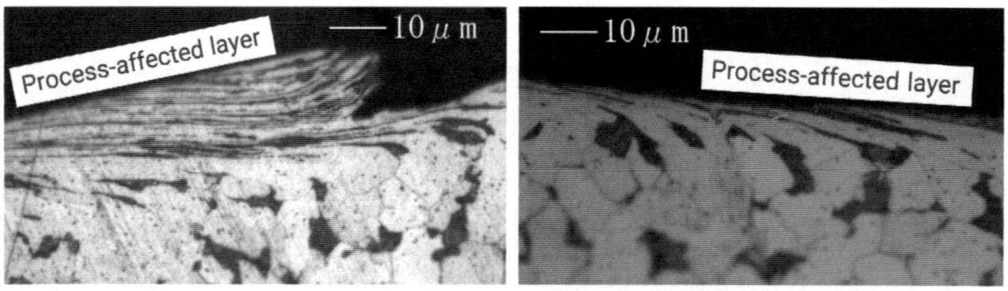

Figure 6: Post-processing structure occurring in processed areas with lathe machining

Figure 7: After heat treatment structure occurring in processed areas with lathe machining

Purpose of this Research

Against this background, we investigated why such deformation phenomena occur even when the shape change at the end of heat treatment is measured.
In this study, as shown in the lower part of Figure 8, we measure the shape change before and after heat treatment and observe how the shape change occurs. We also perform a heat treatment simulation with pre-processing residual stress. By matching the visualization of stress and structural changes inside the object through simulation with the actual observation results of deformation during the treatment, it is possible to confirm the deformation phenomena occurring during heat treatment and to investigate their causes.

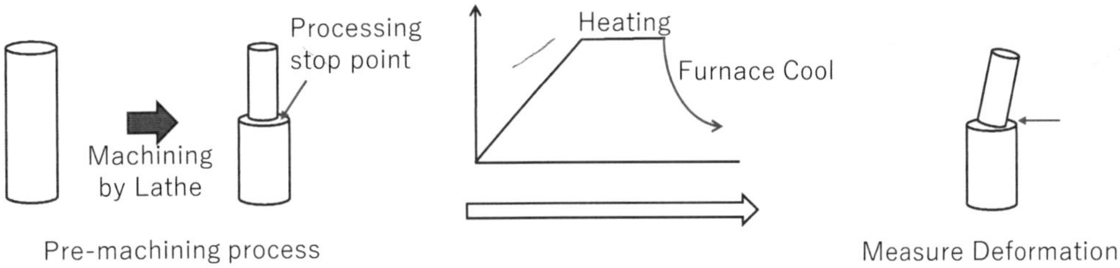

Figure 8: Purpose of this research

Experimental and Simulation Procedure

Experiment
To investigate the deformation caused by heat treatment, a round bar with a diameter of 5 mm and a length of 30 mm was used. After machining the JIS S45C material on a lathe, it was post-conditioned at 600°C for one hour to remove the stress and strain in the test piece. This round bar was cut at a rotation speed of 1020 rpm, a feed rate of 0.610 mm/r, and a cutting depth of 0.5 mm, and machining was stopped when a width of 15 mm had been cut. The point at which machining stopped is defined as the machining stop point. This processing is called stress-adding machining.
After stress-solution processing was completed, the test piece was placed horizontally and heated to 870°C in a vacuum furnace for 30 minutes, and then cooled over 2 hours. This heat treatment is called deformation-imparting heat treatment. The experimental flow is shown in Figure 9.

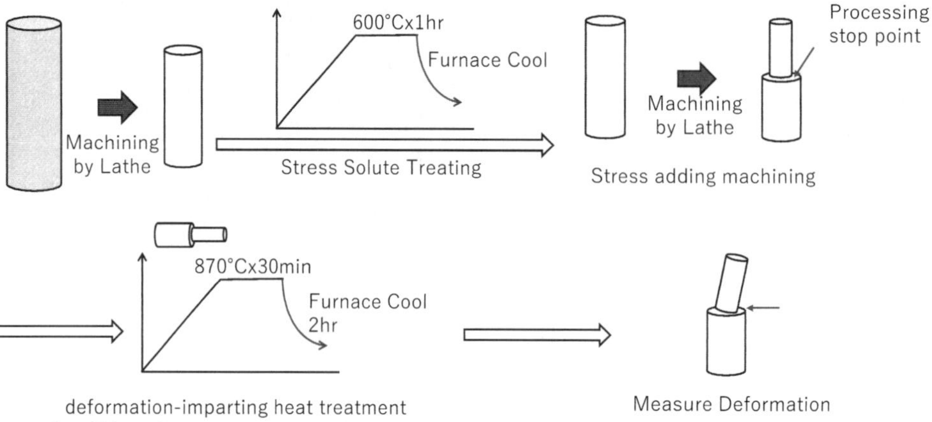

Figure 9: Schematic of Experiment

Simulation

A heat treatment deformation simulation was performed to reproduce the same results as the experiment. SFTC DEFORM-HT® Ver 13.0 was used for the simulation. The shape was modeled after stress adding machining was completed. Inter-machining residual stress was added to the created model. Inter-machining residual stress was based on the research results of Murata et al.[4], and a surface residual stress of 155MPa was added near the completion point of machining, and then the heat treatment conditions shown in Figure 9 were applied to calculate the deformation.

The calculation conditions are shown in Table 1. The applied stresses are shown in Figure 10.

Table 1 Simulation conditions

Solver	SFTC DEFORM-HT Ver.13.0
Number of Elements	27,949
Number of nodes	6,299
Heat Transfer coefficient from atmosphere	0.1kW/m^2/K/s
Material datas	Supplied from SFTC (original datas)

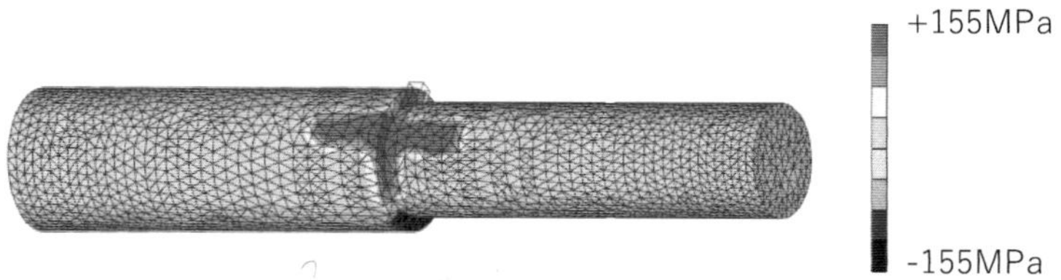

Figure 10: Added mean stress before heat treatment

Results

Experimental results

After the test piece that had been heat-treated after stress machining was heat-treated, the left side of the test piece was fixed to a jig and rotated to measure runout. The results are shown in Figure 11. A test piece that had not been subjected to stress machining or heat treatment was also rotated in the same way and runout was measured. The results are shown in Figure 12. Comparing these results, it can be seen that the test piece that had been subjected to stress machining in Figure 11 had greater heat treatment deformation at the right end of the test piece, and that heat treatment deformation occurred in the direction of the machining stop point between P1 and P2 with about 100 μ m run out.

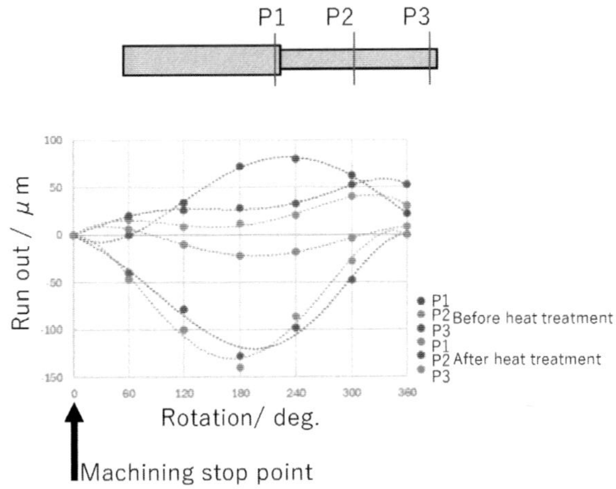

Figure 11: Heat treatment deformation with stress added to machining

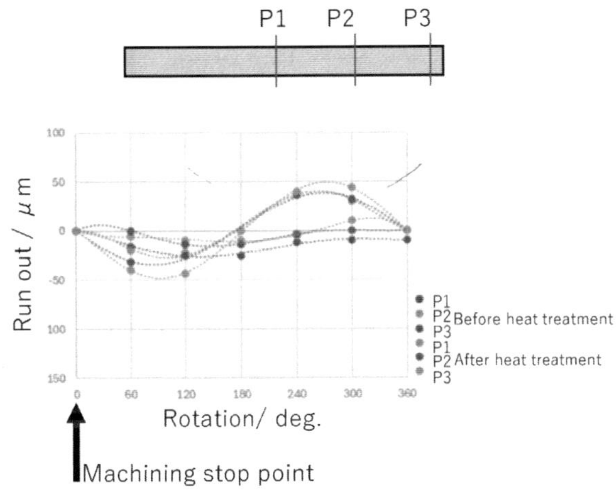

Figure 12: Heat treatment deformation without stress added to machining

Simulation results

A simulation of heat treatment deformation was performed on the process of heating and slow cooling a test piece to which stress had been applied. The final deformed shape and mean stress is shown in Figure 13. Stress is reliefed after machining process. It can be seen that the stress is relieved as heating occurs, and that heat treatment deformation occurs as a result. This demonstrated that when heat treatment is applied to a test piece that has been machined, heat treatment changes occur during heating. It was seen that the stress generated at the completion of machining is relieved as heating occurs, resulting in heat treatment deformation, that large deformation occurs during the austenite transformation, and that deformation remains even in the final work piece.

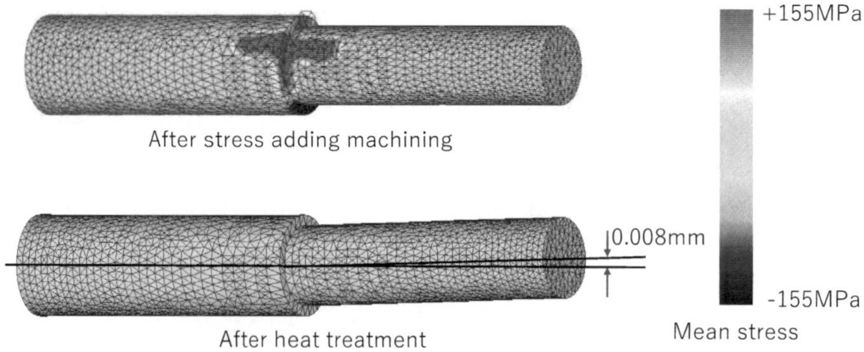

Figure 13: Heat treatment deformation without stress added to machining

Conclusions

It was confirmed that the process-affected layer generated during machining remains in the final product even when heat treatment accompanied by austenite transformation is performed. In addition, when the change in residual stress generated after machining was confirmed by simulation, it was seen that the stress was relieved during subsequent heat treatment when heated. However, it was confirmed that the stress remained even when the heat treatment was completed and the temperature returned to room temperature.

The above results suggest that when machined parts are heat treated, the processing conditions affect the heat treatment deformation. This result was estimated by simulation. However, the current research background is insufficient, including confirmation of the calculation principle, for stress relaxation during heating in the simulation. In the future, it is necessary to verify whether the principle of stress relaxation calculated in the simulation is correct. In addition, there are many conditions in the machining state that are not controlled in the process, such as the position of the machining stop point. It is thought that the variation in heat treatment deformation can be predicted by perturbing the uncontrolled conditions, so we will proceed with verification of the heat treatment deformation variation simulation.

Acknowledgments

This research was partly supported by JSPS KAKENHI Grant JSPS21K14061. We would like to thank all those involved.

References

Literature references are numbered in the order of their appearance in the manuscript and are confined by brackets. They are listed at the end of the manuscript using the "Heat Treat References" format. Include DOI's where applicable. Here is an example of the correct format.

[1] Manufacturing Industry Bureau: Research and development and social implementation direction of the "Decarbonization of thermal processes in the manufacturing sector" project, https://www.meti.go.jp/shingikai/sankoshin/green_innovation/industrial_restructuring/pdf/011_03_00.pdf

[2] Naoya, Kamura, Toshiyuki Miyazaki, Toshihiko Sasaki, Stress measurement in aluminum alloys by Fourier analysis of X-ray diffraction rings, Non-destructive testing, 66(10), pp. 492-497, , 2017

[3] Ruediger Rentsch : Effects of Material Homogeneity and Symmetry Principles in Manufacture on Distortion of Case Hardened Disks, Extended Abstract of QDE2018

[4] Kazuo Murata, Katsumi Yamaguchi, Kazutoshi Adachi, Sakuro Honda, Causes and Features of Residual Stress Generated on Machined Surface, Osaka Industrial Technology Research Institute Report, 15 pp.27-36, 2001

QDE 2025: Proceedings of the 3rd International Conference on
Quenching and Distortion Engineering
May 6–7, 2025, Vancouver, Canada
https://doi.org/10.31399/asm.cp.qde2025p0023

Initial shape optimization of gears using response phase method

Koichiro NAMBU, Osaka Sangyo University
3-1-1, Nakagaito, Daito-city, Osaka, 574-8530, Japan
Knambu.mech@ge.osaka-sandai.ac.jp

Masahiro OKUMIYA, Toyota Technological Institute
2-12-1, Hisakata, Tenpaku-ku, Nagoya-city, Aichi, 468-8511, Japan
okumiya@toyota-ti.ac.jp

Abstract

The miniaturization of automotive components requires higher precision in mechanical elements such as gears. Machining is necessary to achieve high precision, but there are issues such as increased costs and removal of surface modification layers. From this point of view, it is necessary to design the initial shape to achieve the target dimensions after heat treatment, taking thermal strain into consideration. Our group is working on the optimization of the initial shape using the response phase method as a method for this purpose. In our previous research, we have been able to obtain a shape close to the target size after heat treatment of flat plates by optimizing the initial shape. In this study, we applied this method to gear geometry and optimized the initial geometry. To create the response phase, 220 models were created based on the D optimization plan. Finite element analyses were performed on these models. The finite element model is a three-coupled model that considers heat transfer, microstructural change, and stress. The response phase was created from the obtained results, and the initial shape was selected to minimize the strain after heat treatment as the objective function. It was confirmed that the initial shape optimization considering the heat treatment strain reduced the strain after heat treatment compared to the case without considering the heat treatment strain. These results indicate that the response phase method is effective in designing the initial shape to achieve the target dimensions after heat treatment.

Introduction

From the perspective of environmental issues, hybridization and electrification are being promoted in automobiles and other transportation equipment. In addition, the demand for more interior space and longer cruising range requires smaller motors and higher rpm motors. High-strength and high-precision are also important issues for small, high-revolution motors and the reduction gears that are connected to them. Gears used in such reduction gears are often gas carburized and quenched to increase strength. On the other hand, it is known that the quenching treatment causes distortion due to uneven cooling caused by non-uniformity during cooling, in addition to volume expansion caused by martensitic transformation. Compared to nitriding and other treatments, carburizing and quenching treatments are known to cause higher strain. Currently, machining such as grinding is sometimes performed after carburizing and quenching in order to adjust dimensions within predetermined tolerances. In addition, a deep hardening layer is required to account for the grinding allowance due to grinding, which increases the processing time. From this perspective, strain reduction is required not only from a cost standpoint but also to reduce environmental load.
Mist quenching and other methods are available to reduce strain, but they are not yet general-purpose. On the other hand, simulation techniques such as strain distribution analysis after heat treatment are becoming more versatile in addition to improving the analysis system. From this perspective, we propose an optimized shape design method for the initial shape to be the target shape after carburizing and quenching, using simulation-based strain analysis. In this study, as the optimal shape design method, the response phase method was used to optimize the initial shape of gears.

Response phase method

Response phase method is one of the optimization design methods. Design variables, objective functions, and constraints are set as the optimal problem setup. In general optimization problems, the optimal value is obtained through repeated optimization, analysis, and sensitivity calculations for the given settings. In this method, optimization and sensitivity calculations are performed sequentially, which requires computation time and resources. In contrast, the response phase method involves setting parameters and creating multiple models based on experimental design from the optimal problem setting. Then, these models are analyzed to obtain the mean squared error and other information. From the obtained results, a response phase is created and optimization is performed. This method has the advantage that the models are initially set up and the computational load is smaller than that of general calculation methods because the same analysis is performed repeatedly.

Gear model settings and heat treatment analysis conditions

Gear model settings

The 2D model of module 2 shown in Fig. 1 was targeted. In addition, to reduce the computational load, a one-tooth, one-half model was used in the analysis to account for symmetry. A total of 19 design variables were set for the tooth tips, teeth, and tooth bases as shown in Fig. 2. In this study, the edges x1-4 were defined as tooth tips, x5-15 as teeth, and x16-19 as tooth bases. The design variables were set at three levels: -0.03 mm, 0 mm, and +0.03 mm. In addition, the interaction was modeled to consider all variables. From the set design variables, 220 analytical models were created by varying the three levels based on the D-optimal design. Heat treatment analysis was performed on these models, and response surfaces were created.

Analysis of heat treatment

ASTM 4118 steel, a common gear steel, was set as the gear material. The heat pattern was heated to 850°C, held for 1 hour, and then oil cooled. The heat transfer coefficients used are shown in Fig. 3. The heat transfer associations were set to the edges with the design variables shown in Fig. 2. All other edges were set to constrained conditions.

Fig.1 Optimal shape analysis model (gear model)

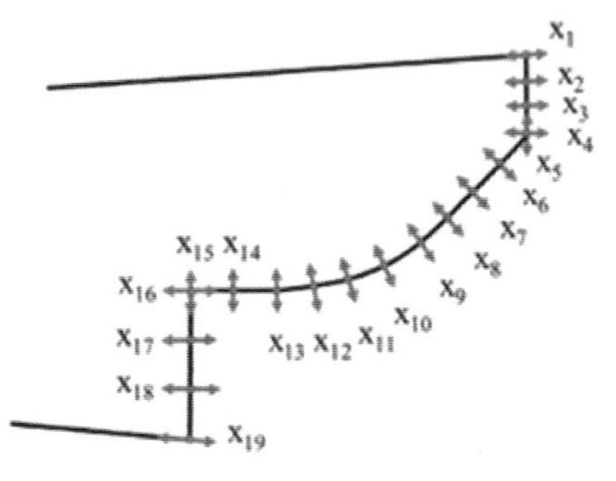

Fig. 2 Points for each design variable

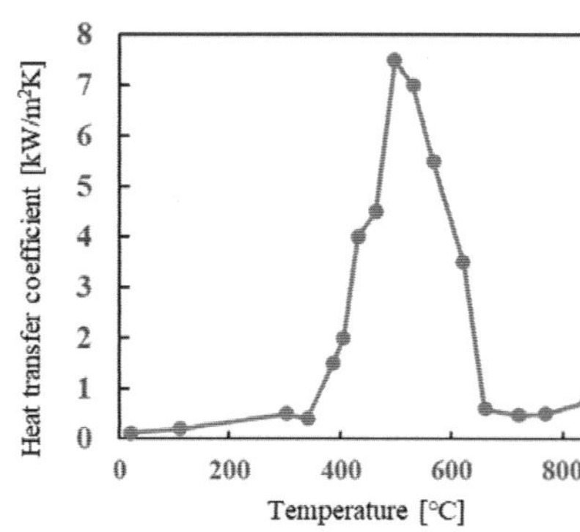

Fig. 3 Relationship between temperature and heat transfer coefficient

Analysis results and optimal shape design results

Analysis of heat treatment

The heat treatment analysis results for the gear model are shown in Fig. 4. The color map in the figure shows the martensite fraction, indicating that the amount of martensite is higher in the tooth tips. It can also be seen that there is a corresponding expansion in the tooth tip direction. Next, Fig. 5 show the mean square error at each point where the design variables were installed for the tooth origin, tooth and tooth tip. From the figures, it can be seen that the errors at each point are large for the tooth tips and the tooth bases.

Fig.4 Martensite fraction and displacement

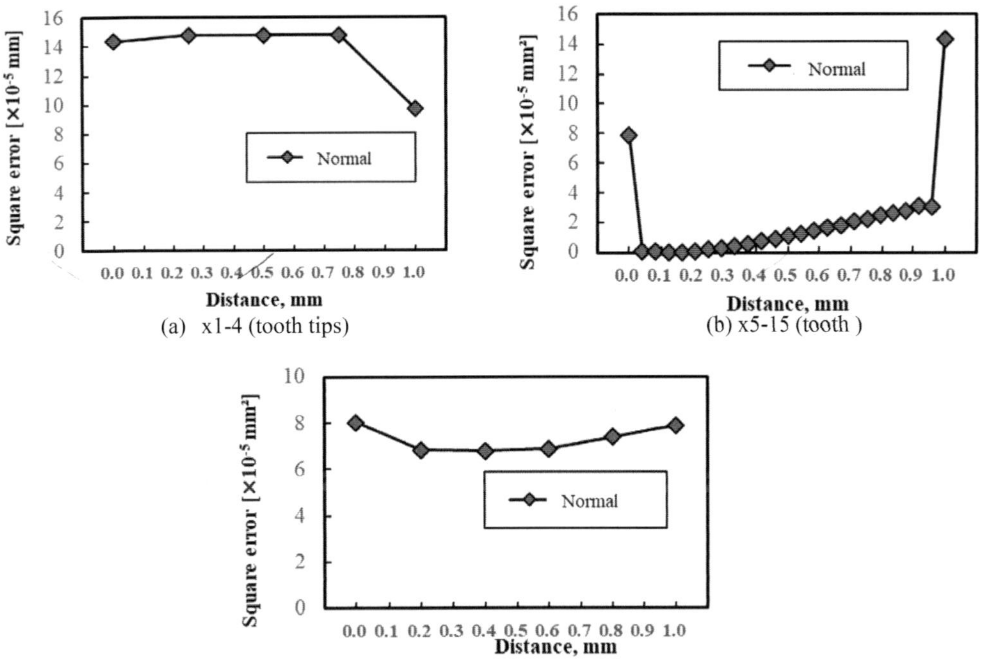

(a) x1-4 (tooth tips)

(b) x5-15 (tooth)

(c) x16-19 (tooth base)

Fig.5 Two Strip Error at Each Edge

Creating a response phase

Fig. 6 shows the mean squared error for all models. The response phase is the maximum value of the mean squared error connected by a line. The coefficient of determination for the response phase is 0.998, confirming that it can be approximated with sufficient accuracy. The response phase was converted into a function, and the value of each design variable with the lowest mean-square error was used as the optimal value.

Initial shape optimization

Table 1 shows each design variable and the mean squared error after shape optimization when the design variables are not changed. The table shows that the mean squared error is reduced by optimization. Next, Fig. 7 show the squared error at each design variable point for the tooth origin, tooth, and tooth tip, respectively. It can be seen that the squared error has been significantly reduced at all points. This indicates that by optimizing the initial shape, it is possible to obtain the desired shape after heat treatment.

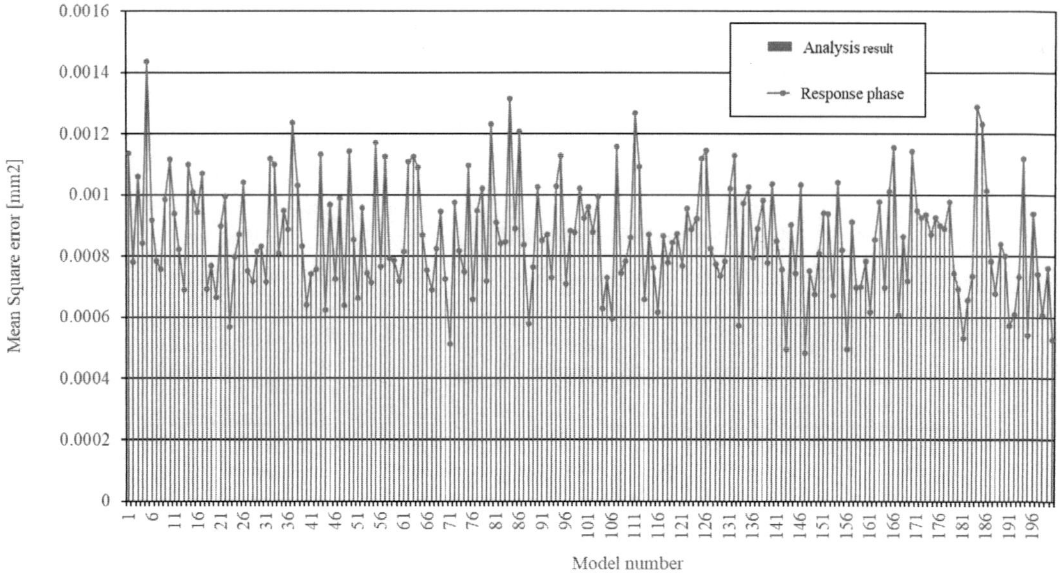

Fig. 6 Mean squared error and response phase for each model

Table 1 Values of each design variable and mean squared error in the optimization model

	x_1	x_2	x_3	x_4	x_5
Normal	0	0	0	0	0
Optimal	-0.00928	-0.01334	-0.01419	-0.01107	-0.00391

	x_6	x_7	x_8	x_9	x_{10}
Normal	0	0	0	0	0
Optimal	-0.00588	-0.00496	-0.00109	-0.00256	-0.00842

	x_{11}	x_{12}	x_{13}	x_{14}	x_{15}
Normal	0	0	0	0	0
Optimal	-0.00202	-0.003	-0.00031	0.00082	-0.00917

	x_{16}	x_{17}	x_{18}	x_{19}	Mean square error
Normal	0	0	0	0	7.62×10^{-5}
Optimal	0.00107	-0.00778	-0.00835	-0.00887	1.67×10^{-6}

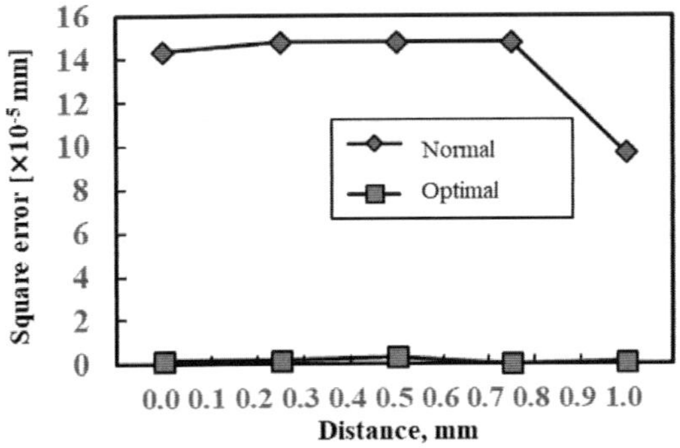

(a) x1-4 (tooth tips)

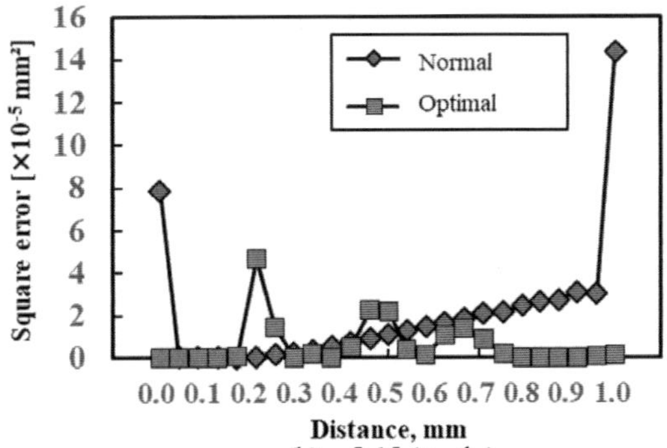

(b) x5-15 (tooth)

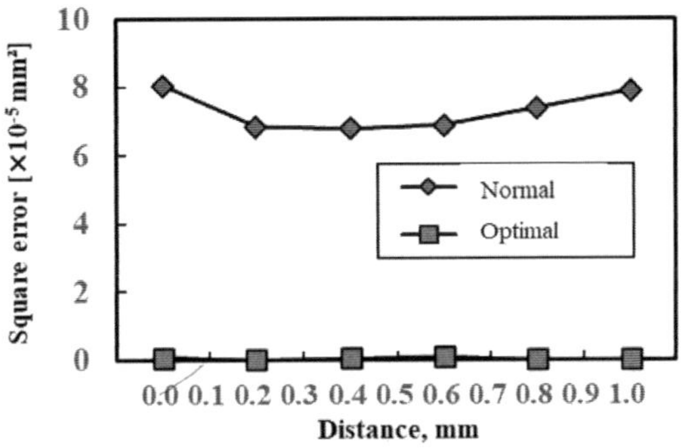

(c) x5-15 (tooth)

Fig. 7 Squared error after heat treatment in the normal and optimized models

Conclusions

1. An optimal initial shape design method using response phase was proposed to reduce heat treatment deformation in gears, and the following conclusions were obtained.
2. It was found that the shape optimization using the response surface can reduce the heat treatment strain in the gear shape, and the shape close to the target shape can be obtained after heat treatment.

QDE 2025: Proceedings of the 3rd International Conference on
Quenching and Distortion Engineering
May 6–7, 2025, Vancouver, Canada
https://doi.org/10.31399/asm.cp.qde2025p0028

Effect of length of vapor blanket duration in quenching oil on mechanical property and distortion of steels

Masahiro Okumiya
Toyota Technological Institute; Nagoya, Japan
okumiya@toyota-ti.ac.jp

Kazuki Nakano
Graduate student of Toyota Technological Institute; Nagoya, Japan
sd22422@tti-j.net

Takuro Aikawa
NIPPON GREASE CO. LTD; Kobe, Japan
takuro.aikawa.3650@nippon-grease.co.jp

Koichiro Nambu
Osaka Sangyo University; Daito, Japan
knambu.mech@ge.osaka-sandai.ac.jp

Abstract

The effect of length of vapor blanket duration when the steels, which were carbon steel JIS-S45C and low-alloy steel JIS-SCM435, quenched in quenching oil on the surface and internal hardness and the distortion of steel was investigated. The heat treatment was accompanied by phase transformation, which caused bending in the direction of the keyway. The quenching oil with the longest length of vapor blanket duration showed the smallest distortion, and the quenching oil with the shortest length of vapor blanket duration showed the largest distortion. It is assumed that this is proportional to the time taken from the temperature of 900°C at the start of quenching to 352°C, the Ms point at which phase transformation from austenite to martensite occurs. The larger the temperature difference between the keyway, which is the low temperature side, and the thick wall, which is the high temperature side, below the Ms point, the larger the distortion. The longer the cooling time from 800°C to the Ms point, the smaller the temperature difference between the low and high temperature sides, and the smaller the distortion.

Introduction

Steel materials are inexpensive and easy to work, and can be alloyed with other metals and heat-treated to have a wide range of properties. In particular, carburizing and quenching are often applied to power transmission parts such as gears and shafts, which are in contact with other parts and require high surface strength.

In the heat treatment of quenching used in this experiment, heat treatment oil is used as a coolant. The purpose of heat treatment oil is to heat and cool metallic materials under specific conditions to satisfy their mechanical properties and to obtain high-quality metallic materials. In recent years, from the viewpoint of environmental protection, there has been a review of heat treatment materials and reduction of heat treatment strain in order to reduce the size and weight of automobiles and other parts and to increase their strength.[1]

In this study, we focused on heat treatment distortion and verified the effects of vapor blanket length on the surface and internal hardness and distortion of steel materials when heat treatment oil is used. S45C, which is widely used for shafts and axes and is called carbon steel for machine structural purposes, was heat-treated, and quenching experiments were also conducted using SCM435 steel to verify the difference in quenching properties.

Experimental method

In this study, four types of heat-treated oils with different vapor blanket stage lengths were used. The cooling curves measured according to JIS K 2242 A are shown in Figure 1. The base oils are all the same, and the additives are adjusted so that the vapor blanket stage lengths are in the order of A oil, B oil, C oil, and D oil from the longest to the longest.

In this experiment, key-grooved cylindrical specimens of S45C and SCM435 with a diameter of 10 mm and a length of 150 mm were used. Their chemical compositions are shown in Table 1. After heat treatment of the specimens, microstructural observation by microscope and SEM, and strain measurement were conducted. The shape of the specimen is shown in Figure 2. The test specimen had a groove on one side that was 2.4 mm deep and 4 mm wide.

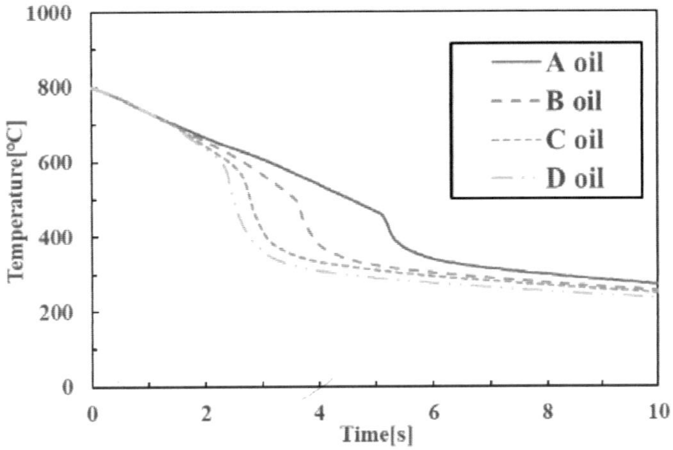

Figure 1 Cooling curves of the heat treatment oils

Table 1 Chemical compositions of the two steel specimens

(Excerpt from JIS G 4051, JIS G 4053) **[mass%]**

	C	Si	Mn	P	S	Ni	Cr	Mo
SCM435	0.33-0.38	0.15-0.35	0.60-0.90	Less than 0.03	Less than 0.03	Less than 0.25	0.90-1.20	0.15-0.30
S45C	0.42-0.48	0.15-0.35	0.60-0.90	Less than 0.03	Less than 0.035	-	-	-

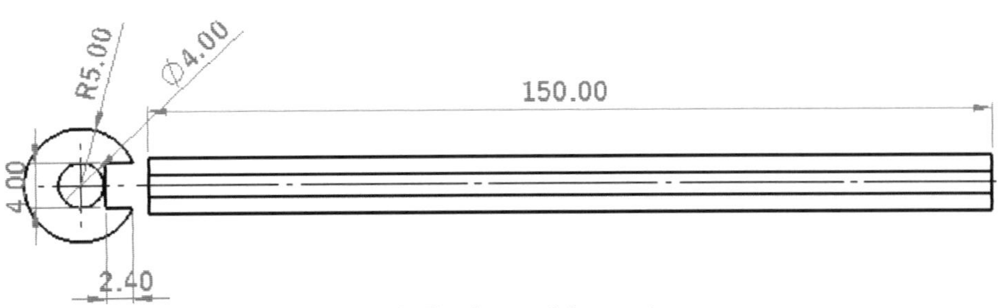

Figure 2 The shape of the specimen

The heat pattern of this study is shown in Figure 3. The treatment temperature was 900°C and was performed in an N_2 atmosphere to prevent decarburization.

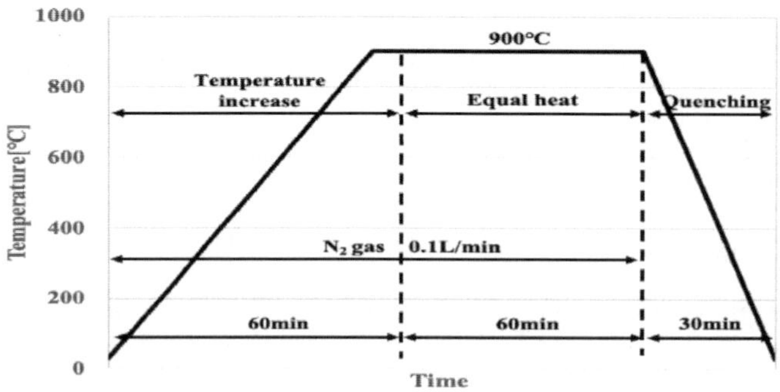

Figure. 3 Heat pattern of quenching experiment

In addition, the quench bend was measured as an indicator of the amount of strain. As shown in Figire 4, the amount of strain was measured as an index of quench bending by holding both ends of the specimen after quenching with a V-shaped block. The runout was measured by a dial gauge as the difference between the maximum and minimum circumferential displacement at the center of the specimen when the specimen held at both ends was rotated around the circumference. This value does not have a positive or negative number, but in this study, the keyway side is assumed to be positive.

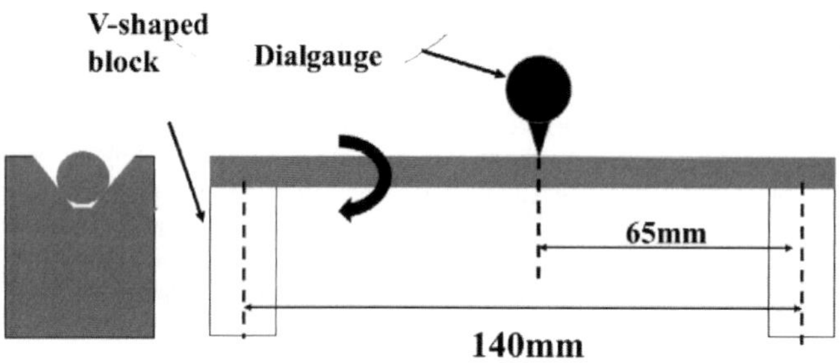

Figure 4 Strain measurement method

Experimental results and discussion

Figure 5 and Figure 6 shows the measured runout (an indicator of burn-in) of the two steels after heat and quench.

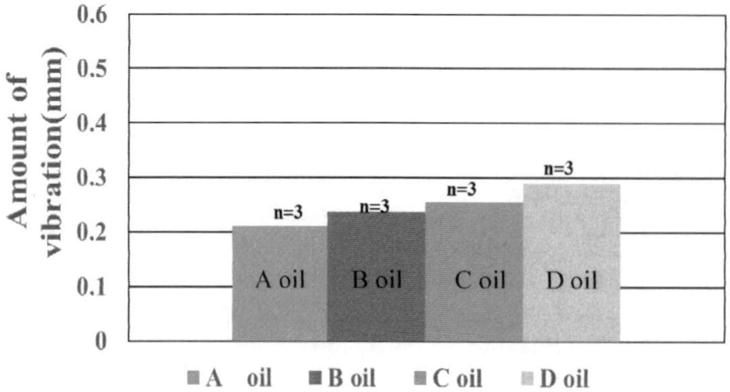

Fig. 5 Amount of vibration after heat treatment of SCM435 specimen

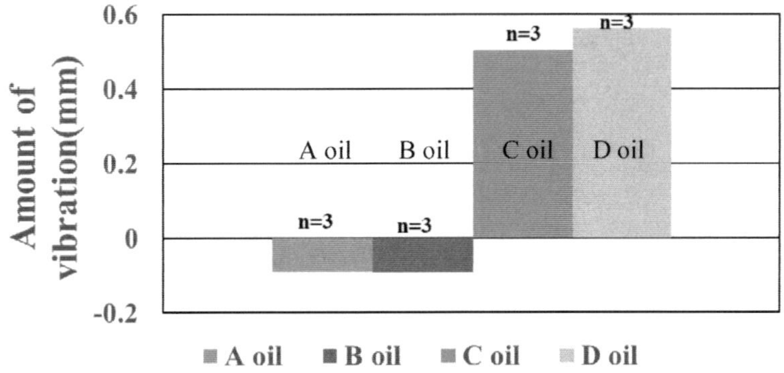

Figure 6 Amount of vibration after heat treatment of S45C specimen

Figure 5 shows that positive distortion occurs in the C and D oils. This is considered to be caused by the fact that the elongation caused by the transformation expansion is treated by the shape change. Figure 5 also shows that the distortion is reduced as the vapor blanket stage length increases. This is considered to be due to the fact that the temperature difference between the thick and thin-walled portions of the steel in the heat-treated oil with a short steam blanket stage length is smaller than that in the heat-treated oil with a long steam blanket stage length, resulting in a smaller time difference for phase transformation to occur between the thin and thick walls of the steel.

Figure 7 shows the internal hardness profile of the S45C test piece when quenched with each oil. No quenching occurred with oil A and oil B, but quenching occurred with oil C and oil D. Therefore, the distortion tendency is different between AB and CD.

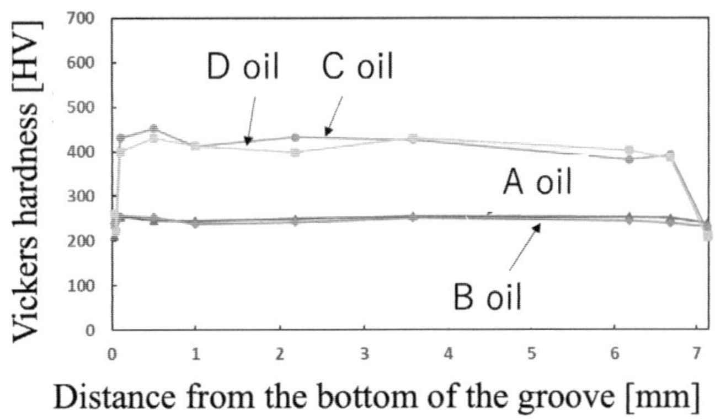

Figure 7 Internal hardness profile of the S45C test piece when quenched with each oil

Summary

The steam blanket stage length has a significant effect on the hardness and distortion of steel, especially with respect to the amount of distortion, which increases as the steam blanket stage length decreases.

References
[1] Nozomu Okumura, Roadmap of Heat Treatment Industry and Current Status of Vacuum Carburizing, 35th Heat Treatment Research Meeting, 2012

QDE 2025: Proceedings of the 3rd International Conference on
Quenching and Distortion Engineering
May 6–7, 2025, Vancouver, Canada
https://doi.org/10.31399/asm.cp.qde2025p0032

Reduction of thermal distortion in the hardening process of a tubular stabilizer through conductive heating and quenching in clamping state

Federico Martinez Guerrero
Martzingenieurbuero, San Luis Potosi, S.L.P.,
Mexicomartz. ingenieurbuero@gmail.com

Ma. Guadalupe Ortiz Lopez
Universidad Politécnica SLP, San Luis Potosí, México
guadalupe.ortiz@upslp.edu.mx

Abstract

In the automotive industry, aggressive weight reduction targets are set to the components supply base. For the suspension, a solution is to substitute the solid bar stabilizers with tubular bars. For these are specified material and hardness to withstand higher stresses and to reach fatigue test cycles similar as solid ones. Another requirement is that they must comply with dimensions and tolerances stated on drawing, often tight due to vehicle package requirements. This is not an easy task; the distorted parts require a time and money consuming process to mechanically reshape them, here undesirable micro-cracks can be originated and lead to premature failure of the part. In this work, it is analytically proposed and physically implemented, alternative manufacturing processes that minimize or avoid distortion. One solution is to heat-up the bars once at a time using an electric conductive heating machine, here the thermal expansion is not restricted by using a moveable fixturing that accompanies the growth of the bar. After the heating, the most severe distortion occurs at the quenching process, this distortion is analyzed in depth and for its minimization a fixturing device is used to clamp the part at defined points and in this clamped state it is controlled immersed in the quench medium, thus restricting the free distortion.

Introduction

With each new vehicle development, weight reduction challenges arise for the component supply focused on lowers the amount of energy needed to move the vehicle, regardless of the efficiency of the propulsion system. For the suspension, a solution is to substitute metallic solid bar elements by tubular ones provided that its higher stresses condition may equal the behavior, in terms of durability or resistance to fatigue, to that of a solid element [1]. The hardness requirement for these tubular elements is in the range of 1600 to 1850 MPa yield tensile strength. Another important requirement is that the parts built with tubulars must preserve their geometric shape and dimensions which is a hard task due that tubular bars which are larger in length compared with its diameter, can easily deform in the heat treatment process applied to them.

Not only the high yield tensile strength of the material but other material / metallurgical requirements such as zero total decarburization (loss of carbon during the heating process), tempered martensitic microstructure and specified grain size contribute to the achievement of the equivalent fatigue test requirements by the less mass element. It is often applied a heat treatment process where there is not possibility that the material losses carbon during the heating process up to 920°C - 950°C. Such process is called controlled atmosphere heat treatment and there, either an inert gas such as nitrogen or an endothermic gas must be introduced into the heating chamber of the furnace in order to protect the work pieces from oxygen which in an open environment will cause carbon loss [2]. Undesirable effects such as an incomplete transformation and a proeutectoid ferrite microstructure is avoided by a fast transport from the heating chamber to the quench tank as well as using a high cooling rate quench oil and a proper agitation for a high heat transfer rate. Conventionally the controlled atmosphere furnace is a batch kind where the tubular pieces are accommodated on a rack with various levels so that the charge of pieces, in the case of stabilizers could be around 100 pieces, depending on their size and geometry. The tubular elements having a very relatively high ratio length versus diameter and being hollow suffer thermal distortion during the quenching process so that the pieces don't fit with the original geometry anymore and a mechanical deformation post-process is required in order to bring it back to the original required state. If continuous furnaces are used, where the process of the parts are in small batches or one by one this problem continues as well [3]. After extensive experimentation with the batch type heating-quenching furnaces is possible to evidence that the distortion problem cannot be substantially reduced [3], therefore an alternative process must be investigated.

To reach the adequate temperature, the bath salt process [4] is not suitable because its application applies to low production and it is very time slow, an individual induction process is difficult to apply in elements with specific geometric shape due to a complex shape induction coil to produce them. Then an option to be considered is a heating process is to use the electrical conductivity of the element itself to be processed by applying current, because it would be a faster process and there would not be any problem when used on a part with complex geometrical shape characteristics inclusive. After the heating process a so called clamped-quench or press-quench process can be applied in order to preserve the shape and the geometric tolerances.

Press Quenching is a specialized quenching technique that is currently applied on parts such as bearing rings, crown and conical, gears, etc., where the tight tolerances on critical dimensional characteristics meaning diameter and planarity must be kept. After this the motivation is that such technique may be used to minimize the distortion of complex, geometrical components during heat treatment [5]. The objective is to hold parts round and flat while they are being cooled, thereby reducing, though typically not drastically eliminating, distortion [6], and is typical applied in large circular or flat parts, gears, bearing races, clutches, saw blades, slitting knives etc.

Based on the causes of distortion, volume changes and deformation, the mechanism for distortion formation in quench press is discussed in [7] as well as the potential for distortion with their precursors. Furthermore, the so-called method of "Distortion Engineering" is explains and presents selected examples to illustrate the connections between distortion and the quenching process. In [8], it is given a widely description about the independent contributing factors to produce distortion in Press Quenching, between them, are cited thermal history and residual stress of the material processes; thermal stresses; temperature, flow rates and properties of quench medium; location and magnitude of contact points on the component to applying external loads as well as quench die tooling. In the technical literature also is possible to found the application of simulation and modeling to improve the accuracy of results. The application of finite element analysis (FEA) is applied as per [9] to establish the calculation model for a face gear die quenching method, as well as the influence of die on gear quenching deformation. The software DANTE is applied in [10] to investigate the sensitivities of several critical process parameters on distortion for a press quenching process of a simplified thin-wall bevel gear. Also, in [11], DANTE software is used to improve the design of quench tooling applied in the press-quenching process of a straight bevel gear. In [12] by COSMAP software is modelled the effect of the not uniform vapor film formed along the surface of a bearing sleeve during the quenching process, that cause its uneven distortion. The simulation results verify the distortion mechanism and solution how to reduce the elliptical distortion. In the case of research and study of the press quenching process applied to bars or shafts, there are very few works carried out in the area, among which it is possible to mention [13], where the distortion mechanisms during carburizing and quenching process of a transmission shaft are analyzed.

The aim of this work is to analyze a treatment method based on the individual processing, said one piece flow, by conductive heating up to austenization temperature of a tubular stabilizer bar and subsequent press quenching process. Subsequently to find out the main process parameters that would determine the behavior where the combination of the two phenomena as well as to propose a fixturing device that could prevent the part from undergoing large deformations that will cause very extensive work post-operations to recover the per drawing specified shape. The remainder of this paper is as follows: In section 2 the geometrical characteristics, the bending and heating process applied to a stabilizer bar is described. Analyzing using FEA software the temperature profile on the part of study after the heating and quenching processes along with the deformation state, in section 3, the critical points are determined to place clamping points on a proposed fixture used to press or hold the part while being quenched to avoid deformation. In section 4, the improvements achieved by the use of the designed fixture are presented. The paper concludes with some final remarks in section 5.

Characteristics and process of the stabilizer bar

Characteristics of analyzed component.
Stabilizer bars are a component of the automotive vehicle suspension whose function is to prevent the vehicle's rollover effect of the vehicle when driving in curves, cornering, and/or moving on bad-condition roads. Its basic construction is a U-bent high-stress steel tubular bar mounted in connection with other suspension components by

links to the vehicle axle and by bushings to the vehicle chassis, as shown in Figure 1. Its position on the vehicle is selected in such a way that it hinders the rotation of the vehicle around its longitudinal axis [7].

The geometry of a bar is given as a function of the packaging space it has in the chassis, i.e. the neighboring components whose path it must avoid, maintaining a free space in relation to these in both static and dynamic states, as well as facilitating assembly and disassembly. The bar under study, shown in Figure 1, is built based on a low alloy high strength steel 26MnB5 tube with 23.8 mm OD, 4.28 mm WT Overall dimensions1080.5 mm L, 171 mm W, 254.3 mm H, and form tolerances of 3 mm maximum at critical zones by engineering specifications as marked the red circle.

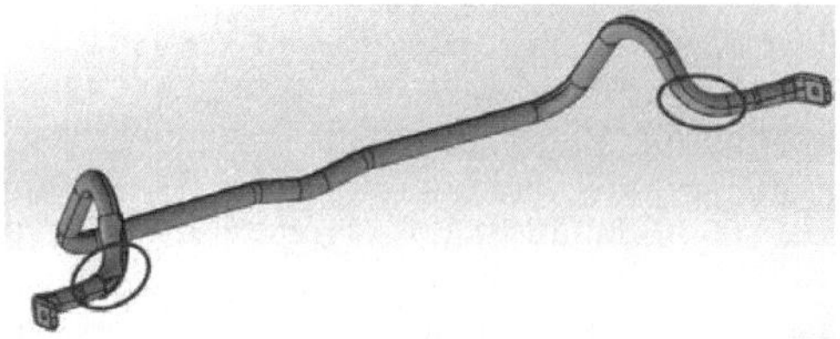

Figure 1. Stabilizer bar

Description of bending process.
In solid stabilizer bars, the geometry is hot formed with a dedicated tooling, the heating process provides a temperature usable for the hot bending and few time after for the quenching process.
On the other hand, tubular stabilizers are primarily CNC cold bent, this to avoid tube collapsing by hot bending. Therefore, the next step is heating treatment. To bend the tubular stabilizers bars, a several axis CNC machines is foreseen. Besides material and mentioned dimensions, the tooling design takes into consideration bents, bent radii and angles. The bending dies are as shown in Figure 2 a). In figure 2b) is shown a coordinate system XY. The Y axis is set with the projection of the bending plane on the cross-section, and the X axis through the geometric center thar divides the tube in two sides. The bending outside far from the bending center and the bending inside near the bending center axis. The full-cross-section profile is described by the curves K_{O1} and K_{O2} on the bending outside, and K_{i1}, K_{i2} on the bending inside.

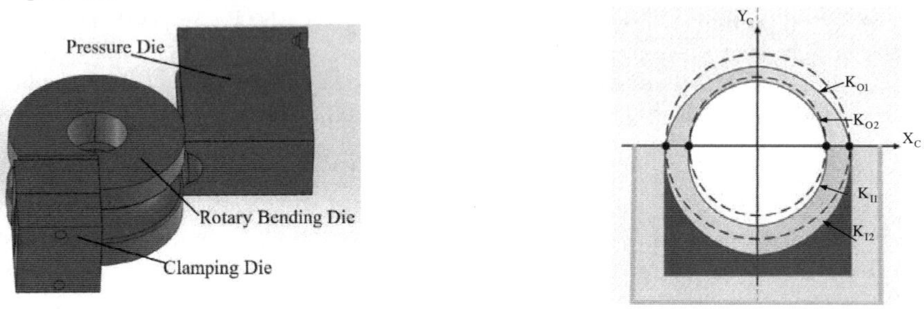

a) *b)*

Figure 2. a) Dies of a CNC bending machine, b) cross section of bending tube.

On the bending inside the material of tube under bending compressive stress produces accumulation. Thus, the wall-thickness would thicken, obviously on the bending innermost. Considering the accumulation within the containing capacity of tube materials themselves, the characteristic curve K_{i1} tends to be a regular curve after tube materials fill the gap between bending die and tube blank because of the restriction of bending die [14]. Therefore, for s_0 representing the gap value and r_0 the initial outer ratio, the curve equation of K_{i1} is given as:

$$ki_1(u) = [r_0 + s_0]Cos(\pi u) - [r_0 + s_0]Sen(\pi u)],\ u \in [0,1] \quad (1)$$

The material accumulation is nonuniform along the radial direction on the cross-section. The closer to bending innermost, the thicker the tube wall is. Therefore, the characteristic curve K_{i2} can be treated as semi-ellipse. With the semi-minor axis of K_{i2} as b_{i2}, and t_1 denoting initial wall-thickness, the curve equation of K_{i2} is:

$$ki_2(u) = [r_0 - t_1]Cos(\pi u) - b_{i2}Sen(\pi u)],\ u \in [0,1] \quad (2)$$

During bending process, the position of the geometric center axis of the bent region is assumed to be consistent, but its length tends to change because of boost and traction. When the drawing load is far larger than the assistant pushing load, the bent region tends to be elongated in a percentage represented by η_e. In the other hand, the accumulation of tube materials causes the tendency that the axial length of the bent region on the bending inside would decrease after bending. The wall-thickness variation is mainly caused by these two factors. It is assumed that the wall-thickness of each tube layer is invariable on the *Xe* axis and each cross-section has the same wall-thickness distribution. According to the principle of metal volume invariance, b_{i2} can be solved by the equation:

$$R_0 \bullet \theta \bullet [r_0^2 - (r_0 - t_1)^2] = (1 - \eta_e) \bullet \frac{(R_0 - r_0) + (R_0 - b_{i2})}{2} \bullet \theta \bullet [r_0^2 - (r_0 - t_1) \bullet b_{i2}] \quad (3)$$

where R_0 represents the bending ratio and θ the bending angle.

Once the bar cold is formed, it must be heated and quenched, the most established process to obtain zero total decarb is a batch furnace with protective gas, here is proposed as an alternative the conductive heating process.

In the case of heating and quenching in a batch furnace, which is shown in Figure 3a), there are two kinds of factors that originate thermal distortion on the geometry of the processed parts. Firstly, is the way the parts are accommodated in the batch load. A rack with number of levels according to part size and often accommodated with alternate directions to maximize the packaging factor, as show Figure 3b). With regards to quenching process, the influence factors are flow rate of quench medium, flow speed needed to promote the most efficient heat removal. Quench medium reduces its flow speed, as it travels from bottom of load towards the upper layers in the load arrangement due to the friction with the parts and the rack that supports them. Besides, variable convective heat transfer values at each level of the load and even between bars at the same level. By these multiple conditions, a Design of the Experiment Method seems not easily to be applied, since the machine-tooling does not have flexibility to variate involved parameters other than rotational speed of quench medium propellers, temperature, time, etc.

Figure 3. Batch Furnace: a) Configuration of furnace, 1-vestible, 2-heating chamber, 2.1- radiant tubes, 3-quench tank. b) Parts loaded on the rack.

Even when there is an advantage regarding the heating way through radiant tubes which transitorily implies that some parts heat up faster than others, it will be that the temperature control of the furnace will ensure that all the points within the volumetric space of the load are within the temperature tolerances.

Analyzing the aforementioned factors, it is found that the geometrical factors on the quenching process (parts distribution, orientation, levels) contribute the most to the distortion on each part of the load at a variable percentage. Due to the critical cooling rate of the respective material must be exceeded in order to produce the required hardness values, this requirement can lead to large temperature gradients which in turn generate thermal stresses which depending on the quenching medium, the dimensions, and the material of the component, producing deformations and distortion [7]. Another fact that produces distortion and large scattering of the dimensional and size changes, is the inhomogeneity of cooling rates in the complete batch volume is much larger than for a single component due to quenching by levels.

Conductive Heating

After extensive experimentation with the batch type heating-quenching furnaces, it is possible to evidence that the distortion problem cannot be reduced by parameters setting-up, quench medium temperature, time, agitation, pieces arrangement, etc. therefore the conductive heating process with posterior press-quenching is studied. The heating process time for a 1.0 to 1.5 ton. load in a batch process is approximately 90 minutes, with an arrangement of 120 bars per load, it results in a cycle time of 45 seconds. The production rate target of the proposed conductive heating process, piece-by-piece, is to reach a similar or less cycle time.

Per cross bending section shown in Figure 2b), cross-sectional area below X_c or $A_{B\ inside}$ is larger than one above X_c or $A_{B\ outside}$: $A_{B\ inside} > A_{B\ outside}$. The electrical resistance for any given shape cross-sectional element is calculated by relationship $R = \rho L / A$, so far, the cross-sectional area is correctly computed, where R is resistance [Ohm], ρ is resistivity [Ohm-m] of the specific material, L is length [m], then:

$$R_{B\ inside} < R_{B\ outside}\ , \qquad I_{B\ inside} > I_{B\ outside} \tag{4}$$

By Joule effect with heating power P [W] and current I [A]: thus $P=I^2R$, $P_{B\ inside} > P_{B\ outside}$ then inner bent section heats faster than outer bent section, metallurgically this represent a problem due inhomogeneous temperature distribution by quenching, therefore in hardness and microstructure results.

To avoid this problem, the conduction heating machine is programmed and set-up so that after having it clamped at its ends by copper jaws, as shown Figure 4a):
-It is supplied with a remarkably high Ampere current, usually with maximum capacity of 12,000 A, until reaches a set-up maximum temperature.
-Switch to current to a very low level for a few seconds, until temperature equalizes all along the part, means a set up temperature difference between IR temperature sensors pointing on sharpest inner bent and less sharp outer bent.
-Finally switch to a third step to reheat and recover the lost temperature during the previous step.
Profile curve electrical current versus time of the described process steps is shown in Figure 4b).

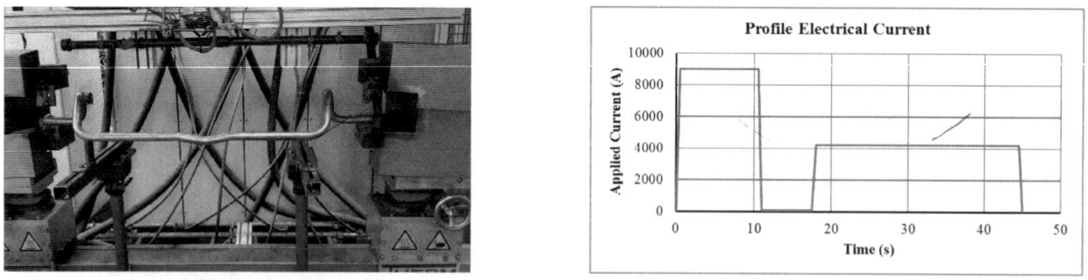

a) b)

Figure 4. a) Stabilizer bar in conductive heating machine, b) Profile curve electrical current versus time

While the bar is being heated, already clamped at the ends, there is thermal expansion and if this is not in some way compensated, it would result in a bow. To avoid such undesirable deformation, the clamping points must travel in synchronicity with the thermal expansion, as show Figure 5. The solution for this is servos programmed at the same speed, calculated used the relation (5), where α_T [W/m-K] is the conduction heat coefficient of material, ΔT [K] the temperature difference between final and initial temperatures T_{sp} [°C] and T_o [°C], and L_f [m] and L_0 [m] the final and initial length while heated up during Δ t [s].

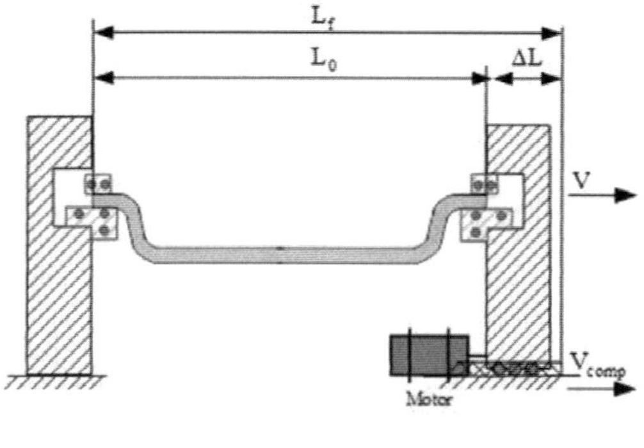

$$\Delta L = \alpha_T \Delta T L_0, \quad V_{comp} = \frac{\Delta L}{\Delta t} = \frac{\alpha_T \Delta T L_0}{\Delta t} \tag{6}$$

Figure 5. Schematic of conductive heating machine.

In Figure 6, thermographic images of a conductive heating process cycle at three different temperature stages are shown. At figures 6a. and 6b. are visible the temperature difference between the outer and inner side of the bents, while at figure 6c the temperature uniformity is reached.

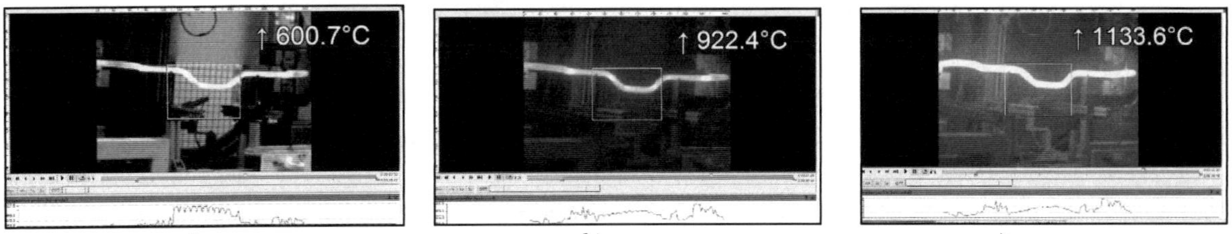

a) b) c)

Figure 6. Thermographic images of conductive process showing the temperature distribution at: a) 600°C, b) 922°C, and c) 1133 °C.

Once the part is heated up to the as the in last step of program set-up Tsp, an industrial robot picks-up the bar and transfers it to the quenching process. At this stage of the process if bar was just let fall in the tank, there would be metallurgical and dimensional huge process variability. To avoid quenching below critical temperature of 730°C for the processed steel and decarburization since the piece is not gas protected heating time and quench delay are critical process parameters. Figure 7 shows the temperature profile obtained by FEA of stabilizer bar with an initial temperature of 920°C, cooled on air with a convective heat transfer coefficient of 35 $W/m^2 \cdot K$ during 30s. The final average temperature is 885°C. To avoid the thermal distortion that would occur if the part was placed in free state in the quench bath, a clamping fixture that holds the bar in certain determined points is proposed.

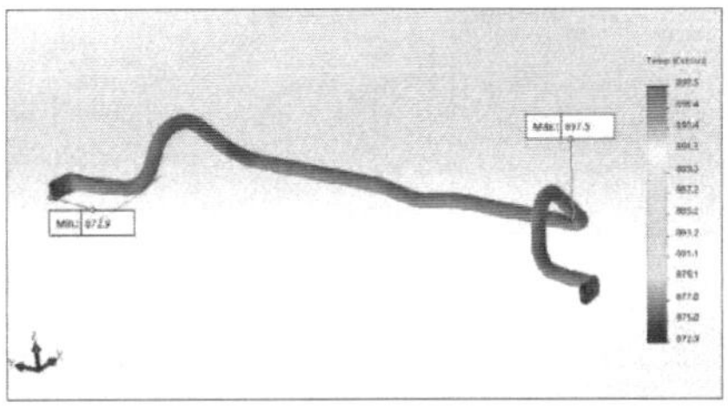

Figure 7. Temperature profile for 30s of air cooling.

Fixture Design

FEA Simulation software is used to study the effect of implementing the clamps in different points, targeting for a deformation of less than 3 mm (surface profile = 2*deformation) at critical zones of the part under study (red circles). The initial temperature is 920 °C, and it is assumed to be air-cooled for 20s, with a convective heat transfer coefficient of 35 $W/m^2 \cdot K$. The first scenario, Figure 8a) has four clamping points: two at the ends of the bar and two on the straight part of the bar where it is normally fixed to the vehicle chassis. As shown in Figure 8b). The critical zone resulted in a total deformation of 3.20 mm, that slightly exceeds the target tolerance. The second scenario has six clamping points, incorporating two more in the critical zone of the stabilizer, as shown in Figure 8c). As result the deformation in the critical zone is 0.30 mm, therefore, the target would be met. However, it presents an undesirable deformation in the central part of the stabilizer bar with a value of 3.32 mm as shown in Figure 8 d).

To avoid deformation of the previous arrangement, a third scenario with seven clamping points, incorporating another one located in the central point of stabilizer bar was proposed, and shown Figure 8 e). As result the deformation in the critical area as well as the deformation in the central part is 0.60 mm Figure 8 f). Finally, an arrangement with eight symmetrical clamping points is proposed and shown in Figure 8g). Due to the clamp of central point is not applied, the deformation in this point persists with a value of 3.2 mm as is depicted in Figure 8h). Based in the aforementioned results, is concluded the scenario of seven clamps is the best option to implement the fixture.

Finally, the scenarios with four and seven clamping points respectively, were analyzed by FEA for the conditions of water quenching process and shown in Figure 9a) and 9b). The initial temperature is 885 °C, and it is assumed to be water-cooled for 120s, with a convective heat transfer coefficient of 3000 W/m2·K. As can be seen, in both cases the deformation conditions of 3 mm in the critical points indicated on the stabilizer bar are met. The option with four clamps allows the bar to be placed inside the fixture in a less complex way than inside the seven-clamp fixture. However, to avoid deformation in the central part of the bar with a value of 1.76 mm, the seven clamps fixture is selected.

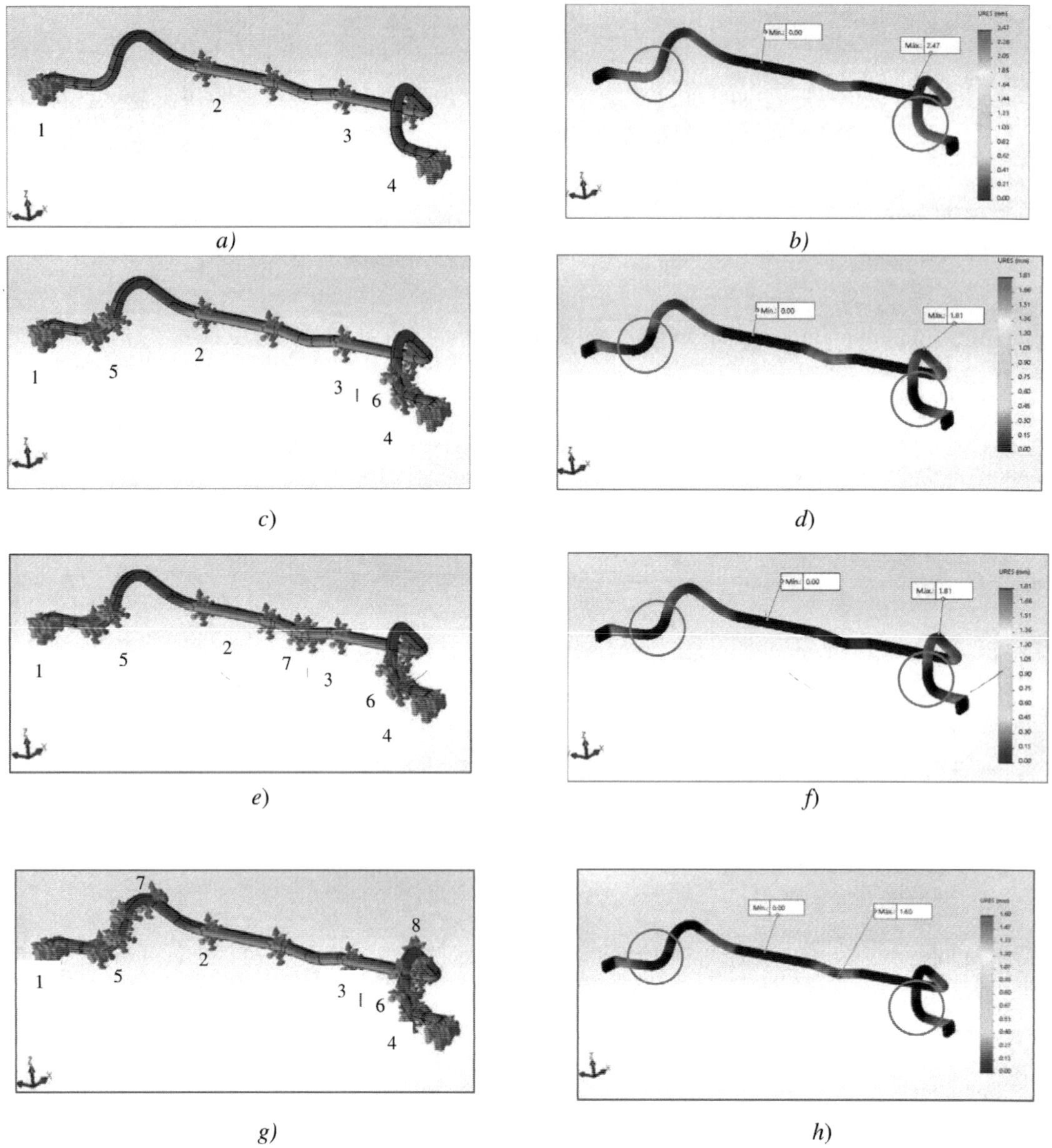

Figure 8. Clamping and deformation diagram using FEA software for four scenarios.

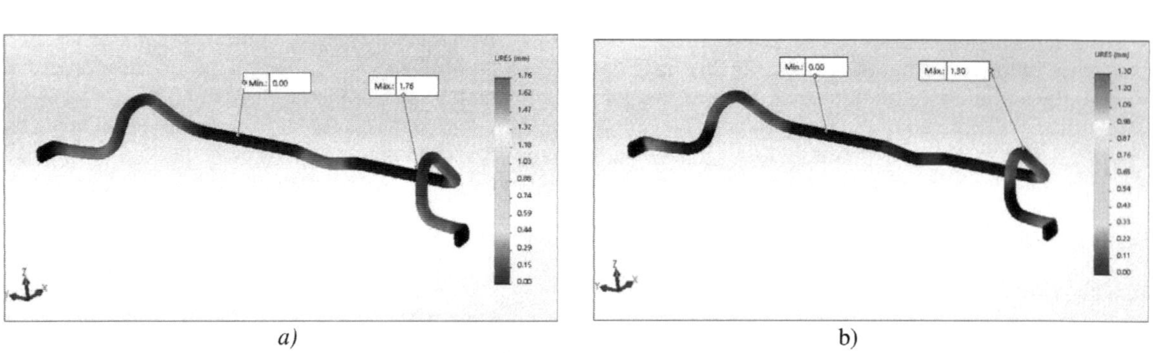

a) b)

Figure 9. Deformation for water press-quenching for: a) four clamps, b) seven clamps.

Implementation of the fixture and improvements achieved

Based on the results obtained through FEA analysis and the dimensions of the stabilizer bar, the fixture with seven clamp points is the one to be used in the press-quenching process and was built according to what is shown in Figure 10, and in Figure 11 its assembly over stabilizer.

a) *b)* *c)*

Figure 10. Stabilizer bar and fixture to press-quenching it: a) stabilizer bar in position to be clamped, b) upper part of clamp-quench fixture, and c) lower part of clamp-quench fixture.

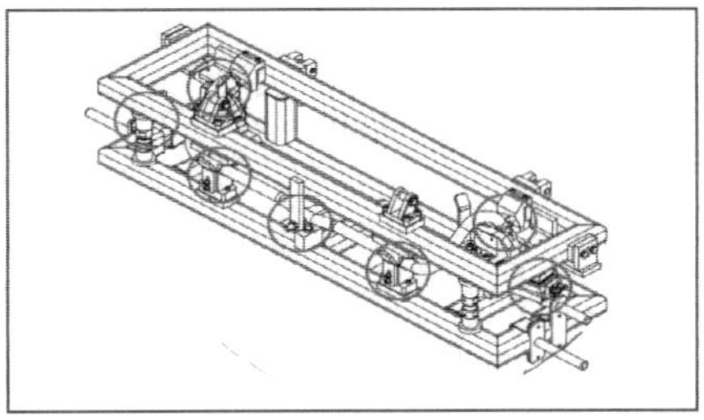

Figure 11. Upper and Lower parts of clamp-quench fixture assembled over stabilizer seven clamp points

To verify the results of deformation, in special in the target critical points, the quality checking fixture shown in Figure 12 was used. Based on the results obtained, a comparative statistical study was developed on a sample of 30 bars as shown in Figure 13a) with quenching without a fixture and 13b) with quenching with a fixture. The lower line of the graph indicates the deviation with respect to the target measurement of the CNC bending process and the upper line with respect to tempering. It is possible to observe a very noticeable improvement in the process using the fixture as seen in the yellow sectors that correspond to the critical zones.

Figure 12. Checking fixture.

Concluding Remarks

Quench press is a very widely used process used in the heat treatment manufacturing of parts and components that require minimal thermal distortion to comply with very tight tolerances regarding planarity, roundness and surface profile such as conical gears, crown gears, transmission shafts and bearing rings with the purpose to assure smooth function, avoid premature wear and reduce noise, therefore in the market exists many quenches press and tooling for quench press manufacturers. Per contrary for suspension parts like the stabilizer bars with more not to tight tolerances regarding surface profile there is no standard process in the industry to systematically reduce or avoid the distortion by quenching process; the parts are all the time quenched in free state and the mechanical rework to reshape them after heat treatment has become a non-standard well accepted process, that is we learned is very high time and cost extensive and a quality detrimental factor for the fatigue life of the part.

There is few to do when the parts are heat treated in batch like heating furnaces and quench tanks as the process conditions vary along the whole batch and the distortion results are random and almost unpredictable. When using a conductive heating machine that runs with one piece flow strategy every single bar is produced with the same parameters and the undesirable effects like overheating of the bents produced by the ovality of the cross section due to tube cold bending process are state of the art controlled by use of local temperature measurement and electrical current heating with special configured recipes that conduce to the bar temperature homogeneity.

The key point is how to design and adequate clamping fixture for quenching that reduce the most the thermal distortion and thus the mechanical rework with press (calibration or straightening). Finite Element Analysis was the tool used to simulate configurations of fixtures with different number of fixing points without performing time and cost extensive prototype trials and CMM or 3D scanner measuring. The final results as evaluated with quality checking fixture (go-no go) are palpable and the calibration process can be reduced from 80-100% percent depending on the bar complexity and the clamping effectiveness of the fixture. Future work appoints for the design of suitable process for solid bars too.

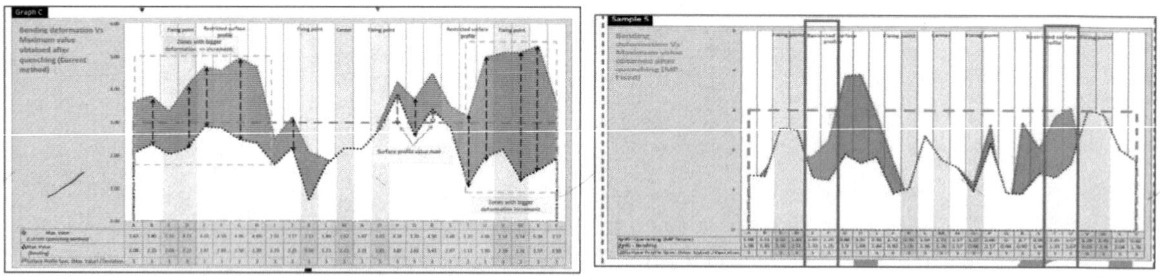

a) b)

Figure 13. Statistical distribution of distortion after CNC bending v.s. after quenching: a) free state, and b) clamp-fixture.

Acknowledgments

Special acknowledgement to TK F& S

References

[1] *Lightweighting Technology Developments, Technical brief No. 6*, March 2017, International Council on Clean Transportation, https://theicct.org>defaults> files>publications, accessed November 2024.

[2] Industrial Heating Equipment Association, *Fundamentals of Industrial Process Heating, 2nd Edition,* 2017, p 19.

[3] Olofsson, A., *Hardening Distortions of Serial Produced Gears*, KTH Royal Institute of Technology, Sweden, 2017, https://kth.diva-portal.org/smash/record.jsf?aq2, accessed October 2024.

[4] Prabhudev, K.H., *Handbook of Heat Treatment of Steels*, Mc Graw-Hill (New Delhi, 1988), p 64-66.

[5] Reardon, A.C., *Press Quenching*, ASM Handbook, Volume 4a, Steel Heat Treating Fundamentals and Processes, J. Dossett and G. Totten, Editors, ASM International, Nov. 2013.

[6] *The Heat Treat Doctor: Fundamentals of Press Quenching by Dan Herring*, Industrial Heating, April 1995, https://secowarwick.com>2018/11 eBook PDF, accessed September 2024.

[7] Fechte-Heinen T. *et al, Quenching and Distortion*, HTM J. Heat Treatm. Mat. Vol. *76* (2021), pp. 390-416, doi:10.1515/htm-2021-0017

[8] *Press Quenching and Effects of Prior Thermal History on Distortion during Heat Treatment,* Reardon A. C., Gear Technology, https://www.geartechnology.com>ext>issues PDF, March, April 2014, accessed November 2024.

[9] Liu, H., *et al, Simulation and Experimental Verification of Die Quenching Deformation of Aviation Carburized Face Gear.* Materials, Vol.16, 690 (2023). doi.org/10.3390/ma16020690

[10] Li, Z, et al, *Distortion Simulation of Bevel Gears Press Quench Hardening Process Using Computer Modeling, Proc 30th ASM Heat Treating Society Conference,* Detroint, MI, October 2019, p 237-244.

[11] *Press quenching tooling design using simulation*, Sims J. Termal Processing, https://thermalprocessing.com/press-quenching-tooling-design-using-simulation/, December 2021, accesed November 2024.

[12] Li, X. *et al*, Analysis of the Distortion of the Bearing Rings after Quenching. Coatings, Vol. 13, 1190 (2023). doi.org/10.3390/

[13] Tewary, U. *et al,* Distortion Mechanisms During Carburizing and Quenching in a Transmission Shaft, ASM International, Vol. 26, (2017), p 4890-4901. doi: 10.1007/s11665-017-2951-9 1059-9495

[14] Fu, M. et al, *Full-cross-section deformation characterization of Cu/Al bimetallic tubes under Rotary-Draw-*Bending based on physics-driven B-spline curves fitting. Materials &Design, Vol. 215, 110493, (2022). doi: 10.1016/j.matdes.2022.220493.

QDE 2025: Proceedings of the 3rd International Conference on
Quenching and Distortion Engineering
May 6–7, 2025, Vancouver, Canada
https://doi.org/10.31399/asm.cp.qde2025p0042

Influence of Tempering Transformation Induced Plasticity (T-TRIP) on the Evolution of Residual Stresses in Laser Heat Treated 50CrMo4 Steel

Niki Nouri, Stefan Dietrich, Volker Schulze
Institute for Applied Materials - Materials Science and Engineering (IAM–WK), Karlsruhe Institute of Technology (KIT), Engelbert-Arnold-Str. 4, 76131 Karlsruhe, Germany

Pierre Lapouge, Justin Dirrenberger
PIMM Lab, Arts et Métiers, CNAM, CNRS UMR 8006, 151 Bd. de l'Hôpital, 75013 Paris, France

Abstract

Laser hardening is an adaptable heat treatment technique recognized for its flexibility, but it also induces significant temperature gradients that result in residual stresses. This study investigates the interaction between tempering processes and the formation of tempering transformation induced plasticity (T-TRIP) in 50CrMo4 steel during laser heat treatment. Various configurations, including single and double laser treatments, were examined along with different initial material states and heat treatment parameters. Comprehensive measurements of hardness and residual stresses were performed, in different directions. The findings demonstrated a notable influence of T-TRIP on the distribution of residual stresses and the mechanical properties of the treated materials. Laser hardening led to the creation of compressive stresses on the surface in the center of the laser, which change to tensile stress at the transition zone to the initial material. The compressive stresses were then shifted to tensile stresses in laser tempered samples because of the intensified volume decrease due to TRIP during cementite precipitation. The findings offer critical insights into optimizing laser heat treatment parameters to control residual stresses and improve material properties, demonstrating the significance of T-TRIP in influencing the outcomes of laser tempering processes in industrial applications. Furthermore, improved simulation would allow for better prediction of the material state after heat treatment, facilitating control over mechanical performance.

Introduction

Tempering Transformation Induced Plasticity

Heat treatments often lead to phase transformations, which may occur intentionally or unintentionally. When phase transformation takes place under the influence of external or residual stresses lower than the yield strength of the softer phase, transformation induced plasticity (TRIP) may occur. There are two known mechanisms leading to this phenomenon: Greenwood-Johnson effect [1] and Magee effect [2]. The first one is based on the volume difference between the existing and forming phase, whereas the latter relates the reorientation of the phases in the stress direction to the plasticity that occurs. While martensitic, bainitic, and austenitic transformations have been extensively studied in relation to TRIP [3], TRIP due to precipitation has only been addressed in recent years [4]. Previous studies have demonstrated the existence of aging and tempering TRIP (T-TRIP) as a result of precipitation of either intermetallic phases in maraging steels [5] or cementite in quenched and tempered (Q&T) steels [6].

Laser Heat Treatment

Laser hardening is a flexible surface heat treatment method that has opened up new possibilities for improving components. It involves rapid heating above the austenitization temperature, followed by a self-quenching process that is fast enough to create martensite on the heat treated surface [7]. These significant temperature gradients lead to the formation of residual stresses. Due to the complexity of predicting the residual stress state, several numerical methods have been developed to simulate them [8–10]. Recently, the complex mechanisms during laser tempering - specifically the influence of different tempering effects, such as T-TRIP, in the heat-affected zone (HAZ) - have also been addressed using the finite element method (FEM) [11].

This Work

This study aims to investigate the influence of transformation-induced plasticity due to precipitation (T-TRIP) on residual stresses following laser heat treatment. The understanding of T-TRIP gained from laboratory-scale investigations will be applied to real-world scenarios. Additionally, the potential for utilizing the residual stresses generated during laser hardening to induce TRIP during the tempering process will be explored. Systematic variation of laser heat treatment parameters will be carried out, followed by comprehensive hardness and residual stress measurements, enabling a thorough analysis of the resulting microstructural and mechanical properties.

Materials and Methods

Material and Sample Geometry

The Q&T steel 50CrMo4 was used in this work. The chemical composition of the material, determined using an emission spectrometer, is listed in Table 1.

The material was initially delivered as rods in the Q&T state before being machined into the required sample geometries. This steel typically undergoes austenitization, followed by rapid quenching to form a martensitic microstructure. A tempering step is then performed to reduce residual stresses and increase ductility through the precipitation of carbides.

Table 1: Chemical composition of the Q&T steel 50CrMo4 in wt-%.

C	Cr	Mo	Mn	Si	Ni	Cu	P	Fe
0.49	1.01	0.20	0.75	0.28	0.20	0.17	0.02	Bal.

The experiments were conducted on steel sheets with the following dimensions: 80 mm in length, 15 mm in width, and 0.6 mm in thickness. The samples were then ground on both sides to remove the oxide layer, resulting in a final thickness of 0.5 mm. A schematic representation of the sample geometry is shown in Figure 1, and a real sample as an example can be found in Figure 2b. The laser heat treated side of the samples is designated as surface A, while the cross section in the middle is labelled as surface B.

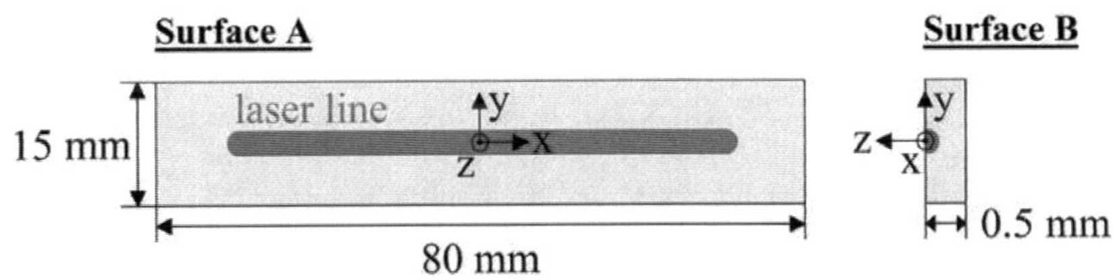

Figure 1: Geometry of the used samples with surface A being the laser heat treated side and surface B the cross section in the middle of the sample.

Heat Treatment

Before the laser heat treatments, two different initial states were created in the Q&T sheet samples using a furnace:

- Furnace hardened (FH): austenitization at 850 °C for 20 min, followed by oil quenching to room temperature.
- Furnace tempered (FT): only tempering at 550 °C for 90 min, followed by air cooling to room temperature.

Two different fixed lasers were used to perform the heat treatments on both sides of the samples. On the front side (surface A), an SPI single-mode fiber laser with a maximum power of 500 W at a wavelength of 1080 nm was used. A DILAS multimode laser diode with a maximum power of 110 W at a wavelength of 980 nm was used on the back surface. A defocused spot size of 2 mm was employed. The sheet samples were positioned at the desired distance from the lasers on a KUKA robotic arm, with movement trajectories programmed for each experiment. A pyrometer was used to track the temperature at the front surface of the sample. Figure 2a shows the setup for laser heat treatments. Table 2 is a list of the parameters, which were varied, along with the corresponding initials used to name each sample.

Mechanical Characterization

The Vickers hardness measurements were performed using a Qness Q10A+ micro hardness tester with an indentation time of 10 s. Due to the size of the areas that needed to be investigated, $HV_{0.1}$ has been used according to the standard DIN EN ISO 6507-1 [12]. All hardness samples were cut in the cross-section perpendicular to the laser line for embedding and polishing prior to the measurements on surface B. Color maps of the hardness were created for a better visualization of the results.

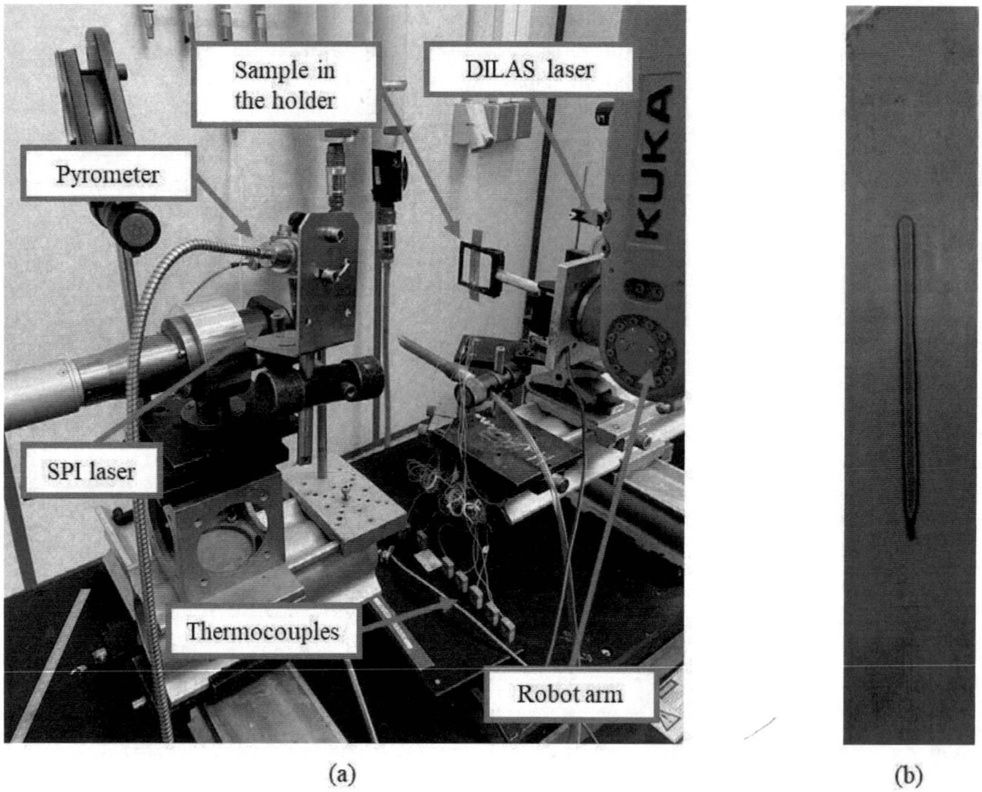

Figure 2: a) The setup of the laser heat treatment including both lasers and the robot arm holding the sample, as well as b) a heat treated sample as an example.

Residual stress measurements were conducted using a DR45 diffractometer (Stresstech GmbH) in the modified χ-configuration according to the DIN EN 15305 standard [13]. The diffractometer was equipped with Cr-Kα radiation at a wavelength of 0.2291 nm and a collimator with a nominal diameter of 1 mm. The ψ-angle range extended from - 45° to + 45°, divided into six equidistant tilt angles in both positive and negative directions. To determine the orientation of the stress state, measurements were taken at ϕ = 0° and 90°, where 0° corresponds to the direction parallel to the laser path (x-axis in Figure 1) and 90° corresponds to the direction perpendicular to the laser path (y-axis in Figure 1). The results were evaluated using the $\sin^2\psi$ method with a cross-correlation peak fitting technique. Residual stresses were measured on heat treated surface A along a path perpendicular to the laser line. Additionally, depth distributions (along z-axis in Figure 1) were obtained by performing electrochemical polishing on surface A after each measurement step, which allowed for material removal without introducing additional residual stresses.

Table 2: Various parameters and the associated initials used during the experiments.

Variables	**Used parameters**
Initial state	Furnace hardened (FH) Furnace tempered (FT)
Laser heat treatment	Laser hardening (LH) Laser tempering (LT) Laser hardening and tempering (LHT)
Laser position	One line in the middle (M) One line at the edge (E) *
Number of lasers	Single laser (SL) Double laser (DL) **
Laser speed	10 mm/s 15 mm/s

* Distance of the center of the laser from the edge was set to 0.5 mm.
** DL heat treatment only in the middle with a laser speed of 15 mm/s.

Results

Determination of laser power

The hardening and tempering temperatures were set to approximately 1000 °C and 500 °C, respectively. For the main studies, it was first essential to determine the laser power required to achieve these temperatures. Due to the influence of various factors, such as laser speed and position, each combination of experimental parameters required a specific laser power to reach the desired heat treatment temperature.

In the pre-studies, several thermocouples were welded to the back surface of the sheets to record temperatures for different laser power settings. The results were compared with a previously developed simulation [11], and the laser power values were optimized accordingly. The necessary laser power for each sample group is listed in Table 3.

Table 3: The laser power used for each sample group, determined during the pre-studies.

Laser heat treatment	Laser position	Number of lasers	Laser speed	Laser Power
LH *	M	DL	15 °C/s	75 W
		SL	10 °C/s	100 W
			15 °C/s	100 W
	E		10 °C/s	100 W
			15 °C/s	100 W
LT *	M	DL	15 °C/s	45 W
		SL	10 °C/s	60 W
			15 °C/s	75 W
	E		10 °C/s	50 W
			15 °C/s	60 W

* LHT samples were heat treated using the corresponding LH and LT laser powers consecutively.

Hardness

The initial hardness of the FH and FT states was measured on at least five samples, yielding results of 816 ± 22 $HV_{0.1}$ for FH and 340 ± 9 $HV_{0.1}$ for FT. All subsequent results are derived from measurements on surface B of the samples. The hardness maps of all SL-M samples are shown in Figure 3.

A laser speed of 10 mm/s results in an almost homogeneous hardness throughout the depth of the sample (Figure 3, upper row). In contrast, a laser speed of 15 mm/s affects only a narrower, triangle-shaped area in approximately the upper half of the sample (Figure 3, lower row) due to the lower laser energy density. The hardened zone of the FT-LH sample reaches a maximum hardness of 895 $HV_{0.1}$ (Figure 3, left column).

After tempering, the hardness in the laser center of the FT-LHT samples decreases to a minimum of 469 $HV_{0.1}$ (Figure 3, middle column). This value is slightly lower than the minimum hardness of 515 $HV_{0.1}$ observed at the center of FH-LT samples after tempering (Figure 3, right column). Notably, FT samples exhibit a distinct transition between the heat treated and initial microstructure due to the hardening process. However, FH samples show a progressive change in the microstructure, with no sharp transition, due to the tempering process.

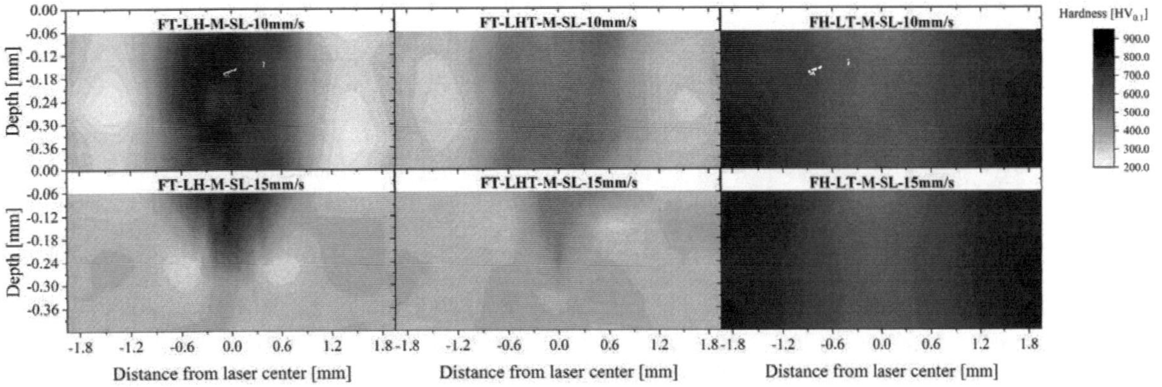

Figure 3: Hardness color maps of the M-SL samples.

Figure 4 shows the hardness maps of all M-DL samples, which exhibit the same transition characteristics as the M-SL samples (Figure 3). Although the laser speeds were set to 15 mm/s, a homogeneous hardness distribution was achieved through the depth of the sample by using both lasers, resulting in comparable outcomes to M-SL samples processed at 10 mm/s.
It is noteworthy that the double-laser FH-LT sample (Figure 4, right column) shows slightly more tempering than the single-laser FH-LT samples (Figure 3, right column), with a minimum hardness of 467 $HV_{0.1}$ at the center of the laser line.

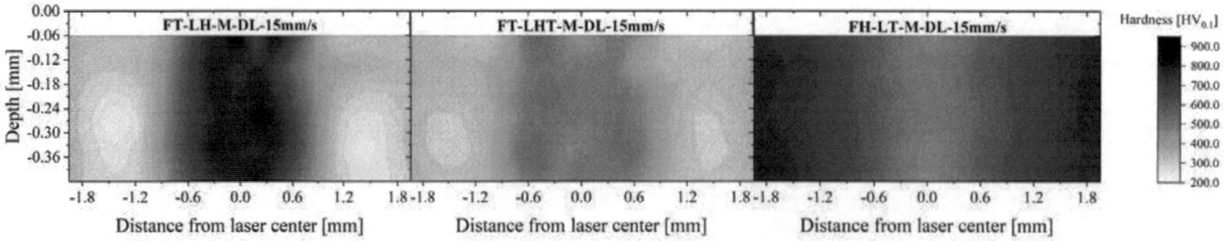

Figure 4: Hardness color maps of the M-DL samples.

Similar to the M samples, hardness maps were also generated for E samples (Figure 5).
All samples exhibit a relatively homogeneous surface at the edge of the side surface. When comparing different scan speeds, it is evident that slower laser speeds result in a larger heat treated zone, extending up to 0.5 mm deeper across the width of the samples.
The transitions between the heat treated and initial zones are similar to those observed in the M samples, with one exception: the FT-LHT-E-SL-10 mm/s sample. In this particular sample, there is a homogeneous tempered zone with a hardness of ~ 450 $HV_{0.1}$ extending up to 0.5 mm from the sample edge. Beyond this point, the hardness increases to ~ 600 $HV_{0.1}$ at 1.25 mm from the edge. After this HAZ, a significant drop in the hardness to the initial value is observed.

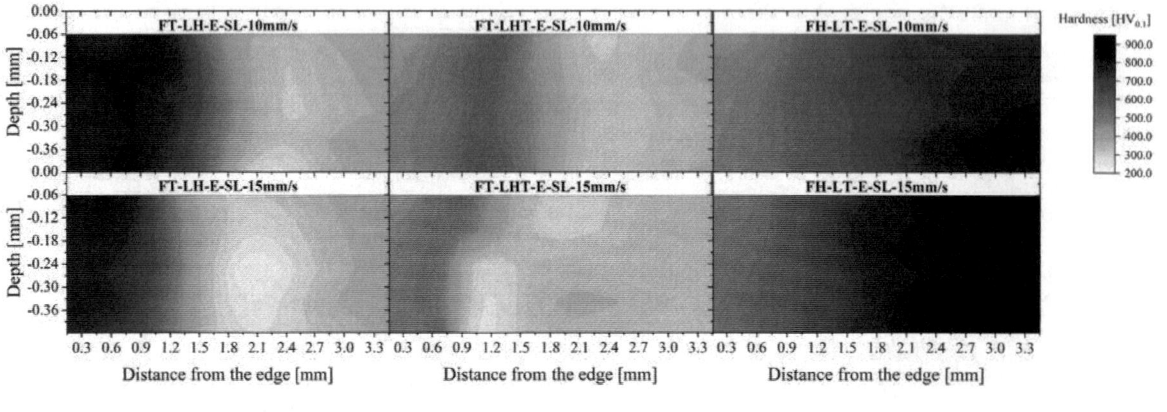

Figure 5: Hardness color maps of the E-SL samples.

Residual stresses
The initial residual stresses of the FH and FT states on surface A were measured. The normal stresses recorded were 31 ± 8 MPa in the 0° direction and 18 ± 11 MPa in the 90° direction, both low enough to be considered negligible. However, furnace hardening resulted in tensile normal stresses of 122 ± 26 MPa in the 0° direction and 140 ± 36 MPa in the 90° direction for the FH samples. Consequently, these samples could not be used as a comparison to the stress-free state, as initially planned.
In the following sections, all curves representing laser-hardened samples are shown in purple, while those for laser-tempered samples are shown in orange. For better visibility, only residual stresses in the 0° direction (parallel to the scan line) are compared, as the stresses in the 90° direction were less significant due to distortion and relaxation effects. The arrows indicate the positions where residual stress depth measurements were performed, which are presented after the surface measurements.

Residual stresses on the laser heat treated surface A

The evolution of normal residual stresses on surface A of M-SL and M-DL samples was measured along a path perpendicular to the laser line. The results are shown in Figure 6 and Figure 7.

A faster laser speed tends to increase stresses in the middle of the laser line due to higher temperature gradients. Compressive stresses were observed only in the FT-LH samples (Figure 6 and Figure 7, left), reaching nearly -200 MPa at the centre of the laser line, regardless of whether the processing was performed with SL-10 mm/s or DL-15 mm/s. However, these compressive stresses transit into tensile stresses, reaching up to 540 MPa in the transition zone between the hardened and initial microstructure (approximately 1 mm from the centre).

The DL sample exhibited a lower maximum stress of up to 430 MPa. Tempering increased these stresses in the FT-LHT samples to 760 MPa for the SL samples (Figure 6, middle) and 640 MPa for the DL sample (Figure 7, middle). The FH-LT samples exhibited a single tensile stress peak in the middle of the laser line, reaching values of ~ 800 MPa (Figure 6 and Figure 7, right).

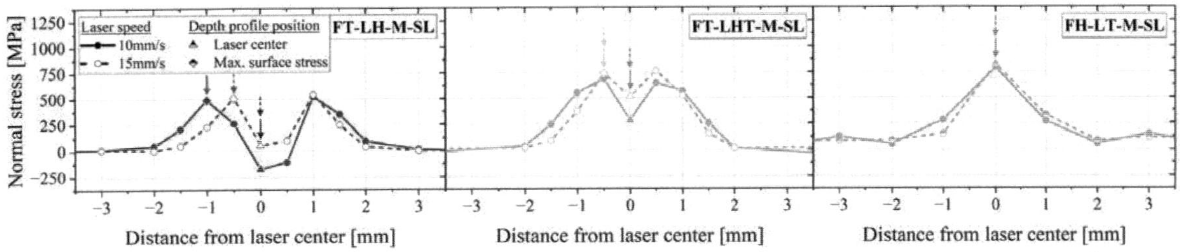

Figure 6: Evolution of the normal residual stresses on the surface A on a path perpendicular to the laser line for M-SL samples in 0° direction.

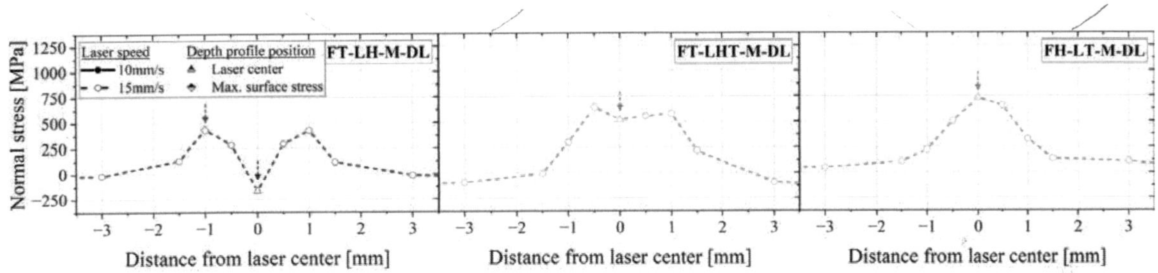

Figure 7: Evolution of the normal residual stresses on the surface A on a path perpendicular to the laser line for M-DL samples in 0° direction.

The same measurements were conducted for surface A of the E-SL samples, with the measurement path starting at the sample edge (Figure 8). The hardened FT-LH samples (Figure 8, left) exhibited compressive stresses at the edge, which transitioned to tensile stresses in the HAZ, reaching up to 490 MPa. The maximum stress was located at 1 mm and 1.5 mm from the edge for the 15 mm/s and 10 mm/s samples, respectively.

Tempering shifted the tensile stress zone closer to the edge, resulting in predominantly tensile stresses within the tempered zone and HAZ. These stresses shifted to compressive values between -100 and -200 MPa in the initial microstructure (Figure 8, middle). The FH-LT samples (Figure 8, right) exhibited a similar progression, but with higher tensile stress values at the edge, reaching ~ 550 MPa, and consisting solely of tensile stresses.

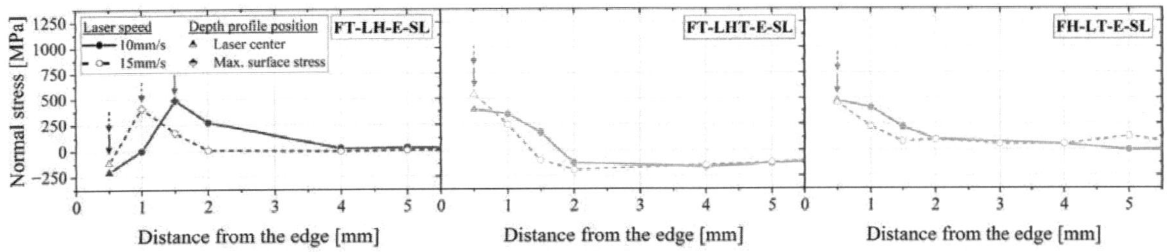

Figure 8: Evolution of the normal residual stresses on the surface A on a path perpendicular to the laser line for E-SL samples in 0° direction.

Residual stresses in the depth
Residual stress measurements in the depth along z-axis were conducted at the center of the laser line, as well as at the location with the maximum surface stress if this was not at the center (positions are marked by arrows in Figure 6, Figure 7 and Figure 8). The depth profiles of the residual stresses for the M-SL and M-DL samples are shown in Figure 9 and Figure 10, respectively.
For the FT-LH-M-SL samples (Figure 9, left), the homogenously hardened sample processed at 10 mm/s exhibits a constant compressive stress through the depth at the laser center. However, for the same sample at the position with the maximum surface stress, as well as the 15 mm/s sample at both positions, the stresses transition to tensile values in the depth, reaching up to 660 MPa. The FT-LH-M-DL sample (Figure 10, left) shows similar behavior to the FT-LH-M-SL-10 mm/s sample.
All FT-LHT samples (Figure 9 and Figure 10, middle) exhibit a similar pattern: the surface residual stresses slightly decrease within the first 0.02 mm of depth, but then significantly increase, reaching up to 930 MPa for the SL samples and 1030 MPa for the DL samples. This increase is much steeper at the laser center compared to the position with the maximum surface stress. Additionally, the maximum stress in the depth occurs at a greater depth for DL samples compared to SL.
FH samples (Figure 9 and Figure 10, right) show a relatively uniform evolution of residual stresses through the depth, maintaining high tensile stress values throughout.

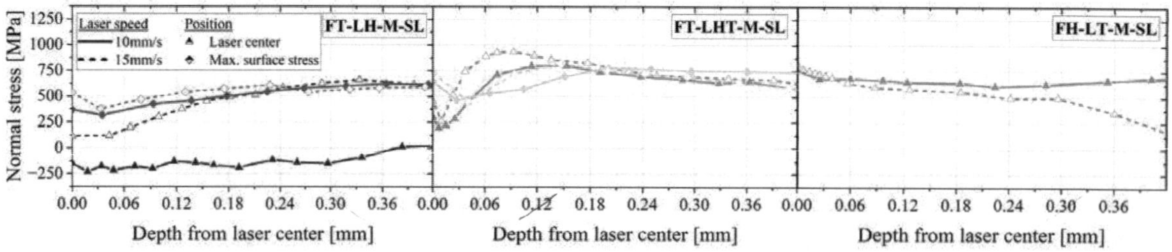

Figure 9: Evolution of the normal residual stresses on the surface B in the depth for M-SL samples in 0° direction.

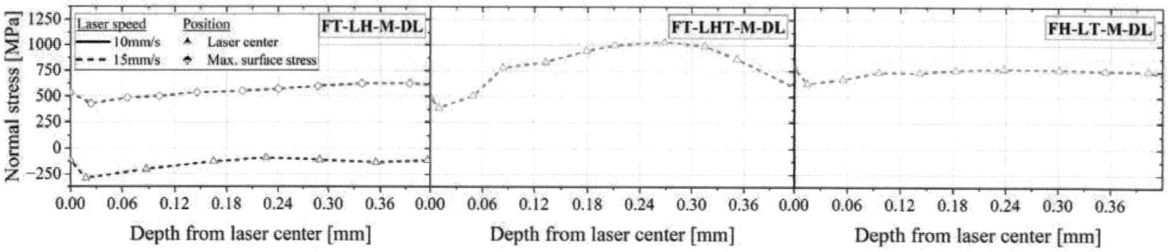

Figure 10: Evolution of the normal residual stresses on the surface B in the depth for M-DL samples in 0° direction.

Figure 11 presents the results of residual stress depth measurements for the E-SL samples. Most samples exhibit relatively constant residual stress values through the depth. An exception is the FT-LHT-E-SL-15 mm/s sample, which initially shifts from tensile to compressive stress, followed by a more pronounced transition back to tensile stresses, reaching up to 430 MPa. Additionally, the FT-LH-E-SL-15 mm/s sample shows a continuous increase in residual stresses with depth, transitioning from compressive to tensile stresses.

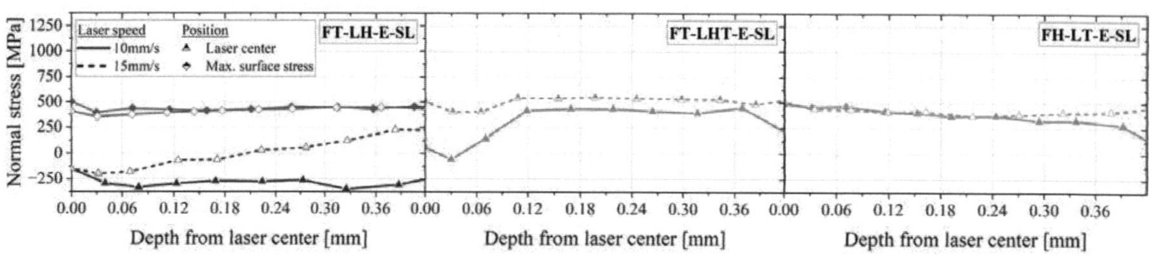

Figure 11: Evolution of the normal residual stresses on the surface B in the depth for E-SL samples in 0° direction.

Discussion

Different laser heat treatments were performed on thin 50CrMo4 sheets, and the evolution of hardness and residual stresses was studied.

Surface hardened samples have led to compressive stress in the center of the laser and tensile stresses at the transition zone to initial material. This corresponds to the typical W-shaped profile over the process zone known from the literature [8]. The evolution of residual stresses leading to this final material state has been described previously [14–16]: At the beginning, thermal expansion causes the surface layer to expand, but the cold surroundings restrict it, creating compressive stresses in the surface and tensile stresses in the surroundings. The stresses rise up to the warm yield strength and are thereby constrained. Upon reaching the austenitizing temperature, the surface transforms to austenite, causing a reduction in volume due to the shift from a BCC to an FCC structure. This contraction lowers compressive stresses and generates tensile stresses. After complete austenite formation, compressive stresses rebuild until the maximum austenitization temperature. During quenching, the surface contracts, but the cold surroundings hinder it, creating tensile stresses limited by the warm yield strength until the martensite start temperature. Martensite formation causes a volume increase and compressive stresses. After martensite formation, final contraction slightly reduces these stresses, and TRIP during martensite formation further reduces compressive stresses.

The samples containing the residual stresses were then tempered at ~ 500 °C, which leads to a shift of the residual stresses from compressive to tensile stresses. The main reason for this change at this tempering temperature is the tempering TRIP, as demonstrated by simulations [11]. In fact, the higher the present residual stresses during the phase transformation, the larger is the occurring tempering TRIP and therefore the total strain [4–6]. During cementite precipitation, a volume decrease takes place, causing the tempered zone to contract. This contraction is restricted by the cold surroundings, resulting in tensile stresses. Given the unfavorable nature of tensile stresses concerning fatigue and crack propagation properties, it is crucial to understand and predict their formation during any heat treatment process.

Further simulation studies and their comparison with experimental results could pave the way for comprehensive and accurate heat treatment simulations, facilitating the prediction of material states under various heat treatment conditions.

Conclusions

Influence of different laser heat treatments on the resulting hardness distribution, as well as the occurring residual stresses in different directions, has been studied in this work. Many heat treatment parameters such as laser speed and number of lasers were varied to gain a better understanding of their impact on the material state. In summary, following key findings were identified:

- The resulting hardness is highly dependent on laser heat treatment parameters, such as processing speed and geometry-dependent thermal distribution.
- For thin sheets, a slower single laser heat treatment yields similar hardness and residual stress results as a faster dual laser heat treatment.
- Surface laser hardening generates compressive stresses at the laser center and tensile stresses near the transition zone to the initial microstructure.
- Laser tempering the same samples promotes the precipitation of cementite under stress, triggering TRIP. This effect increases total strain, causing a pronounced volume decrease and leading to tensile stresses.

For future advancements, understanding the material state after heat treatment is essential. Accurate integration of TRIP effects into simulations will enhance predictive capabilities and process optimization.

Acknowledgments

The authors gratefully acknowledge the financial support of German Research Foundation (DFG) with the project number 428958028. Moreover, we would like to thank the Department of Science and Technology of the French Embassy in Germany for granting the Procope Mobility fund, as well as the Franco-German University for granting the mobility fund of French-German Institute for Industry of the Future, which enabled the necessary research stays for this franco-german research collaboration. Additionally, we express our gratitude to our colleague Christian Krämer and our student research assistants Elena Hillenmeyer and Yifan Lu for their support in conducting the experiences.

References

[1] G. W. Greenwood and R. H. Johnson, "The deformation of metals under small stresses during phase transformations", *Proceedings of the Royal Society of London. Series A. Mathematical and Physical Sciences*, Vol. 283, No. 1394 (1965), p 403–422. https://doi.org/10.1098/rspa.1965.0029

[2] C. L. Magee, "Transformation kinetics, microplasticity and aging of martensite in iron 21 Ni," PhD Thesis, Carnegie Melton University, Pittsburg, (1967).

[3] F. D. Fischer, *et al.*, "Transformation-Induced Plasticity (TRIP)", *Applied Mechanics Reviews*, Vol. 49, No. 6 (1996), p 317–364. https://doi.org/10.1115/1.3101930

[4] D. Kaiser, *et al.*, "A dilatometric study on the influence of compressive stresses on the tempering of martensitic AISI 4140 steel - Evidence of transformation induced plasticity during cementite precipitation", *Materials Science and Engineering: A*, Vol. 705 (2017), p 114–121. https://doi.org/10.1016/j.msea.2017.08.028

[5] N. Nouri, *et al.*, "Transformation Induced Plasticity during Precipitation Strengthening of Maraging Steels," presented at the IFHTSE 2024, Cleveland, Ohio, USA, (2024), p 50–56. https://doi.org/10.31399/asm.cp.ifhtse2024p0050

[6] N. Nouri, *et al.*, "Transformation induced plasticity (TRIP) due to precipitation during continuous tempering of quenched and tempered steels", *Journal of Materials Research and Technology*, (2024), p S223878542402845X. https://doi.org/10.1016/j.jmrt.2024.12.042

[7] P. D. Babu, *et al.*, "Laser surface hardening: a review", *International Journal of Surface Science and Engineering*, Vol. 5, 41398 vols., No. 2/3 (2011), p 131. https://doi.org/10.1504/IJSURFSE.2011.041398

[8] D. Kiefer, *et al.*, "Experimental and Simulative Studies on Residual Stress Formation for Laser-Beam Surface Hardening*", *HTM Journal of Heat Treatment and Materials*, Vol. 74, No. 1 (2019), p 23–35. https://doi.org/10.3139/105.110374

[9] E. Liverani, *et al.*, "A complete residual stress model for laser surface hardening of complex medium carbon steel components", *Surface and Coatings Technology*, Vol. 302 (2016), p 100–106. https://doi.org/10.1016/j.surfcoat.2016.05.066

[10] V. Kostov, "Untersuchungen zur zeitaufgelösten Spannungsentwicklung und Eigenspannungsentstehung beim Laserstrahlstandhärten am Beispiel des Stahls 42CrMo4" (2014), . https://doi.org/10.5445/IR/1000043357

[11] P. Schüßler, *et al.*, "Laser surface hardening: A simulative study of tempering mechanisms on hardness and residual stress", *Computational Materials Science*, Vol. 221 (2023), p 112079. https://doi.org/10.1016/j.commatsci.2023.112079

[12] "DIN EN ISO 6507-1, Metallische Werkstoffe - Härteprüfung nach Vickers - Teil 1: Prüfverfahren," No. DIN EN ISO 6507-1. 2018). https://doi.org/10.31030/2778746

[13] "DIN EN 15305:2009-01, Zerstörungsfreie Prüfung_- Röntgendiffraktometrisches Prüfverfahren zur Ermittlung der Eigenspannungen; Deutsche Fassung EN_15305:2008," DIN Media GmbH. https://doi.org/10.31030/1425472

[14] M. Schwenk, *Numerische Modellierung der induktiven Ein- und zweifrequenzrandschichthartung*. Karlsruher Institut für Technologie, (2014).

[15] M. Habschied, *et al.*, "Fertigung und Eigenspannungen", *HTM Journal of Heat Treatment and Materials*, Vol. 70, No. 3 (2015), p 111–121. https://doi.org/10.3139/105.110261

[16] V. Schulze, *et al.*, "Residual Stresses after Quenching," in *Quenching Theory and Technology*, CRC Press, 0ed., 2010), p 60. https://doi.org/10.1201/9781420009163.

QDE 2025: Proceedings of the 3rd International Conference on
Quenching and Distortion Engineering
May 6–7, 2025, Vancouver, Canada
https://doi.org/10.31399/asm.cp.qde2025p0051

On the Influence of Geometrical Features on Residual Stresses in Additively Manufactured 316L for Lightweight Engineering

Christian Krämer, Volker Schulze, Stefan Dietrich
Institute for Applied Materials – Materials Science and Engineering (IAM-WK), Karlsruhe Institute of Technology (KIT), Karlsruhe, Baden-Württemberg, Germany

Abstract

Additive manufacturing (AM), particularly the Laser Beam Powder Bed Fusion (PBF-LB) process, is increasingly utilized in both scientific research and industrial applications due to its ability to produce complex and customized metal parts. The possibility of AM to create components with high design freedom, offers a considerable potential for lightweight engineering. This potential can be further enhanced by combining the AM process with topology optimization during the design of components. However, the PBF-LB process introduces unique thermal histories to the manufactured parts because of the localized heat input. This results in a distinctive residual stress distribution and complex microstructures. These play a critical role in the fatigue performance of components. This study aims to provide a deeper understanding of the relationship between geometrical features and residual stresses. Therefore, typical geometries, as they often result from topology optimization algorithms, were identified and manufactured from 316L stainless steel with the LB-PBF process. For these geometries, like Y-struts or different node shapes, the near-surface residual stress distributions were characterized. The AM-process also has an impact on the mechanical properties of components. Due to the directed buildup of parts, texture effects may occur. These can also have an effect on the macroscopic mechanical properties of components. Especially in lightweight engineering, these effects need to be considered in the design process. In this study, the texture of the material was determined with non-destructive ultrasound phase spectroscopy as well as with conventional mechanical methods.

Introduction

Additive manufacturing (AM) is a constantly growing field in science and industry. The application ranges from rapid prototyping over individualized parts to serial production. For the manufacturing of metals, PBF-LB is one of the most used techniques. For this process, a metal powder is raked on a build plate with a coater and the contour of the powder layer is consolidated with a laser. Afterwards, the build plate is lowered and a new layer of powder is applied. The process takes place in an inert gas atmosphere to prevent chemical reactions of the molten metal, like oxidation. The dimensions of the melt pool are usually much bigger than the layer height and the distance between two laser lines (hatch distance). Therefore, the material melts and solidifies multiple times until it reaches its final microstructural state **[1], [2]** and even after final solidification, a cyclic annealing is applied to the substrate, dependent on the position of the considered volume.

In general, AM offers a great potential for lightweight engineering. Because of the high design freedom, parts can get manufactured close to the ideal load-adapted shape. This shape is often found by topology optimization algorithms. In **[3]** Liu et al. give an overview over the possibilities and challenges of using topology optimization to optimize parts for AM. They are pointing out that the unique residual stress state of AM parts is a big challenge. This residual stress state is mostly dependent on the local thermal history and the post-treatment of the manufactured parts. Usually, tensile residual stresses occur close to the surface when no post-processing is performed. For 316L, the tensile stresses are in a range of 200 to 500 MPa **[4]**. In the core of additively manufactured parts compressive residual stresses occur **[5]**. The main influence parameters on the residual stress distribution are the structural density and the energy density. However, the structural density is dependent on the energy density and for printing parameter, which leads to a high structural density, the influence of the energy density is negligible **[6]**. Furthermore, the part geometry has an influence on the residual stress state. It is shown for example that interior corners have near-zero residual stresses **[7]**.

In industry, components are usually sandblasted. This process step removes loose powder and increases the surface quality by lowering the roughness. The blasting process itself induces compressive residual stresses in the surface regions. These compression residual stresses overlay the tensile residual stresses of the process **[8]**. Moreover, the blasting process has an influence on the near surface hardness and microstructure, as well as the surface roughness **[9]**. Due to the industrial relevance, this study focuses on sand-blasted material states. The macrostructure of additively manufactured 316L samples is dominated by the cellular structure of the melt pools. On a microstructural level, coarser structures with columnar growth and finer, equiaxed dendritic structures are observable **[10]**. Furthermore, pores are a common characteristic of the microstructure of AM parts. Typically, relative densities of >99% are standard in

industry. But even porosities less than 1 % can have a big influence on the mechanical properties of components. Especially, because of the uneven distribution of the pores **[11] [12]**.
As shown before, there are many examinations about the influence factors on the residual stress distribution in additively manufactured parts. The part geometry is one of which. But there are only a few investigations about the stress distribution in complex parts, which often occur after a topology optimization. Therefore, this study focuses on the residual stress distribution in these geometries and the interaction with mechanical properties.

Materials and Methods

For this study, samples manufactured from 316L (1.4044) powder with a chemical composition according to Table 1 were used. For the definition of the sample shapes, topology optimizations with different boundary conditions were performed for a representative component. Out of the topology optimized geometry, characteristic substructures, which occurred more frequently, were identified. These geometrical features were changes in the build angle, Y-Struts, changes in the diameter and holes. All these geometrical features were isolated and modeled individually, as shown in Figure 1.

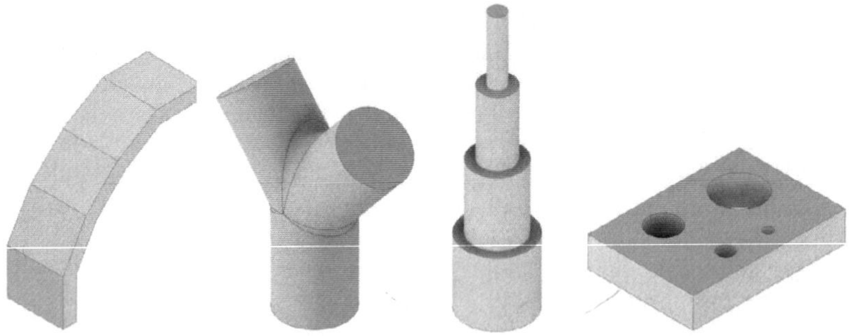

Figure 1 - Isolated geometrical features. From left to right: changes in build angle, Y-strut, changes in Diameter, and holes.

The samples were manufactured with the PBF-LB process. After manufacturing, the samples were sandblasted to remove loose powder and achieve a better surface quality.

Table 1 - Chemical composition

Element	Fe	C	Cr	Mn	Mo	Ni	Si
Content in %	Balance	0.3	16-18	2.0	2.0-3.0	10.5-14	1.0

Ultrasound phase spectroscopy is a non-destructive testing method to determine the propagation speed of ultrasonic waves through a material. A sending and a receiving ultrasonic probe are applied to the sample. By knowing the dimensions of the sample and the wavelength, the phase difference can be measured for a variety of frequencies. By the determination of the slope of the phase difference–frequency plot, the group velocity can be determined. This group velocity is dependent on the elastic constants in the measured direction. The elastic constants C are calculated according to

$$C_{ii} = \rho U^2$$

where ρ is the density and **U** is the group velocity in the considered direction. For a calculation of all elastic constants, measurements of longitudinal and transversal waves are necessary in the investigated direction as well as with a rotation of 45°. From the elastic constants, the elastic moduli can be calculated **[13], [14]**. For a transverse isotropic system, where the z-direction is the symmetry axis, the in-plane and out-of-plane Young modulus is calculated as follows **[15]**:

$$E_x = E_y = (C_{11} - C_{12})(C_{11}C_{33} + C_{12}C_{33} - 2C_{13}C_{13})/(C_{11}C_{33} - C_{13}C_{13})$$

$$E_z = C_{33} - 2C_{13}C_{13}/(C_{11} + C_{12})$$

The 1-direction is equivalent to the x-direction, 2 to y-direction and 3 is equivalent to z, which correspond to the building direction.

The Residual stress analyses were conducted using a DR45 diffractometer from Stresstech GmbH with Mn-Kα-radiation (Wavelength = 0,2103 nm). The device was built up in the modified χ−configuration according to DIN EN 15305 **[16]**. The X-ray beam was masked using a collimator with a nominal diameter of 1 mm. The ψ-range was -45° to +45° divided in six tilt angles for the positive and the negative direction, which were chosen to be equidistant for $\sin^2\psi$. To get information about the orientation of the stress state, measurements for $\phi = 0°$ and 90° were performed, where the 0° describes the measurements parallel to the building direction. For the evaluation of the stresses, the $\sin^2\psi$ method was used with a cross-correlation peak fit. Depth profiles were recorded by performing electrochemical polishing after the stress measurement for each depth step. This allows the removal of material without inducing new residual stresses.

Microstructural analyses were carried out with polished and etched samples using an inverted light microscope from Carl Zeiss Microscopy GmbH (Model: Axiovert 200). For etching, V2A-stain was used. The analysis was performed for thick-walled samples.

Results

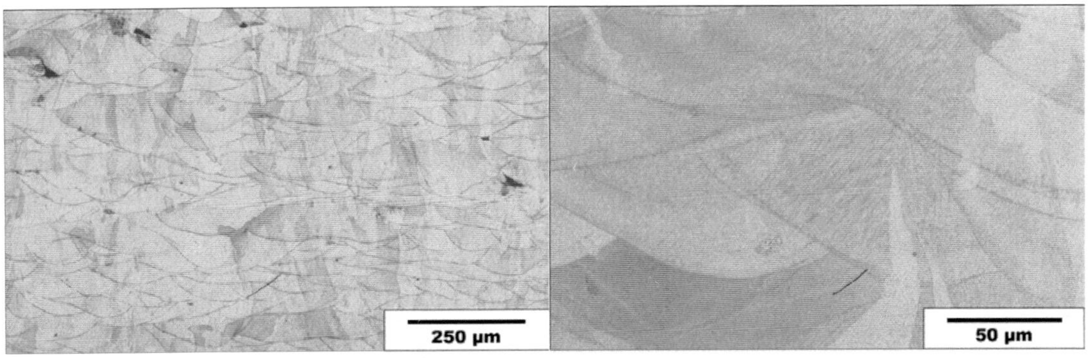

Figure 2 - Macro- (left) and microstructural (right) images of the thick-walled samples.

Microstructural results for thick -walled samples are displayed in Figure 2 at different magnifications. A characteristic cellular macrostructure is visible, which is formed by the boundaries of the melt pools. Because the melt pools are bigger than the layer height and hatch distance, the melt pools are overlapping. Within the cellular structure, fine columnar grains are observable. These seem to be oriented orthogonal to the melt pool boundaries, as also shown in [17].

Because the macro- and the microstructure show a repeating arrangement in building direction, it is also expected that the mechanical properties show an anisotropic behavior. To confirm this assumption, non-destructive ultrasound phase spectroscopy measurements were performed. These enable a direction-dependent determination of the Young Modulus. Ultrasound phase spectroscopy showed a slightly anisotropic material behavior toward the printing direction. The Young Modulus in printing direction is 176 GPa and orthogonal to the printing direction 170 GPa. This represents a difference of 4%. Between the two examined directions orthogonal to the printing direction, no significant differences were measured. This leads to the conclusion, that a transverse isotropic material state is present. The elastic constants are sown in Table 2.

Table 2 - Elastic constants measured with ultrasound phase spectroscopy

Constant	C_{11}	C_{22}	C_{33}	C_{44}	C_{55}	C_{66}	C_{12}	C_{13}
[GPa]	261	260	246	82	84	64	130	122

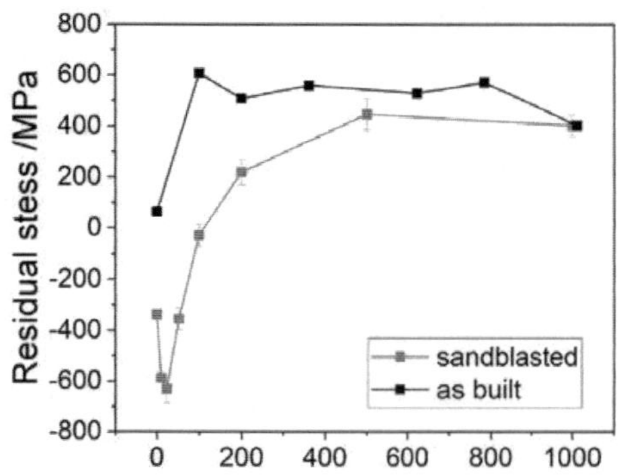

Figure 3 - Residual stress depth profile for a as-built and a sand blasted sample

The near surface residual stresses are mainly influenced by the blasting process, which takes place after the manufacturing. This leads to compression stresses close to the surface, with a maximum around 10-30 µm underneath the surface. In bigger depths, the stress state changes to tensile stress. The transition from compressive to tensile stresses is located around 100-200 µm underneath the surface. A characteristic depth profile is shown in Figure 2. There is also comparison to a part, which is not blasted. The as-built sample is stress-free on the surface and shows tensile stresses up to the measured depth of 1000 µm. Up from a depth of approximately 500 µm the stresses of the as-built sample and the sand blasted sample are nearly identical. So, it is concluded, that the influence of the sand blasting process does not exceed the depth of 500 µm.
As an exemplary evaluation of the surface residual stresses, the mappings on the Y-struts are considered below.

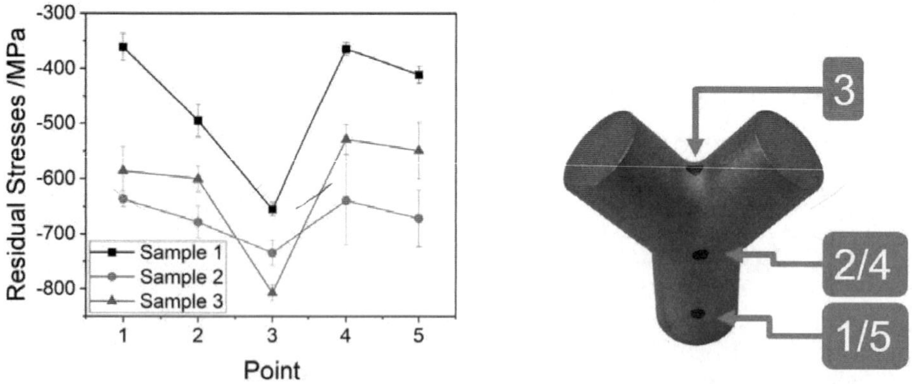

Figure 4 - Residual stress surface mapping on the Y-Strut and a picture of the sample with the measuring points.

The surface residual stresses mapping for the Y-strut is shown in Figure 4. For the three examined samples a similar distribution was measured. For the point directly in the center of the branching (Point 3), the compression residual stresses are higher compared to the stresses on the beam. To get a deeper understanding of the dependence of the geometry on the stress distribution underneath the surface, stress profiles were measured for different build angles.

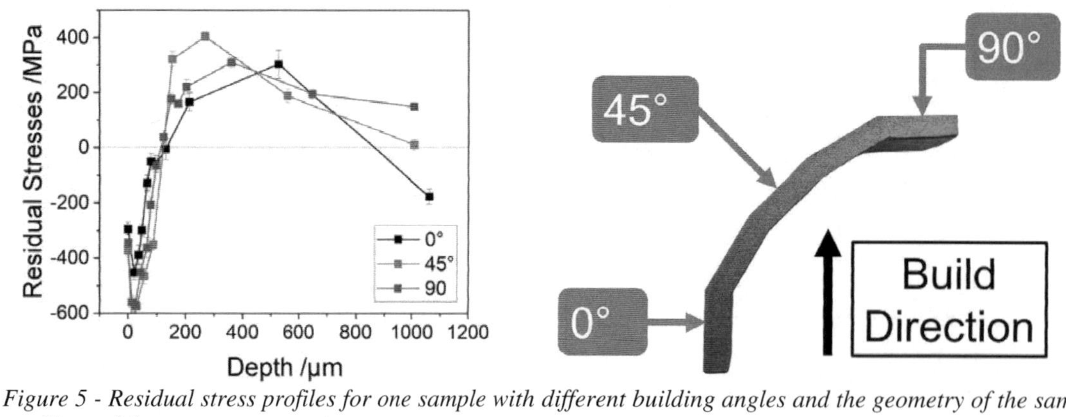

Figure 5 - Residual stress profiles for one sample with different building angles and the geometry of the sample with markings of the measurement points

These measurements of residual stress depth profiles for a sample with different build angles are shown in Figure 5. For measurements closer than 800 µm to the surface, no clear differences are visible. As shown before, for measurements closer than 500 µm to the surface, the stress distribution is mainly dependent on the quality of the blasting process, which should be independent of the build angle. For measurements in a depth of 1000 µm, the three investigated building directions show different behavior. For a vertical sample (0°) compressive stresses are present. The 45° sample is nearly stress-free and on the horizontal sample (90°) tensile stresses were measured.

Discussion

The microstructural examination showed results, that are consistent with the ultrasound measurements. The macrostructure shows a cellular arrangement, which is defined by the melt pool boundaries. On a finer level, columnar grains are visible. These columnar grains are oriented orthogonal to the melt pool boundaries, corresponding to the direction of the maximum temperature gradient within the melt pool. Because a large amount of the solidified melt pool is cyclically remelted, in the end only the melt pool boundaries with an angle close to 90° to the built direction stay in their state. Therefore, most of the columnar grains are oriented close to parallel to the build direction. This orientation of these microstructural features together with the transversal isotropic crystallographic orientation, which forms due to the directional solidification as shown by [18], is strong evidence for the transversal isotropic material behavior with a 4% higher Young Modulus in printing direction.

Regarding the residual stresses, the influence of the blasting process overlays most of the effect from the manufacturing process. The depth profile is similar to the well-known profile for different peening processes. This profile with compression stresses close to the surface leads to an improvement in fatigue life. This is due to the improved surface quality on the one hand and the compression stresses, which inhibit the crack initiation. This usually takes place at or close to the surface. Regarding the depth of the influence of the sand blasting process, a depth of approximately 500 µm was found. From there on, the blasting process has no influence and the residual stress state is mainly influenced by the manufacturing process. In literature there are no measurements of residual stress depth profiles for 316L after sand blasting. In general, a lower depth influence is expected. Similar investigations for shot peening [19] show an analogous profile, but the intensity of the spherical metal shot is much higher. Therefore, the maximum in compression stress is located deeper in the material and the influence reaches further than the influence of the sandblasting process.

With regard to the geometry dependence of the residual stresses, two effects were shown:

Firstly, as analyzed before, the blasting process is mainly influencing the residual stress state on and close to the surface. So, the differences of the surface residual stresses on the Y-strut samples originate probably from the geometry dependence of the blasting process. The tapered form could lead to a focusing effect of blasting particles that hit the surface in the branch center. This could lead to the measured increased compression stresses. In [20] the authors investigated the blasting process for corners and showed that accumulations of blasting material reduce the effect of the process. But the investigated are not comparable, because the Y-strut gives no opportunity for sand to accumulate.

The second notable effect relates to the influence of the build angle on the residual stresses. These were measured in a depth of approximately 1000 µm, where the sandblasting has no effect. As shown in Figure 5, the stresses shift from tensile on a horizontal surface over a stress-free state for 45° to compression stresses for the vertical part of the sample. This could be a result of the dependence of the residual stresses on the length of the laser path within a layer. Because the sample has a rectangular shape (20 x 5 mm), which was rotated stepwise. The average scanning vector is shorter for the upright printed part of the sample. For the horizontal part of the sample the average scanning vector is the longest, because the cross-section projected in the printing direction is the biggest. Moreover, it is nearly quadratic, which also leads to a longer average scanning vector. As shown by Mercelis and Kruth [21], the formation of residual stresses is mainly due to the local shrinkage from cooling effects. This local cooling depends on the length of the scanning vector. For longer vectors it takes much more time for the laser to heat up a spot again by hatching the adjacent path, which also leads to a heating of the material nearby. It was also theoretically shown by Mercelis and Kruth, that these effects lead to compressive stresses in a depth of 1-2 mm for samples printed with shorter building angles and tensile stresses for the samples with longer laser paths. This matches with the results from the measurements shown in Figure 5.

For future investigations, measurements with samples that are not sandblasted could help separate the influences on the formation of the residual stresses. Regarding the influence of the build angle on the residual stresses, samples with equally long scanning vectors for all orientations could be manufactured and measured.

Conclusions

In this study, investigations on the different influential factors in the build-up of residual stresses on samples manufactured from 316L via PBF-LB were presented. The main conclusions are:

- The commonly used sand blasting to clean and increase the surface quality of additively manufactured structures induces a residual stress profile with compression stresses at the surface and tensile stresses in a depth greater than 100-200 µm. The influence of the sandblasting process reaches a depth of 500 µm. Deeper, the thermally induced residual stresses of the manufacturing process are dominant.

- The local geometry of samples and parts can influence the result of the blasting process. Nooks, for example, can have an amplifying effect by concentrating the flow of sand particles.
- Regarding the build angle, an influence of the length of the laser path was shown. Especially in a depth of around 1 mm underneath the surface, shorter average scanning vectors lead to lower residual stresses or compression stresses.

The next steps of this investigation are going to focus on the as-built state. Even if the blasted state is for bigger interest in industry, the as-built state could give more information about the formation mechanisms of stresses. Moreover, the influence of the residual stress state on the mechanical properties and fatigue behavior will be investigated.

Acknowledgments

The authors thanks Axel Meier from PMT GmbH for the manufacturing of the samples as well as Robert Dannecker of Tecosim GmbH and Sven Lenhardt from IPEK, KIT for the identification of the sample geometries via topology optimization. These studies were funded by the "Technologietransfer-Programm Leichtbau" of the German Federal Ministry for Economic Affairs and Climate Action.

References

[1] J. Irwin, E. W. Reutzel, P. Michaleris, J. Keist, and A. R. Nassar, "Predicting Microstructure From Thermal History During Additive Manufacturing for Ti-6Al-4V," *J. Manuf. Sci. Eng.*, vol. 138, no. 11, Nov. 2016, doi: 10.1115/1.4033525.

[2] P. Promoppatum *et al.*, "Numerical modeling and experimental validation of thermal history and microstructure for additive manufacturing of an Inconel 718 product," *Prog. Addit. Manuf.*, vol. 3, no. 1, pp. 15–32, Jun. 2018, doi: 10.1007/s40964-018-0039-1.

[3] J. Liu *et al.*, "Current and future trends in topology optimization for additive manufacturing," *Struct. Multidiscip. Optim.*, vol. 57, no. 6, pp. 2457–2483, Jun. 2018, doi: 10.1007/s00158-018-1994-3.

[4] I. Yadroitsev and I. Yadroitsava, "Evaluation of residual stress in stainless steel 316L and Ti6Al4V samples produced by selective laser melting," *Virtual Phys. Prototyp.*, vol. 10, no. 2, pp. 67–76, Jan. 2015, doi: 10.1080/17452759.2015.1026045.

[5] B. Ahmad, S. O. van der Veen, M. E. Fitzpatrick, and H. Guo, "Residual stress evaluation in selective-laser-melting additively manufactured titanium (Ti-6Al-4V) and inconel 718 using the contour method and numerical simulation," *Addit. Manuf.*, vol. 22, pp. 571–582, Jan. 2018, doi: 10.1016/j.addma.2018.06.002.

[6] T. Simson, A. Emmel, A. Dwars, and J. Böhm, "Residual stress measurements on AISI 316L samples manufactured by selective laser melting," *Addit. Manuf.*, vol. 17, pp. 183–189, Jan. 2017, doi: 10.1016/j.addma.2017.07.007.

[7] A. S. Wu, D. W. Brown, M. Kumar, G. F. Gallegos, and W. E. King, "An Experimental Investigation into Additive Manufacturing-Induced Residual Stresses in 316L Stainless Steel," *Metall. Mater. Trans. A*, vol. 45, no. 13, pp. 6260–6270, Jan. 2014, doi: 10.1007/s11661-014-2549-x.

[8] S. Santa-aho *et al.*, "Additive Manufactured 316L Stainless-Steel Samples: Microstructure, Residual Stress and Corrosion Characteristics after Post-Processing," *Metals*, vol. 11, no. 2, Art. no. 2, Feb. 2021, doi: 10.3390/met11020182.

[9] M. Sugavaneswaran, A. V. Jebaraj, M. D. B. Kumar, K. Lokesh, and A. J. Rajan, "Enhancement of surface characteristics of direct metal laser sintered stainless steel 316L by shot peening," *Surf. Interfaces*, vol. 12, pp. 31–40, Sep. 2018, doi: 10.1016/j.surfin.2018.04.010.

[10] M. Marya, V. Singh, S. Marya, and J. Y. Hascoet, "Microstructural Development and Technical Challenges in Laser Additive Manufacturing: Case Study with a 316L Industrial Part," *Metall. Mater. Trans. B*, vol. 46, no. 4, pp. 1654–1665, Aug. 2015, doi: 10.1007/s11663-015-0310-5.

[11] S. M. Yusuf, Y. Chen, R. Boardman, S. Yang, and N. Gao, "Investigation on Porosity and Microhardness of 316L Stainless Steel Fabricated by Selective Laser Melting," *Metals*, vol. 7, no. 2, Art. no. 2, Feb. 2017, doi: 10.3390/met7020064.

[12] A. Hamada, M. Jaskari, T. Gundgire, and A. Järvenpää, "Enhancement and underlying fatigue mechanisms of laser powder bed fusion additive-manufactured 316L stainless steel," *Mater. Sci. Eng. A*, vol. 873, p. 145021, May 2023, doi: 10.1016/j.msea.2023.145021.

[13] A. Wanner, "Elastic modulus measurements of extremely porous ceramic materials by ultrasonic phase spectroscopy," *Mater. Sci. Eng. A*, vol. 248, no. 1, pp. 35–43, Jun. 1998, doi: 10.1016/S0921-5093(98)00524-3.

[14] S. Czink, S. Dietrich, and V. Schulze, "Ultrasonic evaluation of elastic properties in laser powder bed fusion manufactured AlSi10Mg components," *NDT E Int.*, vol. 132, p. 102729, Jan. 2022, doi: 10.1016/j.ndteint.2022.102729.

[15] W. C. Van Buskirk, S. C. Cowin, and R. N. Ward, "Ultrasonic Measurement of Orthotropic Elastic Constants of Bovine Femoral Bone," *J. Biomech. Eng.*, vol. 103, no. 2, pp. 67–72, May 1981, doi: 10.1115/1.3138262.

[16] "DIN EN 15305:2009-01, Zerstörungsfreie Prüfung_- Röntgendiffraktometrisches Prüfverfahren zur Ermittlung der Eigenspannungen; Deutsche Fassung EN_15305:2008," Beuth Verlag GmbH, Berlin, 2009. doi: 10.31030/1425472.

[17] Y. Zhong, L. Liu, S. Wikman, D. Cui, and Z. Shen, "Intragranular cellular segregation network structure strengthening 316L stainless steel prepared by selective laser melting," *J. Nucl. Mater.*, vol. 470, pp. 170–178, Mar. 2016, doi: 10.1016/j.jnucmat.2015.12.034.

[18] Y.-D. Im, K.-H. Kim, K.-H. Jung, Y.-K. Lee, and K.-H. Song, "Anisotropic Mechanical Behavior of Additive Manufactured AISI 316L Steel," *Metall. Mater. Trans. A*, vol. 50, no. 4, pp. 2014–2021, Jan. 2019, doi: 10.1007/s11661-019-05139-7.

[19] T. Gundgire, T. Jokiaho, S. Santa-aho, T. Rautio, A. Järvenpää, and M. Vippola, "Comparative study of additively manufactured and reference 316 L stainless steel samples – Effect of severe shot peening on microstructure and residual stresses," *Mater. Charact.*, vol. 191, p. 112162, Sep. 2022, doi: 10.1016/j.matchar.2022.112162.

[20] J. Badreddine, E. Rouhaud, M. Micoulaut, and S. Remy, "Simulation of shot dynamics for ultrasonic shot peening: Effects of process parameters," *Int. J. Mech. Sci.*, vol. 82, pp. 179–190, May 2014, doi: 10.1016/j.ijmecsci.2014.03.006.

[21] P. Mercelis and J. Kruth, "Residual stresses in selective laser sintering and selective laser melting," *Rapid Prototyp. J.*, vol. 12, no. 5, pp. 254–265, Jan. 2006, doi: 10.1108/13552540610707013.

QDE 2025: Proceedings of the 3rd International Conference on
Quenching and Distortion Engineering
May 6–7, 2025, Vancouver, Canada
https://doi.org/10.31399/asm.cp.qde2025p0058

Characterization of the sintered steel Astaloy CrA for the simulation of induction surface hardening of small-modulus gear wheels

Moritz Klug, Sandipan Sen, Stefan Dietrich and Volker Schulze
Institute for Applied Materials – Materials Science and Engineering (IAM-WK) / Karlsruhe Institute of Technology (KIT), Karlsruhe, Baden-Württemberg, Germany

Abstract

Sintered steel components allow for near-net-shape manufacturing and lightweight design, positioning them as a key material for modern industrial applications. As the inherent porosity leads to reduced load-bearing capabilities, the component can be subjected to a post-sintering heat treatment process such as induction surface hardening to enhance its application limits. Numerical methods such as finite-element (FE) simulations have proven to be valuable tools in streamlining process development, reducing experimental effort for process optimization and refining process understanding. This study focuses on evaluating the transferability of an established induction heat treatment simulation model to the sintered steel Fe-1.8%Cr-0.6%C (Astaloy CrA). As the porosity affects the electromagnetic, thermal and metallurgical material behavior during induction hardening, these material properties were experimentally determined as a function of temperature across all relevant phases. The material data was integrated into a coupled electromagnetic-thermo-metallurgical finite-element simulation. The simulation was validated using a small-modulus gear wheel subjected to induction hardening. The results demonstrate that close to contour hardening profiles can be achieved through induction hardening, confirming that the process is applicable for small modulus sintered steel gear wheels. The FE-model has shown high accuracy in predicting surface layer conditions, including microstructure and hardness distribution, providing a valuable tool for optimizing induction hardening processes.

Introduction

Powder-metallurgical (PM) manufactured components allow for near-net-shape manufacturing, minimizing machining costs and material waste [1]. Due to their versatility, they are receiving growing recognition for industrial applications, particularly for complex shaped parts such as gears used in transmissions in the automotive industry [2]. PM steel components are characterized by a defined level of porosity [3]. When subjected to mechanical loading, the inherent porosity of PM steels can result in local stress concentrations, which may reduce their load-bearing capacity compared to conventional materials [4, 5]. To counteract this drawback and ensure sufficient fatigue and wear resistance, appropriate surface layer heat treatments are required. Induction surface hardening is particularly advantageous, as it offers a fast process with localized heating, close-to-contour hardening and minimal distortion [6]. Induction hardening can significantly improve the mechanical performance by enhancing surface hardness and inducing beneficial compressive residual stresses in the surface layer [7].

For conventionally manufactured components, induction hardening is a well-established process [8]. However, when applied to PM steels, additional complexities arise as the inherent porosity of PM components influences their thermal, mechanical and electromagnetic behaviour during the heat treatment process [9, 10]. Unlike conventional steels, the presence of pores in PM materials affects heat conduction, alters the distribution of induced currents and can lead to altered kinetics of phase transformations [9, 11, 12]. These differences lead to deviations in microstructure and hardness profiles, posing significant challenges for process control and optimization.

The optimization of heat treatment processes typically involves extensive experimental effort. With the availability of advanced simulation tools, numerical methods such as finite-element simulations are becoming a time- and resource-efficient alternative. To obtain accurate simulation results, detailed material models that reflect on the electromagnetic, metallurgic as well as thermomechanical behaviour of the alloy used are required [13]. This study aims to address these challenges by experimentally characterizing the electromagnetic, thermal and metallurgical properties of the PM steel Fe-1.8%Cr-0.6%C (Astaloy CrA). These properties are integrated into a finite-element framework to simulate the induction hardening process. The developed simulation is validated against experimental data, demonstrating its capability to predict the surface layer state with high accuracy. This work provides a foundation for an FE-based optimization for the induction hardening processes of PM steels, enabling broader industrial application of these materials.

Material

The experiments were carried out using the water atomized, pre-alloyed powder Fe-1.8%Cr (Astaloy CrA) as a base material. Graphite was added to adjust the carbon content of the base material to approximately 0.6 %. The chemical composition of the alloy used is listed in table 1. The material was then pressed and sintered to discs as well as gear wheels with modulus 2 in an atmosphere of 90 % N_2 and 10 % H_2 at 1120 °C for 30 min with subsequent cooling, leading to a part density of 7.2 g/cm^3. The sintering process results in a fine predominantly pearlitic microstructure with an initial hardness of 260 HV 0.1. The geometry of the initial parts is depicted in figure 1.

Table 1: *Chemical composition of the used batch of Astaloy CrA in wt.%.*

C	Cr	Mn	Fe.
0.58	1.76	0.17	bal.

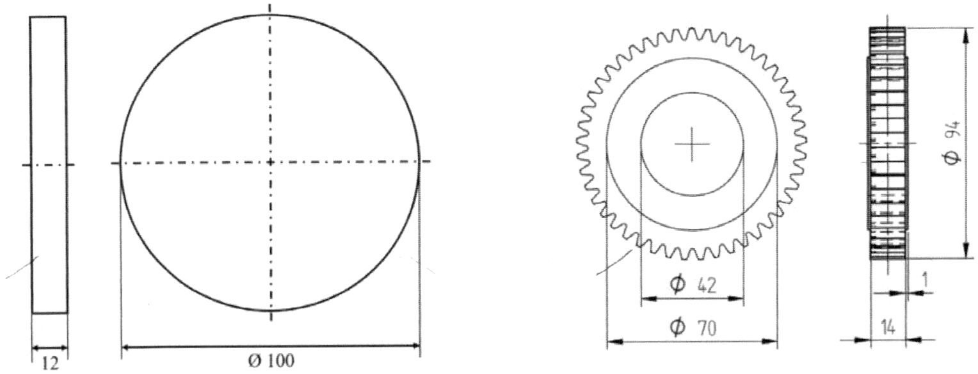

a) *Cylindrical base geometry* b) *Gear wheel*

Figure 1: *Initial geometries of the sintered steel specimen. All dimensions are given in millimeters.*

Methods

Material analysis

The electromagnetic characteristics of the material were measured using a toroid sample with an inner diameter of 66 mm, outer diameter of 82 mm and a height of 8 mm. The toroid was wound with a primary and secondary winding, both containing 28 loops, hereby enabling to use the specimen as a transformer. The wires were insulated to allow for measurements up to the Curie temperature of the material. To measure the electromagnetic properties, the toroid was heated inductively with a thermocouple for temperature monitoring. For the measurements, the primary winding is supplied with an excitation current inducing a current into the secondary winding. The magnetic field strength (H) in the primary winding can be calculated from the excitation current using Ampere's law. According to Faraday's law of induction, the integration of the voltage measured on the secondary side leads to the magnetic flux (B), which in turn enables the derivation of the B-H curve.

The specific heat capacity of the material was measured by differential scanning calorimetry (DSC) using a Netzsch DSC 404 F1 Pegasus including an A54 sample holder equipped with a type S-thermocouple and a high temperature Pt-furnace. A cylindrical sample with 6 mm in diameter and 1 mm in height was machined from the base material. The specimen was placed in a Pt-Rh sample holder inlayed with an Al_2O_3 liner. This sample-containing Pt-Rh crucible along with an empty but identical Pt-Rh crucible was placed in the sample holder and tested under argon atmosphere. The DSC experiment was conducted with a constant heating rate of 20 K/min until the maximum temperature of 1250 °C was reached. During the heating phase, DSC signals (μW/mg) was measured based on the differential heat flow between the sample and the empty reference crucible. This experiment was also repeated for a standard sapphire sample provided by Netzsch. The specific heat capacity (dH/dT) was evaluated according to DIN 51007 [14] using the DSC signals obtained for the samples and those obtained for the standard sapphire sample.

The thermal diffusivity of the material was measured through laser-flash-analysis using a Netzsch LFA 427. Cylindrical specimen with a diameter of 12.7 mm and a thickness of 2.2 mm were tested at different temperatures ranging from room temperature to 900 °C. The thermal conductivity was then calculated as the product of thermal conductivity, specific heat capacity and density.

Phase transformations were characterized using a quenching dilatometer DIL 805 A from TA Instruments. Hollow cylinders with an outer diameter of 4 mm, length of 10 mm and wall thickness of 0,5 mm were machined from the base material. To prevent oxidation or decarburization, the chamber was evacuated to a pressure of $< 10^{-4}$ mbar prior to the experiment. The specimens were heated inductively with a type S thermocouple spot-welded onto the surface for process control and monitoring. During the experiment the length change of the sample was measured by a linear variable differential transformer (LVDT). To characterize the austenitization, the samples were heated with a constant heating rate of 300 K/s to a maximum temperature of 1100 °C. For martensitic hardening, the sample was first heated with a constant heating rate of 100 K/s to an austenitization temperature of 900 °C with a subsequent annealing step of 10 s. The samples were then cooled to room temperature using helium gas as the quenchant with a constant flow rate of 52 l/h.

Induction hardening

An induction hardening machine Fiand KHM 750 equipped with a dual frequency converter IDEA SMS 850 was used for the induction hardening experiments. The generator allows for simultaneous dual frequency heating providing up to 400 kW in the medium frequency range (10 – 30 kHz) and 450 kW in the high frequency range (150 – 450 kHz). An optimized ring inductor particularly designed to match the gear wheel geometry was used for the experiments. The process was controlled by specifying the heating time as well as the power distribution between the medium and the high frequency range. To avoid influences from the inductor feed line, the part was set to rotation with 300 rpm prior to starting the heating cycle. The part was quenched using a water-polymer-mixture. During the hardening process, temperatures at the gear tip and root were measured using a spot-welded thermocouple type K connected to a rotating datalogger. To allow for a close to contour hardening pattern, the process parameters were set to a heating time of 140 ms using 60 % power in the medium frequency as well as 70 % in the high frequency band. After the experiment was completed, a segment of the gear wheel was cut, embedded, ground and polished up to 1 µm. The hardness depth profiles at the tooth root and tip were measured on the polished surface using a Vickers hardness tester Qness Q10A+ with a testing load of 1 N (HV 0.1). To visualize the microstructure, the sample was etched in 2 % alcoholic nitric acid (Nital) for 10 s and analyzed using a light microscope.

***Figure 2:** Experimental setup for the induction hardening experiments: (a) tailstock, (b) inductor, (c) quenching shower, (d) gear wheel, (e) thermocouple with connection to datalogger*

FE-Analysis

The simulation framework used in this work is based on the model developed by Schwenk [15], Kaiser [16] and further extended by Dollhofer [17]. The model involves an electromagnetic as well as a thermometallurgic model and is integrated in Abaqus 2019. An indirect coupling between the two models was incorporated using a python interface. First, the eddy current losses are calculated using Maxwell's equations. In case of ferromagnetic materials there is no linear relationship between the magnetic field strength and the magnetic flux density below the Curie temperature. To allow for the use of a time-harmonic approach, enhancing the model's computational efficiency, Labridis [18] proposed a model to approximate the nonlinear behavior using a fictious material with constant permeability μ_r^{f*}. The fictive permeability is determined based on an energetic analysis, utilizing both the linearized and integrated areas of the magnetization curve. To integrate the area under the curve, Labridis suggests using the Fröhlich equation (Eq. 1) due to its simplicity. Under the assumption of constant thermal properties, the approximation of μ_r^{f*} is repeated iteratively until a predefined convergence criterion is met.

$$B = \frac{H}{\alpha + \beta|H|} \tag{1}$$

In the electromagnetic model, both the initial ferromagnetic pearlitic phase as well as the paramagnetic austenite with their respective phase fractions are considered. For modelling the field concentrator, air and cooling water, constant properties were assumed. To allow for the simulation of the hardening process, the Joule heat calculated from the electromagnetic model is then transferred to the thermometallurgic simulation. The thermometallurgic model is based on the heat conduction equation, also taking phase transformations into account. The thermal conductivity and specific heat capacity were implemented as temperature-dependent properties, with distinctions made between the face-centered-cubic (fcc) and body-centered-cubic (bcc) structures. The temperature dependency was implemented using the logistic function (Eq. 2) with parameters k_0, k_1, T_0 and p.

$$f(T) = k_1 + \frac{k_0 - k_1}{\left(1 + \left(\frac{T}{T_0}\right)^p\right)} \tag{2}$$

The kinetics of the austenitization are based on the Avrami equation (Eq. 3) [19, 20]. For non-isothermal conditions, the growth rate parameter b must be considered as temperature-dependent incorporating the velocity constant C, the activation enthalpy ΔH, Boltzmann constant k_B as well as the Avrami exponent n.

$$w_A = 1 - exp(-(bt)^n),\ b(T) = C \cdot exp\left(\frac{-\Delta H}{k_B T}\right) \tag{3}$$

The martensitic phase transformation is described by the model of Koistinen-Marburger (Eq. 4), consisting of the martensite start temperature M_s as well as a coefficient α_{KM} [21].

$$w_M = 1 - exp\big(-\alpha_{KM} \cdot (M_s - T)\big) \tag{4}$$

The hardness was calculated using a linear mixing rule, accounting for contributions from both the martensitic and the initial ferritic-pearlitic phase. The temperature profile as well as the phase fraction are transferred back to the electromagnetic simulation and the next iteration is started with a predefined step time. These iterations are repeated until the inductive heating phase is completed. Then the quenching of the part is simulated considering both radiational and convectional heat losses. The cooling of the part during the movement of the shower was implemented using a heat transfer coefficient of α_{air} = 35 W/m²K and an emission coefficient of ε = 0.8. The quenching process with the water-polymer solution was implemented by a convective boundary condition with a heat transfer coefficient of α_{quench}= 5000 W/m²K.

To reduce computational time, the geometry of the gear wheel was reduced to a 2,5D model of half a tooth (Fig. 3). For a more precise description of the surface layer condition, an element bias was implemented to the surface of the gear tooth. The electromagnetic model accounts for the reduced gear wheel, inductor, inductor coolant as well as the surrounding air, whereas in the thermometallurgic model, only the gear wheel is considered.

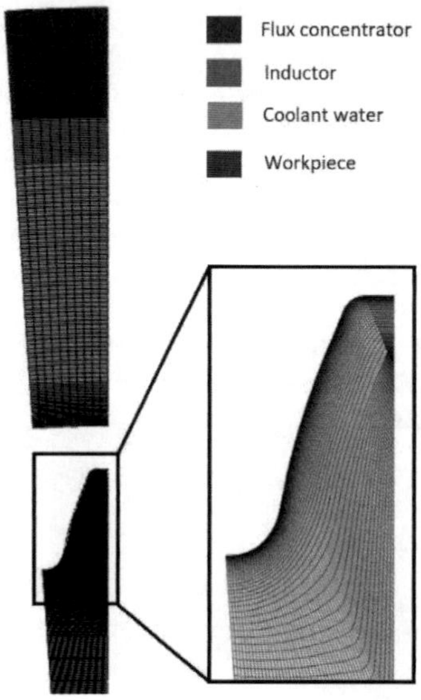

Figure 3: *2,5D model used for the simulation of the induction hardening process*

Results

Electromagnetic properties

The results from the electromagnetic measurements are presented in figure 4. From the temperature dependent hysteresis curves depicted in figure 4 (a) it can be observed that the saturation polarization, coercive field strength, and magnetic remanence decrease with increasing temperature. These changes are less pronounced in the temperature range between room temperature to 300 °C than at higher temperatures. Above 300 °C these properties decline with further temperature rise. The paramagnetic transition occurs at a temperature of approximately 764 °C.
The magnetization curves, representing the materials behavior during initial magnetization can be derived from the hysteresis loops by averaging the measured values of positive magnetic flux density and are depicted in figure 4 (b). Using temperature-dependent parameters α and β in the Fröhlich equation (Eq. 1), a good correlation between the experimentally obtained data and the model predictions can be achieved.

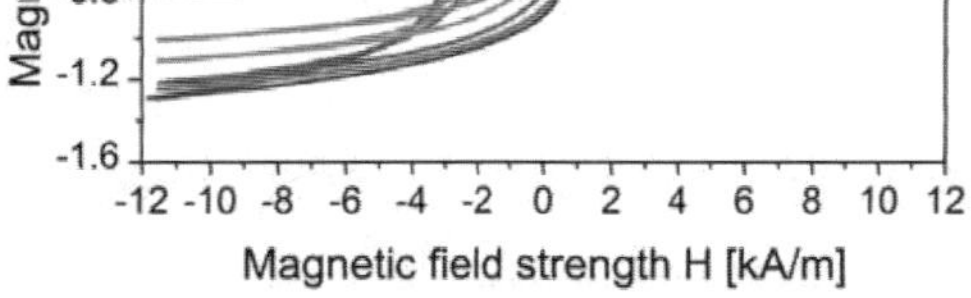

a) Hysteresis curves measured at different temperatures

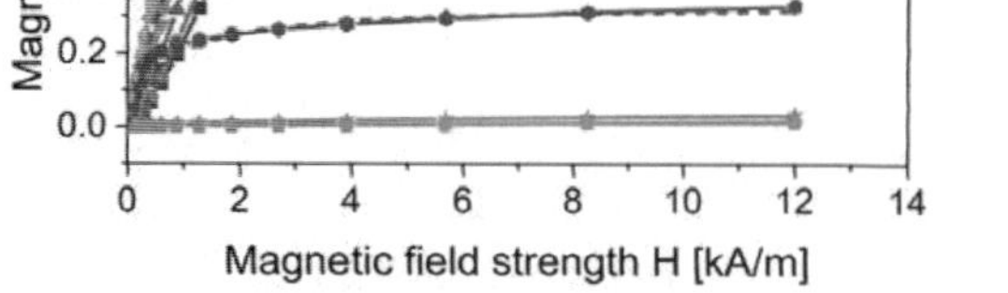

b) Magnetization curves derived from the hysteresis curves including the fitted Fröhlich model

Figure 4: *Electromagnetic properties of the sintered steel Astaloy CrA*

Thermophysical properties

The specific heat capacity c_p determined from DSC-measurements is presented in figure 5 (a). Between 200 °C and 700 °C the specific heat capacity increases steadily with increasing temperature. Upon reaching the Curie temperature, the specific heat capacity reaches its peak due to the material's transition to a paramagnetic state and the overlapping phase transformation to austenite. Beyond this point, the specific heat capacity decreases rapidly. Once the austenitization is completed, the specific heat capacity stabilizes at a nearly constant level with almost no increase with further temperature rise. The thermal conductivity calculated from the measured thermal diffusivity, specific heat capacity and density is shown in figure 5 (b). The results indicate a decrease in thermal conductivity with increasing temperature, stabilizing at nearly constant values upon reaching the austenitization temperature.

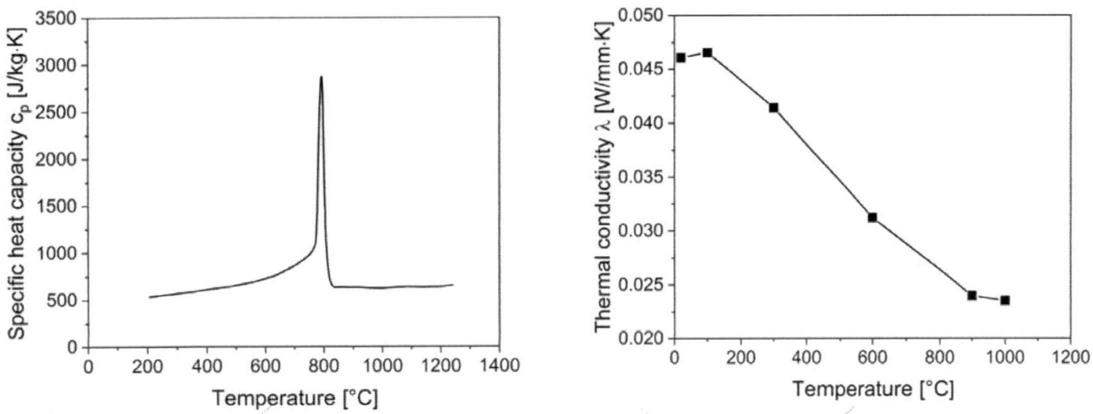

(a) Specific heat capacity *(b) Thermal conductivity*

Figure 5: *Thermophysical properties of the sintered steel Astaloy CrA*

Metallurgic properties and kinetics of phase transformation

The strain derived from the dilatometer experiments during continuous heating and quenching is presented in figure 6. In both cases, the measurements show excellent reproducibility. During continuous heating, an initial linear increase in strain is observed, corresponding to the thermal expansion coefficient of the material. Upon surpassing the Curie temperature (T_C), a sudden increase in strain occurs, which can be attributed to the transition to the paramagnetic state. Due to this transition, a significant increase in inductor power is needed to continue heating. The transition initially occurs in the middle of the sample where the temperature is monitored. Due to temperature gradients in the sample, the increased power input results in an enhanced expansion of the outer ferromagnetic regions, leading to a strain peak. Following this, the strain increases linearly again until the onset of austenitization at the Ac_1 temperature is reached. The formation of austenite continues until completion of the transformation is reached at the Ac_3 temperature. The volume fraction of the phase transformed during continuous heating and cooling can be determined using the lever law (Eq. 5), taking the measured strain $\varepsilon(T,t)$ as well as the linearized strain of the initial mother $\varepsilon_m(T)$ and resulting daughter phase $\varepsilon_d(T)$ into account [22].

$$w(T,t) = \frac{\varepsilon(T,t) - \varepsilon_m(T)}{\varepsilon_d(T) - \varepsilon_m(T)} \tag{5}$$

The austenite fraction formed is illustrated in figure 6 (a). At a heating rate of 300 K/s, the austenitization starts at approximately 817 °C and ends at 892 °C. Upon reaching the austenitization temperature, the austenite volume fraction initially rises steeply before gradually leveling off and reaching saturation upon complete austenitization. This behavior can be attributed to the rapid transformation of pearlitic regions into austenite at Ac_1. With increasing temperature, the transformation of ferritic regions occurs. The measured strain signal is compensated by the thermal expansion of the austenite already transformed, resulting in a decrease in slope. A Nelder-Mead-algorithm with multiple startvalues was used to fit the Avrami model to the dilatometer experiments. By setting the parameters of the Avrami equation to $\Delta H = 1.20$ eV, $C = 3.09 \cdot 10^7$ s^{-1} and $n = 2.58$ a good correlation between the model and the experimental data can be achieved.

Figure 6 (b) presents the strain curves obtained from dilatometry during cooling from the austenite region. The initial linear decline of the strain curve represents the thermal contraction of the supercooled austenite. Due to the rapid cooling rate, only martensitic transformation occurs, initiating at a martensite start temperature of 284 °C leading to an average hardness of 920 HV 0.1 of the dilatometer specimen. By setting the coefficient α_{KM} to 0.026 °C^{-1}, a good agreement between the experimental data and the Koistinen-Marburger model can be achieved.

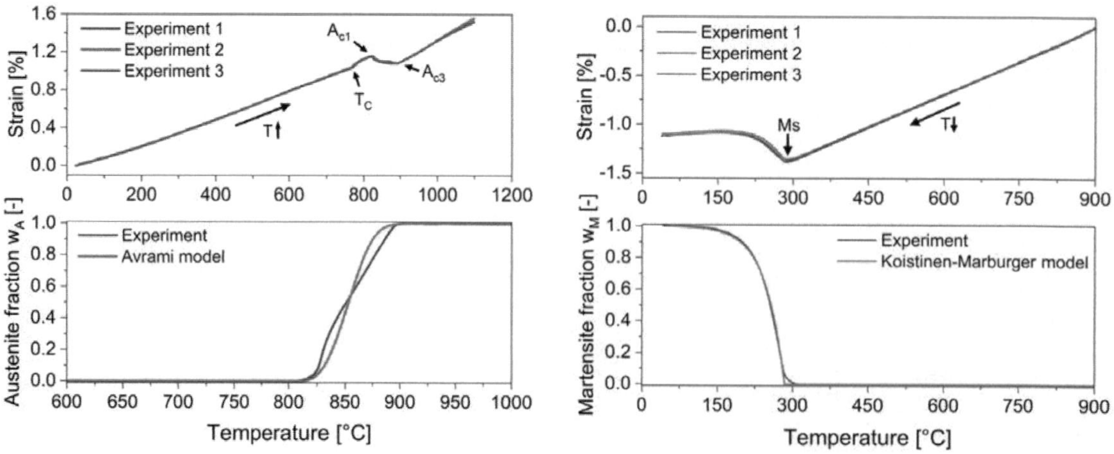

(a) Measured strain and derived volume fraction during austenitization compared to the the Avrami model

(b) Measured strain and derived volume fraction during martensitic transformation compared to the Koistinen-Marburger model

Figure 6: *Metallurgic properties and kinetics of phase transformation*

Validation

To validate the FE-simulation, the frequencies recorded during the process were incorporated into the electromagnetic simulation for each time step. Since the data was recorded at a frequency of 100 Hz, the time step for the EM simulation was set to 0.01 s. The induction hardening process led to a resonance frequency of 12.69 kHz in the MF-circuit, while the HF circuit operated at 169 kHz. During the process, only the currents at the generator could be measured. As losses in the oscillating circuit likely result in a reduced current at the inductor, the current densities required as input parameters for the electromagnetic simulation were calibrated to align with the microstructural profile obtained from the metallographic analysis. Given that the proposed austenite kinetic model was calibrated for a heating rate of 300 K/s, whereas in the actual experiment heating rates exceeding 7000 K/s were achieved, an austenite kinetic model developed by Miokovic [23], which is suitable for heating rates of up to 10.000 K/s, was integrated into the thermo-metallurgical FE-model.

Figure 7 (a) illustrates the comparison between the simulated martensite fraction and the micrograph of the gear tooth, while figure 7 (b) presents the resulting hardness depth profile at the tooth root and tip. The micrograph reveals that the induction hardening process leads to a hardening depth of approximately 400 μm which is homogeneously distributed along the contour of the gear tooth. Furthermore, no macroscopic cracking occurred during inductive surface hardening. When comparing the simulation results to the experimental data, a good agreement regarding the martensite distribution as well as the hardening depth is observed. At the tooth root, both the experiment and simulation lead to a surface hardening depth of 410 μm with hardness values exceeding 900 HV 0.1 at the surface layer of the gear. While the hardening depth at the tooth root and tip closely matches between experiment and simulation, the simulation slightly underestimation the hardening depth at the tooth flank. The maximum temperature from the simulation in the root and tip of the gear were calculated to approximately 1000 °C and 900 °C respectively and are in good agreement with the temperatures measured during the experiment.

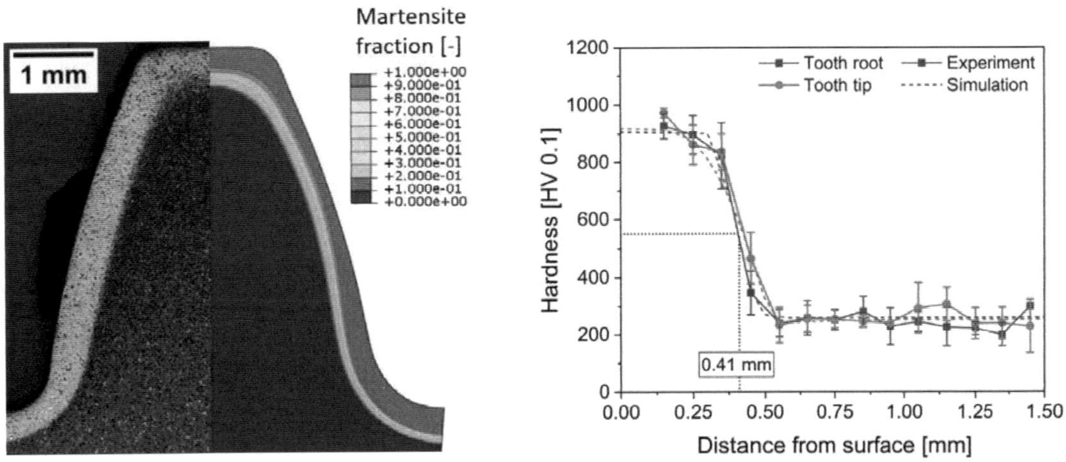

(a) Comparison between simulated martensite distribution and the micrograph

(b) Comparison between the measured and simulated hardening depth profiles

Figure 7: *Comparison between the experiment and the FE-simulation*

Discussion

While a simulation-based optimization approach for the induction hardening process may have advantages in reducing experimental effort and extend process understanding, it can clearly be seen that correct material and process data is a critical factor for obtaining precise results. Although the model already describes the induction hardening process very well, there are some aspects that might influence the accuracy of the simulation.

Regarding the measurement accuracy of electromagnetic properties, reliable and repeatable measurements of the temperature is key for determining correct hysteresis loops. Due to an issue with temperature monitoring the measurements could not be performed during the first heating cycle. After the first measurement cycle was completed, the specimen cooled from approximately 860 °C to room temperature through radiation and convection and then entered a second heating in a heat-treated condition, likely containing martensitic or bainitic structures. During reheating, additional tempering may have occurred. The effect of tempering on the electromagnetic properties of a quenched steel 42CrMoS4 has been analysed by Akujärvi [24]. It was found out that at a tempering temperature of 560 °C hysteresis loss, coercivity and remanence are reduced with increasing tempering time. Both coercivity and hysteresis loss decreased by almost a factor of two from the hardened state after 30 minutes of tempering. These changes were mainly attributed to a reduction of residual stresses and dislocation density caused by tempering. Similarly, a comparison of the first and second measurement cycles in this study revealed a reduction in coercivity in the second cycle. While hysteresis losses remained approximately constant, the second cycle exhibited higher saturation polarization, resulting in an increased relative permeability. Slightly altered electromagnetic behaviour can impact the skin depth, potentially leading to an underestimation in the FE-model due to an overestimation of the relative permeability. Consequently, the electromagnetic data used in the simulation likely represents the behaviour of a tempered martensitic-bainitic material during the initial stages of the induction heating process. The discrepancy to the predominantly pearlitic base material of the gear wheels could affect the accuracy of the electromagnetic model, particularly in predicting the distribution of induced currents, associated Eddy current losses and the resulting Joule heat.

Another important aspect to consider is the austenitization kinetics. Due to the limited power of the dilatometer's generator and the temperature gradient in the dilatometer specimen, reproducible results were restricted to heating rates less than 1000 K/s. However, during the actual induction hardening process, heating rates close to 10.000 K/s can be achieved. The austenitization behaviour of the tempered steel 42CrMo4 at heating rates up to 10.000 K/s has been studied by Miokovic [23], who observed that higher heating rates significantly accelerate the austenitization process. For heating rates between 1000 K/s and 10.000 K/s, Miokovic determined Avrami parameters independent from the heating rating rate. Compared to the results of this work, especially the velocity constant C is significantly increased. This suggests that the kinetic model proposed in this work is likely too slow to accurately represent the rapid austenitization occurring during induction heating. If the proposed model from this work is used in the simulation, the temperatures archived in the tooth root during induction heating are high enough to ensure that austenitization is fully completed. In contrast lower temperatures occur at the tooth tip. In this region, heat dissipation during the dwell time of the quenching shower is primarily governed by radiation and convection to the surrounding atmosphere, resulting in a thermal accumulation. As a result of the lower but almost constant temperature level, delayed but complete austenitization is caused. At the tooth flank, temperatures similar to those at the tooth tip are reached. However, due to the higher mass of the surrounding base material, heat is dissipated more efficiently due to conduction, leading to a faster temperature loss and resulting in an incomplete austenitization due to the slow kinetics. The integration of the obtained material data into the coupled electromagnetic thermo-metallurgical FE-simulation model has shown accurate results regarding the temperature distribution, martensite fraction and hardness profiles, particularly at critical regions such as the tooth root. These results underline the value of using accurate, phase-dependent electromagnetic, thermophysical and metallurgical material data, not only for improving simulation accuracy but also for optimizing process parameters in future application and enhancing process understanding.

Conclusion

This study evaluated the transferability of an inductive heat treatment simulation to the sintered steel Astaloy CrA. Comprehensive material data was obtained to account for electromagnetic, thermophysical and metallurgical properties of Astaloy CrA. The material data was then integrated into a coupled electromagnetic-thermo-metallurgical finite-element simulation. The parameter used for the induction hardening process led to a close-to-contour hardening profile along the gear tooth with a hardening depth of approximately 400 µm. It has proven to be effective in avoiding the formation of cracks during the induction hardening process which is particularly noteworthy, as it demonstrates the suitability of induction hardening for gear wheels manufactured from the sintered steel Astaloy CrA, considering the presence of very high heating rates and steep thermal gradients. When validated against data from an induction-hardened small-modulus gear wheel, the simulation demonstrated a high degree of accuracy in predicting surface layer conditions, including martensite and hardness distribution. While the model performed well at the tooth root and tip, minor deviations were observed at the tooth flank, where hardening depths were slightly underestimated. These discrepancies highlight the sensitivity of the process to localized variations in material properties as well as process data, underscoring the importance of accurate modelling in FE-models.
Overall, the findings illustrate the potential of numerical methods as a valuable tool for optimizing induction hardening processes of sintered steels and to deepen the understanding of the process. Furthermore, the potential to extend the application limits of sintered steel gear wheels through induction surface hardening has been shown. The close alignment between experimental results and FE-based predictions validates the applicability of the model to these materials and provides a foundation for further refinements. Future work could focus on adding a mechanical submodel to the FE-simulation to account for the formation of residual stresses in the component.

Acknowledgments

The authors gratefully acknowledge the financial support from the German Research Foundation (DFG) – project number 458684492. The authors would furthermore like to thank Höganäs AB for providing the material.

References

[1] T. Frech, *et al.*, "Design for PM Challenges and Opportunities for Powder Metal Components in Transmission Technology", *Procedia CIRP*, Vol. 70 (2018), p 186–191. https://doi.org/10.1016/j.procir.2018.03.267
[2] Oğuz, Erden, "The Development and Applications of Powder Metallurgy Manufacturing Methods in Automotive Industry", *International Journal of Engineering Research and Development*, Vol. 9 (2017), p 100–112. https://doi.org/10.29137/umagd.349955
[3] M. Dlapka, *et al.*, "Defining the pores in PM components", *Metal Powder Report*, Vol. 65, No. 2 (2010), p 30–33. https://doi.org/10.1016/S0026-0657(10)70093-X
[4] H. Danninger and C. Gierl, "Powder metallurgy steels for highly loaded precision parts", *International Journal of Materials and Product Technology*, Vol. 28, No. 3/4 (2007), p 338–360. https://doi.org/10.1504/IJMPT.2007.013084
[5] A. Hadrboletz and B. Weiss, "Fatigue behaviour of iron based sintered material: a review", *International Materials Reviews*, Vol. 42, No. 1 (1997), p 1–44. https://doi.org/10.1179/imr.1997.42.1.1
[6] V. Rudnev, *et al.*, *Handbook of Induction Heating*, 2nd Edition, CRC Press (2017). https://doi.org/10.1201/9781315117485

[7] V. Rudnev, "Induction Hardening of Gears and Critical Components", *Gear Technol*ogy (2008), p 58–63.
[8] H.-W. Zoch, "Randschichtverfestigung - Verfahren und Bauteileigenschaften", *HTM Journal of Heat Treatment and Materials*, Vol. 50, No. 5 (1995), p 287–294. https://doi.org/10.1515/htm-1995-500507
[9] F. Ternero, *et al.*, "Influence of the Total Porosity on the Properties of Sintered Materials—A Review", *Metals*, Vol. 11, No. 5 (2021), Art.-no 730. https://doi.org/10.3390/met11050730
[10] H. Danninger and M. Dlapka, "Heat Treatment of Sintered Steels – what is different?", *HTM Journal of Heat Treatment and Materials*, Vol. 73, No. 3 (2018), p 117–130. https://doi.org/10.3139/105.110353
[11] V. S. Warke, *et al.*, "The effect of porosity on the austenite to bainite transformation in powder metallurgy steels", *Journal of Materials Research*, Vol. 24, No. 10 (2009), p 3213–3219. https://doi.org/10.1557/jmr.2009.0379
[12] P. Nusskern, *et al.*, "Austenite-Bainite Transformation Kinetic Model for the Powder-Metallurgical Steel Astaloy 85 Mo", *Materials Science Forum*, Vol. 706–709 (2012), p 1485–1490. https://doi.org/10.4028/www.scientific.net/MSF.706-709.1485
[13] F. Mühl, *et al.*, "Simulation of induction hardening: Simulative sensitivity analysis with respect to material parameters and the surface layer state", *Computational Materials Science*, Vol. 184 (2020), p 109916. https://doi.org/10.1016/j.commatsci.2020.109916
[14] "DIN 51007-1, Thermische Analyse (TA) - Differenz-Thermoanalyse (DTA) und Dynamische Differenzkalorimetrie (DSC) - Teil 1: Allgemeine Grundlagen" (2024). https://doi.org/10.31030/3542721
[15] M. Schwenk, *et al.*, "Process Simulation of Single and Dual Frequency Induction Surface Hardening Considering Magnetic Nonlinearity", *Materials Performance and Characterization*, Vol. 1, No. 1 (2012), p 1–20. https://doi.org/10.1520/MPC104374
[16] D. Kaiser, "Experimentelle Untersuchung und Simulation des Kurzzeitanlassens unter Berücksichtigung thermisch randschichtgehärteter Zustände am Beispiel von 42CrMo4", PhD thesis, Karlsruhe Institute of Technology, (2019).
[17] B. Dollhofer, *et al.*, "Improvement of an Electromagnetic-Thermal-Mechanical Coupled Simulation for the Optimization of Complex Processes in Induction Hardening", *Netsu-Shori, Journal of the Japan Society for Heat Treatment*, Vol. 64, No. Special Issue of 28th IFHTSE Congress (2024), p 66–71. https://doi.org/10.14940/netsushori.64.ex66
[18] D. Labridis and P. Dokopoulos, "Calculation of eddy current losses in nonlinear ferromagnetic materials", *IEEE Transactions on Magnetics*, Vol. 25, No. 3 (1989), p 2665–2669. https://doi.org/10.1109/20.24506
[19] M. Avrami, "Kinetics of Phase Change. I General Theory", *The Journal of Chemical Physics*, Vol. 7, No. 12 (1939), p 1103–1112. https://doi.org/10.1063/1.1750380
[20] M. Avrami, "Kinetics of Phase Change. II Transformation-Time Relations for Random Distribution of Nuclei", *The Journal of Chemical Physics*, Vol. 8, No. 2 (1940), p 212–224. https://doi.org/10.1063/1.1750631
[21] D. P. Koistinen and R. E. Marburger, "A general equation prescribing the extent of the austenite-martensite transformation in pure iron-carbon alloys and plain carbon steels", *Acta Metallurgica*, Vol. 7, No. 1 (1959), p 59–60. https://doi.org/10.1016/0001-6160(59)90170-1
[22] M. Hunkel, *et al.*, "Dilatometry," in *Handbook of Thermal Analysis and Calorimetry*, Elsevier, Vol. 6 (2018), p 103–129. https://doi.org/10.1016/B978-0-444-64062-8.00019-X
[23] T. Mioković, *et al.*, "Description of short time phase transformations during the heating of steels based on high-rate experimental data", *Journal de Physique IV*, Vol. 120 (2004), p 591–598. https://doi.org/10.1051/jp4:2004120068
[24] V. Akujärvi, *et al.*, "Evolution of magnetic properties during tempering", *The International Journal of Advanced Manufacturing Technology*, Vol. 119, No. 3–4 (2022), p 2329–2339. https://doi.org/10.1007/s00170-021-08464-7

QDE 2025: Proceedings of the 3rd International Conference on
Quenching and Distortion Engineering
May 6–7, 2025, Vancouver, Canada
https://doi.org/10.31399/asm.cp.qde2025p0068

Influence of As-Built Microstructure and Thermal Post-Processing on the Fatigue Strength of AISI 4140 Manufactured by Laser Powder Bed Fusion

Philipp Schüßler, Volker Schulze, Stefan Dietrich
Institute for Applied Materials – Materials Science and Engineering (IAM-WK), Karlsruhe Institute of Technology (KIT), Karlsruhe, Baden-Württemberg, Germany

Abstract

This study examines the impact of thermal post processing, specifically induction hardening and tempering, on the fatigue performance of laser powder bed fusion (PBF-LB) manufactured AISI 4140 steel. PBF-LB specimens exhibit significantly lower fatigue strength ($S_{e,50\%}$ = 363 MPa) compared to conventionally manufactured ones ($S_{e,50\%}$ = 631 MPa), primarily due to process-induced porosity acting as stress concentrators. Induction hardening introduced compressive residual stresses of up to 650 MPa and increased surface Vickers hardness to about 675. Therefore, it shifts the crack initiation sites from near-surface pores to the region of maximum tensile residual stress at the boundary between the hardened zone and base material. Thermal post processing mitigated the adverse effects of tempered martensite embrittlement and resulted in comparable low-cycle and high-cycle fatigue strengths ($S_{e,50\%}$ from 348 to 375 MPa) across batches, despite differences in microstructure and hardness of the as-built states. These results underscore the importance of porosity and microstructure control in enhancing fatigue performance. Induction hardening shows promise for addressing stress concentrations from near-surface porosity, particularly in non-machined parts.

Introduction

Laser powder bed fusion (PBF-LB) has become a key additive manufacturing (AM) technology for producing metallic components with complex geometries. Its versatility has driven widespread adoption across various industries. However, challenges such as process-induced residual stresses, high surface roughness, and inherent porosity can significantly compromise the load-bearing capacity compared to conventionally cast or shaped materials [1–4]. Additionally, the microstructure formed during the PBF-LB process plays a crucial role in determining fatigue performance, much like in traditionally manufactured components [1, 3, 5, 6].

The formation of residual stresses in additive manufacturing (AM) components arises from the complex thermal history and temperature gradients in the melt pool and component [7, 8]. These stresses depend on process parameters and component geometry, with elevated tensile residual stresses typically observed near the surface [4, 7]. Surface defects such as porosity and partially melted powder significantly reduce fatigue strength [1, 2, 9, 10]. For high-carbon steels like AISI 4140, martensitic transformations during cooling introduce additional challenges, such as localized stresses and cold cracking due to volume changes [3, 11–15]. Optimizing parameters like laser power and scan speed can prevent micro cracking but often narrows the process window [16]. Preheating the base plate has proven effective in reducing crack susceptibility, slightly widening the process window and reduces unfavorable tensile residual stresses [8, 17, 18]. Furthermore, the complex thermal history during AM influences the resulting microstructure, with variations in quenching rates leading to bainitic or martensitic phases depending on the process strategy [19].

To mitigate these negative effects, post processing strategies like mechanical surface treatments and heat treatments are commonly employed to reduce porosity or improve mechanical properties [4, 9, 20]. Thermal surface treatments showed positive effects on fatigue performance for conventionally manufactured specimens. One established process is the induction hardening in combination with a low temperature tempering process [21–23]. Generally, the induction hardening uses a coil to generate an alternating magnetic field inducing resistive eddy current losses in the work piece. By controlling the current and frequency, the hardening depth can be adapted. For low alloyed steels, the localized heat input, combined with the rapid quenching, leads to the formation of martensite, additionally inducing favorable compressive residual stresses in the hardened zone.

This publication builds on a previous publication and aims to answer the question of the effectiveness of thermal surface hardening methods to improve the fatigue performance of specimens manufactured by the PBF-LB process [6]. To address these questions, an investigation was conducted on three distinct groups of PBF-LB samples, each subjected to a different base plate temperature. In addition, a control group was included, produced using conventional manufacturing methods. Half of each batch was further processed by a combination of inductive surface hardening and oven tempering. Rotating bending tests were then conducted on the six PBF-LB and two conventionally manufactured batches, half of each in the as-built state, the other half after thermal post processing. Several metallographic examinations methods were applied to characterize each material state. The fatigue performance was correlated with the measured material states, offering valuable insights into the process dependent fatigue behavior of additively manufactured components with additional thermal post processing.

Material and Methods

Material

Specimens were produced using the PBF-LB process with inert gas-atomized pre-alloyed AISI 4140 powder (equivalent to the German grade 42CrMo4). For comparison, control specimens were fabricated from AISI 4140 rods. The chemical compositions of the powder, the additively manufactured parts, and the control specimens are provided in Table 1. The compositions were determined using inductively coupled plasma optical emission spectroscopy (ICP-OES) for Mn, Cr, Si, Mo, and Fe, and an elemental CS analyzer for C.

Table 1: Chemical composition of the powder, as-build PBF-LB batches and the conventionally manufactured control batch in wt.%.

	C	Mn	Cr	Si	Mo	Fe
Powder	0.39	0.76	1.09	0.34	0.25	Bal.
PBF-LB	0.38	0.75	1.17	0.40	0.29	Bal.
Control	0.43	0.77	1.15	0.27	0.21	Bal.

Specimen processing

A total of six batches, each consisting of 36 specimens, were manufactured using the PBF-LB process on a ConceptLaser M2 (GE Additive, USA) by Jell GmbH & Co. KG (Germany). The as-built specimens had a diameter of 17 mm and a length of 104 mm. The samples were machined to their final shape with a gauge diameter of 10 mm and a gauge length of 20 mm. The laser parameters and scan strategy were kept constant across all batches: 350 W laser power, 175 µm spot size, 800 mm/s scan speed, 110 µm hatch distance, and 50 µm layer thickness. A stripe strategy was employed, featuring a bi-directional scan path, 5 mm stripe width, 72° hatch rotation per layer, and a single perimeter line. The base plate preheating temperatures were set to 150 °C, 300 °C, or 450 °C, and all specimens were machined after the PBF-LB process.

Two batches were produced for each base plate preheating temperature—one retained in the as-built state and the other subjected to induction hardening and oven tempering. Additionally, two comparison batches were machined from conventional AISI 4140 rods. These batches were divided to match the as-built specimens printed with a 150 °C preheating temperature and those undergoing the same PBF-LB process followed by thermal post processing. To achieve the same hardness as the as-built specimens, heat treatment parameters were determined using the Hollomon-Jaffe equation, involving austenitization at 850 °C for 30 minutes, oil quenching, and tempering at 450 °C for 2 hours. All batches were machined to their final shape after the initial heat treatment. The thermal post processing was selected as a combination of a short-time induction surface hardening in combination with an oven based low temperature tempering.

The induction surface hardening was conducted on a FIAND KHM 750 hardening machine using an IDEA SMS850 dual-frequency converter. The parameters were optimized to 400 ms heating time with the power of 153 kW. The inductance of the setup resulted in a resonance frequency of 295 kHz for all experiments. The heating step was followed by water quenching to room temperature resulting in a mean surface hardening depth (SHD) of 1200 µm. An oven tempering step for 2 h at 200 °C was added as final processing step.

Metallographic characterization

The density of each specimen was measured using Archimedes' principle, employing a ME503TE precision scale (Mettler Toledo, USA) with water as the measurement medium. A dispersive agent was added to minimize surface tension. All density measurements were performed after the specimens had been machined to their final geometry. To analyze the pore distribution and structure, one sample was examined using micro-tomographic (µCT) imaging. The µCT scan was conducted with a YXLON Precision µCT system (YXLON International GmbH, Germany) at an acceleration voltage of 200 kV and a target current of 0.24 mA. A total of 3000 projections were captured over a full 360° rotation using a Perkin Elmer XRD1620 AN flat-panel detector with a resolution of 2048 × 2048 pixels and a pixel pitch of 0.2 mm. The imaging setup included a focus object distance (FOD) of 36.25 mm and a focus detector distance (FDD) of 900 mm, achieving a voxel size of 8 µm. To minimize noise, each projection used a 1-second detector integration time and frame binning of two images. Image reconstruction was performed with the filtered back-projection (FBP) algorithm in VGStudioMAX 2022.4 (Volume Graphics International GmbH, Germany), with a median filter applied for noise reduction. Porosity analysis was conducted using the VGEasyPore module. To analyze the hardness profiles and microstructures, specimens were sectioned perpendicular to the axial direction (XY plane). Hardness testing was conducted using a Qness Q10A hardness tester (ATM Qness GmbH, Austria) with a 10 N load and a 10 s dwell time. The average hardness for each batch was determined from three depth profiles measured across five specimens. For microstructural characterization, specimens were sectioned, embedded, ground, polished, and etched with a solution of 99% ethanol and 1% nitric acid. Microstructures were examined using a LEO Gemini 1530 scanning electron microscope (SEM; Carl Zeiss AG, Germany) with an acceleration voltage of 10 kV and a secondary electron (SE) detector. Fracture surfaces were captured using an Aristomet light microscope (Leica Microsystems GmbH, Germany).

X-ray diffraction

The residual stresses in axial and tangential direction were determined using the X-ray diffraction method $\sin^2(\psi)$ with the material parameters $E_{211} = 211$ GPa and $\nu_{211} = 0.3$. The Ψ-diffractometer DR45 (Stresstech GmbH, Germany) was used with Cr-Kα radiation (30 kV and 9 mA) to measure the {211} α-Fe diffraction line at 11 Ψ-angles between −45° and 45°. For all measurements, an exposure time of 10 s and a collimator with a beam diameter of 1 mm was used.

Mechanical testing

Fatigue testing was carried out using a moment-controlled rotating bending machine. The experiments were conducted with a constant frequency of 50 Hz and a load ratio of $R = -1$. The SN-curve was determined using a combination of the staircase method (at least 15 samples per batch) and the horizon method (at least 4 samples per load level).

Results

Density and surface roughness

The density of components fabricated through AM is a crucial factor that is closely connected to the selected process strategy. Although the base plate preheating temperatures altered cooling rates, there were no notable differences in the measured densities across the various batches. Additionally, the average densities of all AM parts were comparable to those of the control batch made using traditional subtractive manufacturing methods (Table 2). The lowest mean density was recorded for batch AM300-M with a value of 7.814 g cm^{-3}, while the highest was observed for batch AM450-M with 7.835 g cm^{-3}. This represents a small deviation, indicating that these process variations had minimal effect on the final density of the parts. Therefore, no clear trend or correlation between baseplate preheat temperatures and part densities could be identified. The µCT scan of the specimen identified a small number of spherical pores, most likely gas-induced (Figure 1 (a)). The average pore sphericity was determined to be 0.84 with a standard deviation of 0.03. No indications of lack-of-fusion pores were observed, confirming the high density and non-critical morphology of the produced parts. Pore frequency analysis across the specimen radius showed an almost uniform distribution, with fewer pores detected close to the surface, as only pores not connected to the surface were included in the count (Figure 1 (b)). Following the machining process, the surface roughness was found to be similar across all batches, independent on the parameter variation during the PBF-LB manufacturing process (Table 2). A minimal increase of surface roughness was noted after the thermal post processing step for all four batches, likely related to a small oxide layer.

Table 2: Measured density and surface roughness in axial direction after all processing steps. Results are listed as mean values with the standard deviation.

	Batch	Density / g cm^{-3}	Surface roughness R_z / µm	Ref.
As-built	Control-QT-M	7.827 ± 0.003	5.53 ± 0.99	[6]
	AM150-M	7.826 ± 0.003	6.60 ± 0.62	[6]
	AM300-M	7.831 ± 0.006	5.05 ± 0.76	[6]
	AM450-M	7.835 ± 0.006	6.48 ± 0.97	[6]
As-built & post processing	Control-QT-M-IHT	7.827 ± 0.003	7.90 ± 0.73	This work
	AM150-M-IHT	7.823 ± 0.005	7.64 ± 0.49	This work
	AM300-M-IHT	7.814 ± 0.005	6.82 ± 0.42	This work
	AM450-M-IHT	7.825 ± 0.004	8.64 ± 0.54	This work

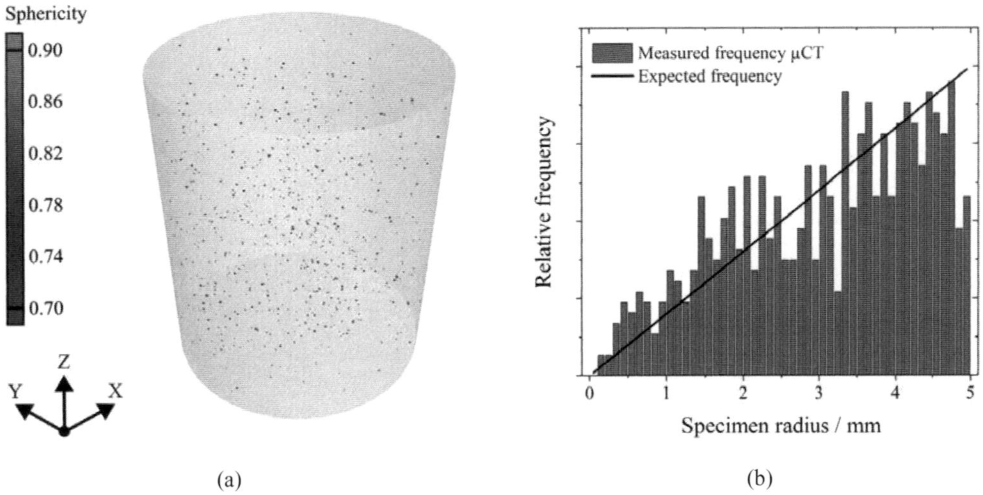

Figure 1: Representative reconstructed µCT scan of a specimen of the AM150-M batch, visualizing (a) the 3D volume of the specimen with highlighted detected pores and (b) statistical analysis of the detected pores dependent on the radius.

Hardness and residual stresses

An analysis of the as-built specimens' hardness and residual stress depth profile was previously reported by Schüßler et al. [6]. The hardness depth profile was examined along the XY-plane, perpendicular to the build direction. A reduction in hardness is noted as the base plate preheating temperature is increased from 150°C to 450 °C (Figure 2 (a)). The Vickers hardness values for the Control-QT-M and AM150-M specimens were found to be similar, with an average Vickers hardness HV1 of (416 ± 3) and (420 ± 5) respectively. No discernible trend was observed across the diameter of the samples, indicating a homogeneous distribution of hardness. With the additional thermal post processing (induction surface hardening and oven tempering), the surface hardness was increased to a Vickers hardness HV1 of 675 for all four different batches (Figure 2 (b)). A minimal difference for the surface hardening depth SHD of about 100 µm (SHD of 1.2 to 1.3 mm) was noted. The induction surface hardening led to a heat affected zone of up to 500 µm, with a slightly lower Vickers hardness compared to the bulk material. The heat affected zone is located between the hardening zone and the base material. Comparing the two batches with the same PBF-LB process parameters, no difference in hardness was noted for the base material.

Residual stress profiles were examined in both the axial (parallel to the build direction) and circumferential orientations from the surface to a depth of up to 1.5 mm. The measurements for the as-built batches revealed no significant differences in both measurement directions (Figure 2 (c)). The residual stresses were significantly influenced by the machining process, which induced compressive stresses near the surface, reaching up to 650 MPa. Following a distance of 100 µm to 200 µm, the compressive residual stresses transition to mildly tensile residual stresses, extending into the core of the specimen. The batches with additional thermal post processing also reach compressive residual stresses of up to 650 MPa near the surface but the transition towards tensile residual stresses was noted near the measured SHD of about 1.2 mm (Figure 2 (d)). The measurements indicate that the peak tensile residual stresses were located near the 1.5 mm depth. Measurements at a higher depth were not conducted due to the increasing measurement error [6].

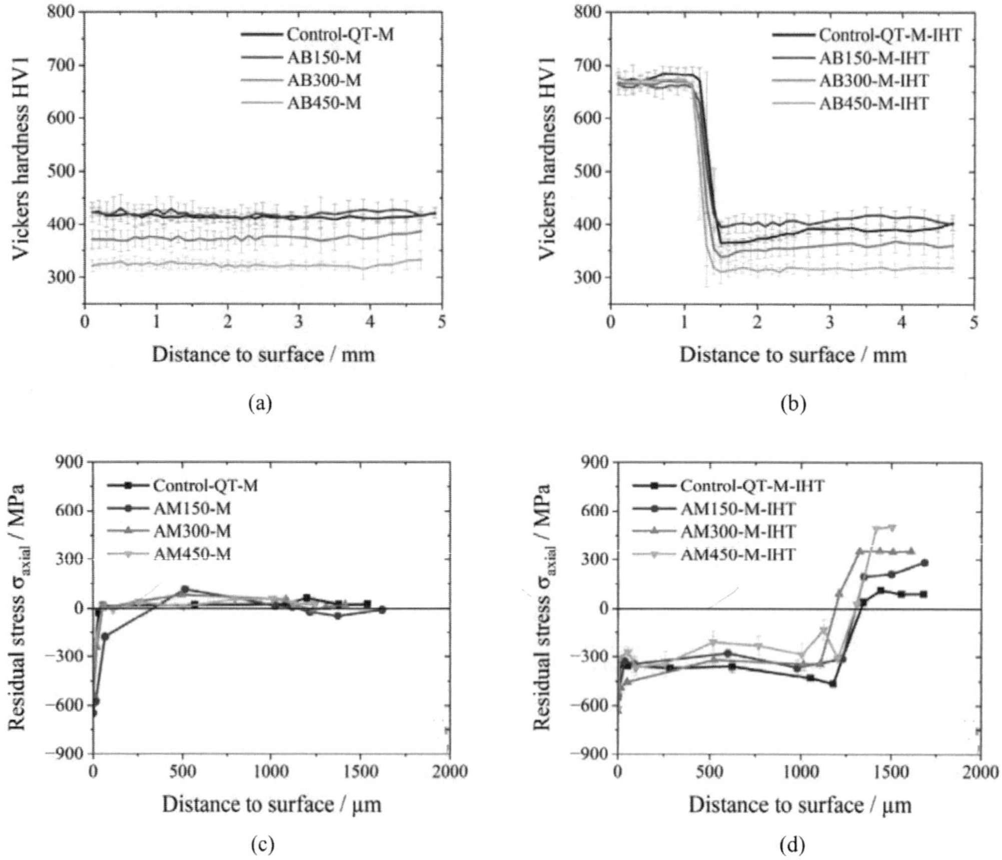

Figure 2: Measured hardness depth profile for (a) all three as-built PBF-LB and the conventionally manufactured batches based on [6] and (b) these batches after additional thermal post processing. The hardness was measured on the XY cross section perpendicular to the build direction. As well as the measured residual stresses in axial direction for (c) all three as-built PBF-LB and the conventionally manufactured batches and (d) these batches after additional thermal post processing.

Microstructure

A comprehensive analysis of the as-built specimens' microstructure was previously reported by Schüßler et al. [6]. In brief, variations in base plate preheating temperature resulted in distinct microstructural differences. Preheating at 150 °C produced a fully martensitic structure with small intra lath carbide precipitates and islands of retained austenite (Figure 3 (b)). Increasing the temperature to 300 °C yielded a mixed bainitic-martensitic microstructure with coarser carbide precipitates (Figure 3 (c)). The bainite laths were characterized by the characteristic 57° orientation of the intra lath carbides [6]. No significant amounts of retained austenite were noted for this batch. The 450 °C base plate preheating temperature led to a fully bainitic microstructure, characterized by coarse carbide precipitates along the boundaries (Figure 3 (d)). For the conventionally manufactured batch, the processing route created a tempered martensitic structure characterized by well-defined martensite laths with dispersed intra lath precipitates (Figure 3 (a)).

Thermal post processing caused notable changes in the near-surface microstructure. Induction hardening induced austenitization followed by martensitic transformation, forming martensite laths near the surface (Figure 3 (e)). Subsequent tempering at 200 °C promoted the formation of fine intra lath epsilon carbide precipitates. Notably, the microstructure after post processing was consistent across all four batches, regardless of their initial base material state.

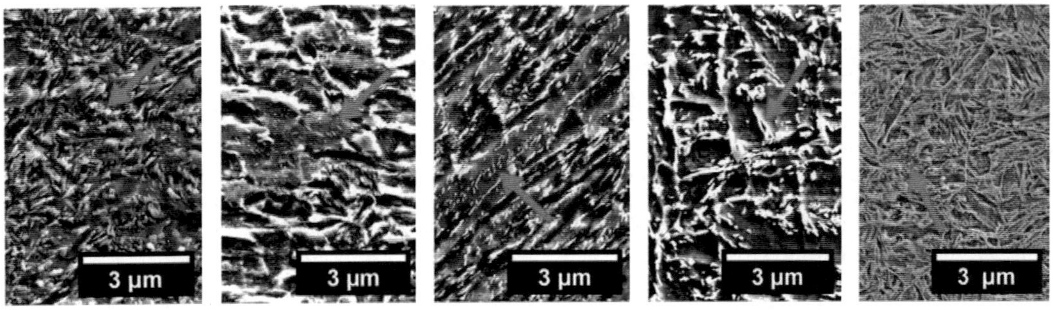

(a) (b) (c) (d) (e)

Figure 3: SEM-SE micrograph images of a cut section parallel to the build direction for the batches (a) Control-QT-M, (b) AM150-M, (c) AM300-M, (d) AM450-M and (e) representative the near surface microstructure of the Control-QT-M-IHT batch after the thermal post processing. Specimens were ground, polished and etched with nital etchant. The representative carbide precipitates described in the section were marked with red arrows. Figures (a)-(d) are based on [6].

Fatigue properties and fractography

The SN-curves for the as-built samples (Figure 4 (a)) were discussed in detail in the previous publication by Schüßler et al. [6]. The PBF-LB process with a base plate preheating temperature of 150 °C led to a significant reduction of HCF fatigue strength ($S_{e,50\%}$ = 363 MPa) compared to the conventionally manufactured Control-QT-M batch ($S_{e,50\%}$ = 631 MPa), despite an equal hardness for both batches. While the LCF fatigue strength was comparable for all three as-built PBF-LB batches, differences were noted for the HCF strength. The increase of the base plate preheating temperature to 300 °C led to a minimum of HCF fatigue strength ($S_{e,50\%}$ = 301 MPa), while a further increase to 450 °C resulted in an intermediate HCF fatigue strength of $S_{e,50\%}$ = 329 MPa.

Induction hardening led to an significant increase of the LCF and HCF fatigue strength for the conventionally manufactured Control-QT-M-IHT batch (from $S_{e,50\%}$ = 631 MPa to $S_{e,50\%}$ = 890 MPa, Figure 4 (b)). An increase of about 41 %. In contrast, a change of 0 % (AM150-M-IHT), 6 % (AM450-M-IHT) and 25 % (AM300-M-IHT) was noted for the batches manufactured with the PBF-LB process. The lower the fatigue strength of the as-built PBF-LB batch, the greater was the noted increase of HCF strength. The induction hardening process also led to the equalization of the HCF fatigue strength for the PBF-LB batches, despite different base material microstructures and hardness. A HCF fatigue strength of 348 to 375 MPa was measured. A significant increase of fatigue strength in the LCF regime was noted. The average cycles to failure increased by about one decade for all PBF-LB batches with the additional thermal post processing.

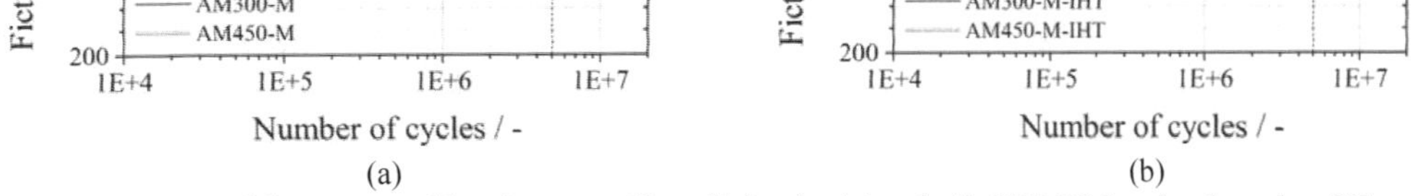

(a) (b)

Figure 4: SN-curves of the rotational bending tests (R = -1) for the (a) as-built PBF-LB batches based on [6] and (b) the batches with the additional thermal post processing.

The analysis of the fracture surfaces revealed a switch of the limiting mechanisms and crack initiation sites for the resulting service life. While the failures of the Control-QT-M batch were mainly initiated at surface defects (Figure 5 (a)), the failure mode switched to a subsurface failure mode likely initiated at non-metallic inclusions or gas porosity in the base material near the boundary of the hardened zone (Figure 5 (c)).
The failure mode of the PBF-LB batches in both the as-built and induction hardened batches was dominated by the residual porosity linked to the manufacturing process (Figure 5 (b)). While the failure mechanism was the same, the location of the crack initiation site changed with the added induction hardening process step. While the as-built specimens crack initiation sites were located near the surface, the crack initiation sites after the induction hardening were found to be located near the boundary of the hardened zone (Figure 5 (d)). The mean distance of the crack initiation site from the surface was calculated to 1.42 mm with a standard deviation of 0.19 mm. This coincides with the location of the maximum tensile residual stresses.

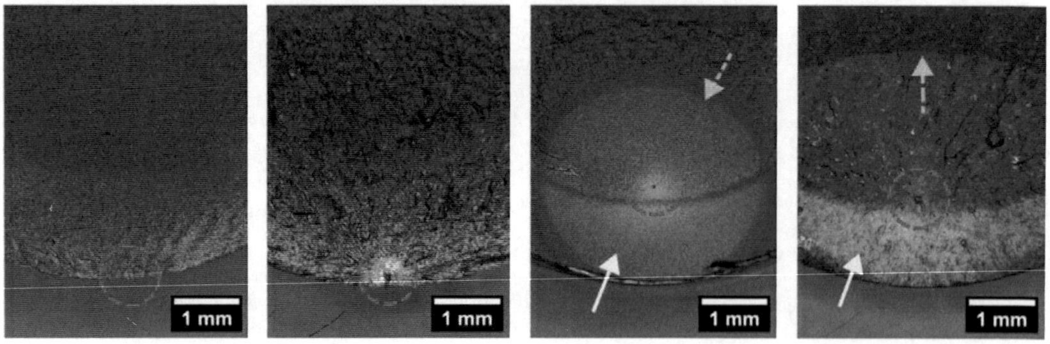

(a) (b) (c) (d)

Figure 5: Representative light microscope images of the fracture surfaces for (a) the conventionally manufactured Control-QT-M batch, (b) the PBF-LB as-built batch AM150-M as well as the same states ((c) Control-QT-M-IHT and (d) AM150-M-IHT) after the thermal post processing step. The initial crack sites were marked with red circles, the hardened zones with yellow arrows and the noted fisheye boundaries with dashed blue arrows. Figures (a) and (b) are based on [6].

Discussion

The comparison between AM150-M and Control-QT-M reveals that the reduction in fatigue strength is primarily influenced by local porosity, which acts as a stress concentrator (notch factor). Despite identical hardness values in both cases, minor differences in microstructure, such as retained austenite and the size of carbide precipitates, were observed. The variation of base plate preheating temperatures significantly affects the microstructure, transitioning from fully martensitic (AM150-M) to mixed bainitic-martensitic (AM300-M), and eventually to fully bainitic (AM450-M). This change influences high-cycle fatigue (HCF) strength. The minimum HCF fatigue strength observed at a preheating temperature of 300 °C is associated with the decomposition of retained austenite into coarse inter lath cementite, a phenomenon known as tempered martensite embrittlement (TME) [24–27]. This embrittlement reduces material toughness and promotes brittle failure modes. In summary, the combination of high mechanical load (rotating bending tests), porosity acting as stress concentrators, and slightly tensile residual stresses contribute to the lower HCF fatigue strength of the PBF-LB batches compared to the conventionally manufactured Control-QT-M batch.

The results for PBF-LB batches with thermal post processing suggest that fatigue strength in both the LCF and HCF regimes is significantly more influenced by tensile residual stresses and porosity near the hardened zone boundary than by variations in microstructure from the PBF-LB process itself. The similarity in fatigue strength across these batches can be attributed to the similar porosity and residual stress depth profiles for all three post processed batches. Additionally, the induced compressive residual stresses near the surface shifted the crack initiation sites towards the base material of the specimens, effectively countering the stress concentrations of the near-surface porosity. Compressive residual stresses, which align with the crack initiation site, exacerbate the detrimental effects of local porosity, ultimately facilitating crack formation.

In summary, induced compressive residual stresses enhance the local fatigue strength near the surface, while tensile residual stresses decrease the fatigue strength in the heat-affected zone. Therefore, thermal post processing methods should be tailored to the specific pore distribution of each specimen for optimal results. For the PBF-LB batches in this study, near-spherical pores were uniformly distributed (See Figure 1). However, non-machined specimens often exhibit higher porosity between the outer contour lines and inner hatching lines [28]. These hotspots of porosity arises from collapsing keyholes caused by the laser deceleration at the end of scan lines [28–30]. While thermal post processing cannot eliminate this process-induced porosity, the findings strongly suggest that the induced compressive residual stresses can effectively counter the notch effect of near-surface pore clusters. Short-time thermal post processing methods could therefore be a more economical alternative to processes like hot isostatic pressing (HIP).

Conclusions

This study explored the influence of a thermal post processing method (induction hardening and tempering) on the low-cycle fatigue (LCF) and high-cycle fatigue (HCF) strength of specimens manufactured using the laser powder bed fusion (PBF-LB) additive manufacturing process with quench and tempering steel AISI 4140. The following key findings provide insights into process optimization and fatigue performance:

- Direct comparison of batches with similar hardness revealed a significant reduction in fatigue strength for PBF-LB specimens, from $S_{e,50\%} = 631$ MPa in conventionally manufactured specimens to $S_{e,50\%} = 363$ MPa. This reduction was primarily driven by process-induced porosity, which acted as stress concentrators. Small microstructural differences, such as those caused by the localized energy input during the PBF-LB process, also contributed to the reduced fatigue performance.
- Induction hardening induced compressive residual stresses of up to 650 MPa and increased the Vickers hardness near the surface to 675. This shifted crack initiation sites from near-surface pores to pores located near the maximum tensile residual stress region, which is located at the boundary between the hardened zone and the base material. Although the specimens in this study featured uniformly distributed pores due to machining, non-machined specimens often exhibit higher porosity near the transition zones between outer contour lines and inner hatching lines. The observed shift of crack initiation sites indicates that induction hardening has the potential to enhance fatigue strength in batches with significant near-surface porosity.
- Induction hardening resulted in similar HCF fatigue strength ($S_{e,50\%}$ between 348 and 375 MPa), regardless of the microstructural and hardness variations retained in the base material of the specimens. The LCF fatigue strength was increased by about one decade compared to the as-built batches. Thermal post processing effectively mitigated the adverse effects of non-ideal microstructural properties, such as tempered martensite embrittlement (TME), improving the mechanical properties of the as-built state.

These findings underscore the importance of controlling both porosity and microstructure to improve fatigue performance. Thermal post processing, particularly induction hardening, demonstrated significant potential for countering the stress concentrations caused by near-surface porosity in PBF-LB specimens. Future research should focus on refining thermal post processing methods to further enhance fatigue strength and explore their application to specimens with higher porosity gradients, especially in non-machined parts. As well as reducing residual porosity by the means of process parameter optimization of further post processing methods such as hot isostatic pressing, shot peening or deep rolling.

Acknowledgments

The authors acknowledge the chemical analysis by ICP-OES conducted at the Institute for Applied Materials (IAM-AWP), Karlsruhe Institute of Technology (KIT). This study was funded by the Deutsche Forschungsgemeinschaft (DFG, German Research Foundation) - 516837935.

References

[1] A. Yadollahi and N. Shamsaei, "Additive manufacturing of fatigue resistant materials: Challenges and opportunities", *International Journal of Fatigue*, Vol. 98 (2017), p 14–31. https://doi.org/10.1016/j.ijfatigue.2017.01.001

[2] G. Meneghetti, *et al.*, "An analysis of defects influence on axial fatigue strength of maraging steel specimens produced by additive manufacturing", *International Journal of Fatigue*, Vol. 118 (2019), p 54–64. https://doi.org/10.1016/j.ijfatigue.2018.08.034
[3] J. Damon, *et al.*, "Process porosity and mechanical performance of fused filament fabricated 316L stainless steel", *Rapid Prototyping Journal*, Vol. 25, No. 7 (2019), p 1319–1327. https://doi.org/10.1108/RPJ-01-2019-0002
[4] J. Damon, *et al.*, "Mechanical surface treatment of EBM Ti6Al4V components: Effects of the resulting surface layer state on fatigue mechanisms and service life", *Materials Science and Engineering: A*, Vol. 849 (2022), p 143422. https://doi.org/10.1016/j.msea.2022.143422
[5] A. Yadollahi, *et al.*, "Effects of process time interval and heat treatment on the mechanical and microstructural properties of direct laser deposited 316L stainless steel", *Materials Science and Engineering: A*, Vol. 644 (2015), p 171–183. https://doi.org/10.1016/j.msea.2015.07.056
[6] P. Schüßler, *et al.*, "Influence of base plate preheating temperature on fatigue strength of AISI 4140 manufactured by laser powder bed fusion", *Journal of Materials Research and Technology*, Vol. 35 (2025), p 1358–1368. https://doi.org/10.1016/j.jmrt.2025.01.073
[7] J. L. Bartlett and X. Li, "An overview of residual stresses in metal powder bed fusion", *Additive Manufacturing*, Vol. 27 (2019), p 131–149. https://doi.org/10.1016/j.addma.2019.02.020
[8] P. Mercelis and J. Kruth, "Residual stresses in selective laser sintering and selective laser melting", *Rapid Prototyping Journal*, Vol. 12, No. 5 (2006), p 254–265. https://doi.org/10.1108/13552540610707013
[9] H. Soyama and F. Takeo, "Effect of Various Peening Methods on the Fatigue Properties of Titanium Alloy Ti6Al4V Manufactured by Direct Metal Laser Sintering and Electron Beam Melting", *Materials*, Vol. 13, No. 10 (2020), p 2216. https://doi.org/10.3390/ma13102216
[10] H. Fayazfar, *et al.*, "A critical review of powder-based additive manufacturing of ferrous alloys: Process parameters, microstructure and mechanical properties", *Materials & Design*, Vol. 144 (2018), p 98–128. https://doi.org/10.1016/j.matdes.2018.02.018
[11] W. Hearn and E. Hryha, "Effect of Carbon Content on the Processability of Fe-C Alloys Produced by Laser Based Powder Bed Fusion", *Frontiers in Materials*, Vol. 8, 800021 vols. (2022), p 547. https://doi.org/10.3389/fmats.2021.800021
[12] W. Hearn, *et al.*, "Development of powder bed fusion – laser beam process for AISI 4140, 4340 and 8620 low-alloy steel", *Powder Metallurgy*, (2022), p 1–13. https://doi.org/10.1080/00325899.2022.2134083
[13] C. Shi, *et al.*, "Parameter optimization and mechanical properties of 42CrMo4 manufactured by laser powder bed fusion", *The International Journal of Advanced Manufacturing Technology*, Vol. 121, No. 3–4 (2022), p 1899–1913. https://doi.org/10.1007/s00170-022-09474-9
[14] C. Shi, *et al.*, "High cycle fatigue behaviour of AISI 4140 steel manufactured by laser-powder bed fusion", *International Journal of Fatigue*, Vol. 168 (2023), p 107469. https://doi.org/10.1016/j.ijfatigue.2022.107469
[15] C. Shi, *et al.*, "Influences of laser remelting on mechanical performances of AISI4140 steel", *Materials Science and Technology*, (2024), . https://doi.org/10.1177/02670836231212614
[16] P. Schüßler, *et al.*, "A novel multiscale process simulation to predict the impact of intrinsic heat treatment on local microstructure gradients and bulk hardness of AISI 4140 manufactured by laser powder bed fusion", *Virtual and Physical Prototyping*, Vol. 18, No. 1 (2023), p e2271455. https://doi.org/10.1080/17452759.2023.2271455
[17] R. Mertens, *et al.*, "Influence of Powder Bed Preheating on Microstructure and Mechanical Properties of H13 Tool Steel SLM Parts", *Physics Procedia*, Vol. 83 (2016), p 882–890. https://doi.org/10.1016/j.phpro.2016.08.092
[18] R. Mertens, *et al.*, "Application of base plate preheating during selective laser melting," in Procedia CIRP, 2018), Vol. 74, p 5–11. https://doi.org/10.1016/j.procir.2018.08.002
[19] W. Wang and S. Kelly, "A Metallurgical Evaluation of the Powder-Bed Laser Additive Manufactured 4140 Steel Material", *JOM*, Vol. 68, No. 3 (2016), p 869–875. https://doi.org/10.1007/s11837-015-1804-y
[20] R. Seede, *et al.*, "Effect of heat treatments on the microstructure and mechanical properties of an ultra-high strength martensitic steel fabricated via laser powder bed fusion additive manufacturing", *Additive Manufacturing*, Vol. 47 (2021), p 102255. https://doi.org/10.1016/j.addma.2021.102255
[21] D. Kaiser, *et al.*, "Experimental investigation and finite-element modeling of the short-time induction quench-and-temper process of AISI 4140", *Journal of Materials Processing Technology*, Vol. 279 (2020), p 116485. https://doi.org/10.1016/j.jmatprotec.2019.116485
[22] F. Mühl, *et al.*, "Tailored bainitic-martensitic microstructures by means of inductive surface hardening for AISI 4140", *Materials & Design*, Vol. 195 (2020), p 108964. https://doi.org/10.1016/j.matdes.2020.108964
[23] T. Oevermann, *et al.*, "Fatigue Properties of Steels SAE 1045 and SAE 4140 upon Integrated Inductive Heat Treatment and Deep Rolling at Elevated Temperature," presented at the 13th International Conference on Shot Peening, Montréal (Canada), 2017).

[24] E. B. Kula and A. A. Anctil, "Tempered Martensite Embrittlement And Fracture Toughness In 4340 Steel," Defense Technical Information Center, 1967. https://doi.org/10.21236/AD0651066
[25] R. O. Ritchie, "Near-Threshold Fatigue Crack Propagation in Ultra-High Strength Steel: Influence of Load Ratio and Cyclic Strength", *Journal of Engineering Materials and Technology*, Vol. 99, No. 3 (1977), p 195–204. https://doi.org/10.1115/1.3443519
[26] F. A. Darwish, *et al.*, "On the tempered martensite embrittlement in AISI 4140 low alloy steel", *Materials Science and Engineering: A*, Vol. 132 (1991), p L5–L9. https://doi.org/10.1016/0921-5093(91)90388-4
[27] G. Krauss, "Tempering of Lath Martensite in Low and Medium Carbon Steels: Assessment and Challenges", *steel research international*, Vol. 88, No. 10 (2017), p 1700038. https://doi.org/10.1002/srin.201700038
[28] L. Englert, *et al.*, "How defects depend on geometry and scanning strategy in additively manufactured AlSi10Mg", *Journal of Materials Processing Technology*, Vol. 299 (2022), p 117331. https://doi.org/10.1016/j.jmatprotec.2021.117331
[29] A. A. Martin, *et al.*, "Dynamics of pore formation during laser powder bed fusion additive manufacturing", *Nature Communications*, Vol. 10, No. 1 (2019), p 1987. https://doi.org/10.1038/s41467-019-10009-2
[30] S. A. Khairallah, *et al.*, "Laser powder-bed fusion additive manufacturing: Physics of complex melt flow and formation mechanisms of pores, spatter, and denudation zones", *Acta Materialia*, Vol. 108 (2016), p 36–45. https://doi.org/10.1016/j.actamat.2016.02.014

QDE 2025: Proceedings of the 3rd International Conference on
Quenching and Distortion Engineering
May 6–7, 2025, Vancouver, Canada
https://doi.org/10.31399/asm.cp.qde2025p0078

Thermo-metallo-mechanical modeling of case-hardened steel parts for prediction of distortions during intermittent gas quenching

Guillaume Troutet, Denis Delagnes, Vanessa Vidal, Vincent Velay
Institut Clément Ader ICA ; Université de Toulouse ; CNRS, IMT Mines Albi, INSA, ISAE- SUPAERO, UPS ; Campus Jarlard, F-81013 France

Charlie Poulat, René Billardon, Pierre Heugue
Safran Transmission Systems, Colombes, France

Abstract

Prediction of the residual distortions induced by quenching requires determining the thermomechanical properties of the material at different stages of the process. For case-hardened steels, with the appearance of a phase transformation, the impact of the phases must be taken into account, as well as the impact of the carbon content gradient. To develop a behavior model, properties such as Young's modulus, viscous stress, kinematic hardening, isotropic hardening, yield strength and transformation-induced plasticity parameter (TRIP) for austenite and martensite were determined using a specially developed experimental set-up.

Introduction

The aeronautical industry is developing turbofan engines with high bypass ratios to increase efficiency. This requires the use of reduction gearboxes whose role is to lower the rotation speed of the fan in relation to the rotation speed of the turbine. These gearboxes transmit high power. These systems are made up of large gears that require perfect geometric quality and functional surfaces with very high mechanical properties. Case-hardened martensitic steels provide a good solution to this specific application.

The steel investigated is M50NiL or 13MoCrNiV42-16-14 grade. Its composition (Table 1) [1] was developed to ensure high toughness. Its low carbon content (0.13%) also prevents the formation of eutectic carbides during solidification. The addition of nickel, and therefore the stabilization of the austenitic phase, also compensates for the loss of hardenability caused by the lower carbon content.

C	Mn	Si	Cr	Ni	Mo	V
0,13	0,25	0,25	4,20	3,40	4,25	1,20

Table 1: Nominal chemical composition of M50NiL steel (weight %)

In our case, the parts are case-hardened and then heat-treated by gas quenching, cryogenic treatment and several tempering. This treatment, provides the expected properties, but causes geometric distortions of the functional surfaces, mainly during quenching.
Therefore, the industry needs are to understand and anticipate these distortions through numerical modeling. Academic studies have addressed internal stresses genesis during martensitic quenching and their consequences on distortions and residual stresses; the phenomenon is fairly well understood [2], [3]. While, models for the quantitative prediction of stress states and distortions have been developed by various research teams, models able to accurately predict distortion for complex parts and complex cooling path are much rarer.

The objective of this study is to develop a model describing stress and strain evolutions during heat treatment. This model must take into account the effect of the composition gradient on the metallurgical mechanisms, in particular martensitic transformations and related phenomena like transformation plasticity, retained austenite stabilization and self-tempering. The approach first involves characterizing the thermomechanical and metallurgical behavior of the material. Then, behavior laws will be proposed to describe the overall metallurgical and mechanical behavior of the steel during quenching.

Experimental approach

A specific experimental set-up (Figure 1) was developed to determine martensite and austenite behavior laws, across the heat treatment temperature range (from 1090°C to -75°C) and for the various carbon contents involved in case hardening (from 0.13% to 0.8%).
The device simulates the heat treatment conditions. It consists of an induction heating system and a compressed air cooling system. Argon flow minimizes oxidation and decarburization. An electrohydraulique testing machine applies and controls loads and deformations at selected rates.

The thermo-elastoviscoplastic behavior of each phase (martensite/bainite and austenite) is identified through various isothermal mechanical tests carried out in the temperature range where these phases exist independently. The behavior is then extrapolated where the phases coexist. In this context, specimens with homogeneous carbon enrichment are used to evaluate the effect of the composition gradient and the properties throughout the depth of the cemented layer. Specifically, during these characterization tests (Figure 2), the specimen is austenitized under industrial conditions, to produce a full austenitic phase with initial characteristics (austenitic grain size, chemical homogeneity, dissolution of carbides) similar to the industrial part. Cooling is subsequently performed and stopped at the desired characterization temperature (550°C in Figure 2).

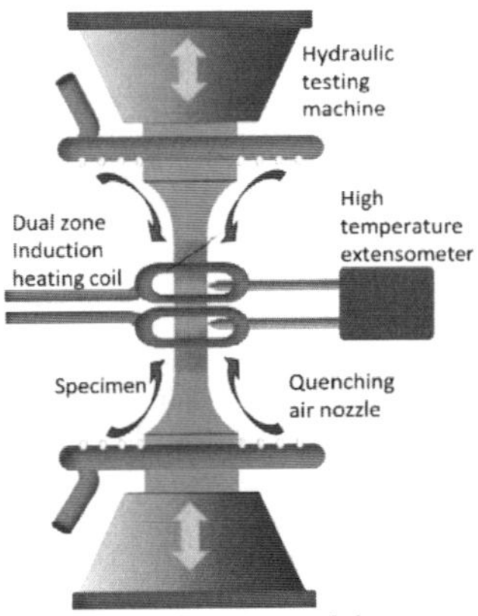

Figure 1: Diagram of the experimental set-up

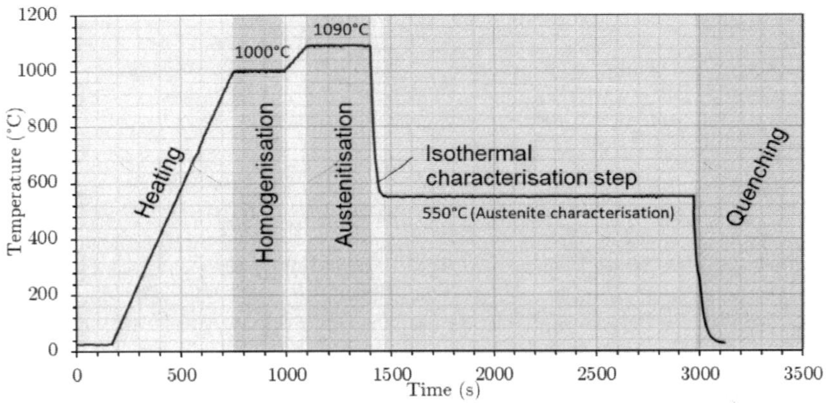

Figure 2: Example of a thermal cycle performed on the hydraulic testing machine (shown in fig.1) for a thermomechanical characterisation test of the austenite (here at 550°C).

During this isothermal interruption, a set of mechanical tests are carried out (Illustration Figure 3). First, a tensile load is applied at a constant strain rate to identify the Young modulus and yield strength. The strain rate is then increased to reveal the viscous behavior (strain rate sensitivity). Following this, a relaxation phase is then conducted by maintaining the strain which is used to identify the material's viscosity. The relaxation test offers a wide range of strain rates (from 10^{-3} s^{-1} to slow rates of around 10^{-8} s^{-1}). The test concludes with cyclic loadings (tension/compression) at a constant strain rate. This cyclic strain-hardening test is used to identify kinematic component and isotropic component.

The aim is to carry out this testing sequence for each of the two phases. For austenite, the experimental plan includes tests conducted at 9 different temperatures in the austenitic range (1050°C, 1000°C, 900°C, 800°C, 700°C, 600°C, 500°C, 400°C, 370°C). To evaluate the effect of carbon content, these tests are repeated on multiple batches of specimens with carbon compositions representative of the carburized layer (0.13%, 0.4%, 0.6% and 0.8%).

For martensite, characterization of its thermo-mechanical behavior follows a similar approach, i.e., testing at several temperatures for different compositions. However, achieving a fully martensitic microstructure throughout the sample requires prior heat treatment (quenching and cryogenic treatment) conducted outside the characterization device. The tests are then performed isothermally, after controlled and interrupted reheating of the specimen on the testing machine. The experimental plan includes tests at 4 different temperatures (20°C, 100°C, 250°C, 350°C). To examine the effect of carbon content, the martensite characterization tests are repeated in the same manner as for the austenite using specimen with identical carbon compositions (0.13%, 0.4%, 0.6% and 0.8%).

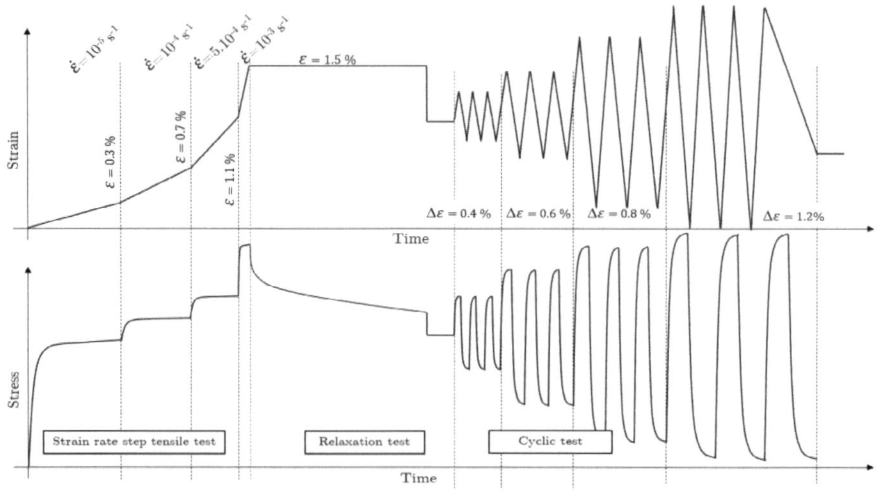

Figure 3: Schematic illustration of a thermomechanical characterization test.

Additional characterization is carried out using dilatometry tests on the same experimental set-up. The tests involve monitoring the dimensional changes of specimens during thermal austenitization and quenching cycles. This method provides information on transformation kinetics, start and end transformation temperatures and thermal expansion coefficients for each phase.

The dilatometry experimental plan includes quenching tests (continuous cooling from 1090°C to 20 °C or -75°C) for various carbon contents (0.13%, 0.4%, 0.6% and 0.8%) and cooling rates (50°C/s, 5°C/s, 0.5°C/s). Variations in cooling rate reveal the effects of self-tempering or austenite stabilization. These phenomena can influence transformation temperatures and kinetics [4-9]. Dilatometry tests under external loading are also planned to identify the transformation-induced plasticity effect. Tests with isothermal interruption (TTT) are also planned in order to identify and characterize potential phase transformations above martensite start temperature, particularly the bainitic transformation. Similarly to the phenomena of self-tempering and stabilization, a bainitic transformation prior to the martensitic transformation will have an impact on transformation temperature and kinetics [7], [10-12]. These tests will be supplemented by phase ratios quantification after treatment (using X-ray diffraction, electron and optical microscopy).

Thermo-metallo-mechanical modeling

A global behavior model was developed to predict the distortions resulting from the quenching process. The model, inspired by the work of Roux [13], assumes a material composed of two phases, austenite (the parent phase) and martensite (the daughter phase). The volume fractions and properties change throughout the heat treatment. The mechanical response of the material during quenching is governed by the local response of the two phases. Assuming small perturbations, the total strain is express as an additive decomposition of various contributions: elastic strain, thermal strain, viscoplastic strain, strain due to volume change from phase transformation and transformation induced plasticity strain (TRIP). For the two-phase material, the same decomposition is introduced for each phase.

A self-consistent scheme as proposed by Kröner [14] is employed to incorporate the behavior of these two phases into the calculation. The approach models the mechanical state of each phase in the heterogeneous medium as equivalent to that of a spherical inclusion in an isotropic, infinite homogeneous medium based on the Eshelby inclusion [15]. The properties of the equivalent homogeneous medium are determined using a volume-weighted mixture law.

The thermomechanical behavior model also incorporates metallurgical mechanisms to describe the evolution of the phase fractions during quenching. The Koistinen & Marburger equation [16] is currently used to describe the progress of the martensitic transformation as a function of temperature. This equation depends on coefficients experimentally determined (including Ms the temperature at which the martensitic transformation starts and a kinetic parameter).

Results and Discussions

A part of the characterization work focuses on the evolution of phase fractions during cooling. Figure 4 shows the results of rapid quench dilatometry tests conducted at 50°C/s. The tests were carried out on M50NiL specimens in the as-received condition and on specimens enriched with 0.8% carbon by mass.
These curves show a drop in austenitization temperatures, with Ac1 dropping from 780°C at 0.13% carbon to 720°C at 0.8% carbon. Additionally, the enriched sample exhibits faster Austenitisation kinetics.

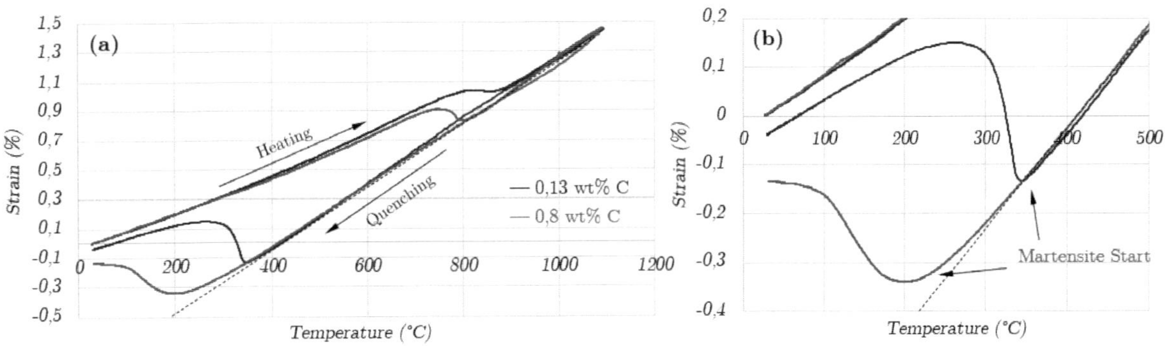

Figure 4: Dilatometric curves for rapid quenching (50°C/s) carried out on the thermomechanical device for a specimen not enriched with carbon (0.13 wt% carbon) and for an enriched specimen (0.8 wt%). (a) Full treatment curves, (b Detail of martensitic transformation expansion)

In agreement with the literature [17], the Ms temperature is significantly reduced by the addition of carbon. It decreasing from 365°C to 250°C for the sample with 0.8% carbon. The martensitic transformation kinetics are also slowed for the enriched sample, with a change in the transformation rate at around 120°C, followed by an additional slowdown. We also note that for this enriched sample, the transformation does not appear to be complete at room temperature, requiring further cooling to cryogenic levels to achieve full transformation.

Figure 5 shows the preliminary results of the thermomechanical characterisation of austenite at high temperature. These curves show the results of tensile tests with strain rate jumps at 700°C, 800°C, 900°C and 1000°C. As the temperature increases, the yield strength decreases significantly, of 65 MPa at 700°C and 18 MPa at 1000°C.
At 900°C and 1000°C, the strain rate sensitivity appears to increase with temperature. In contrast, at 700°C and 800°C, the strain rate sensitivity appears to be very low. However, as shown in figure 6 with an example of austenite relaxation tests at 800°C, the stress decreases significantly and rapidly when the strain level is maintained. A similar phenomenon is also observed at 700°C.

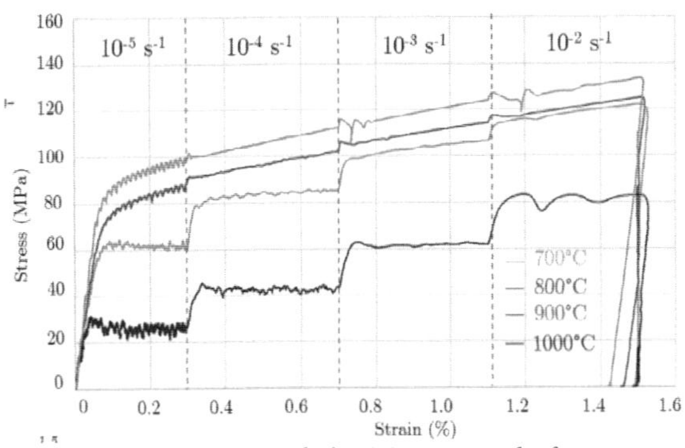

Figure 5: Viscosity and plasticity test results for austenite (at 700°C, 800°C, 900°C and 1000°C) established using tensile tests with a strain rate step.

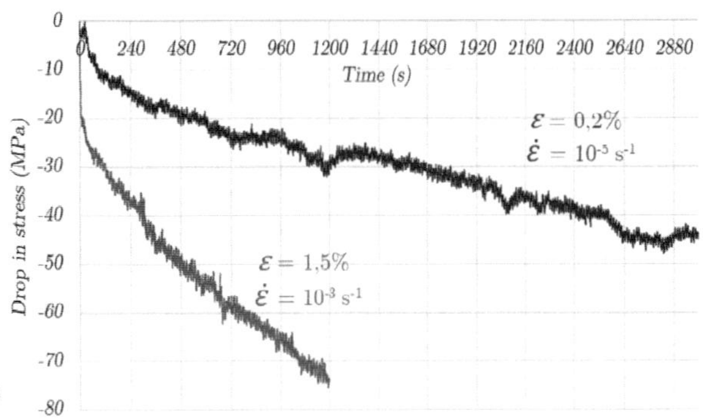

Figure 6: Results of austenite relaxation tests at 800°C for two different initial deformations:

Free dilatometry tests at the same temperatures are conducted out in parallel with these thermomechanical tests to confirm that no phase transformation occurs during characterization and that only austenite is present.

Conclusions

The experimental set-up developed for this study gives promising initial results. The first experiments demonstrate that it is possible to gain reliable and repeatable access to the overall behavior of austenite as a function of temperature. These initial fundings reveal complex behavior for the austenite, with significant variations as a function of temperature, particularly for the viscous effect.

The device also proves its capacity as a dilatometer, with successful initial experiments in free dilatometry (and under stress) during martensitic quenching. In this context, the results show good repeatability and values very close to those found in the literature.

Model parameters are currently being identified for unenriched steel, and the first calculations on homogeneous volume elements will be carried out shortly.

References

[1] LATROBE SPECIALITY STEEL COMPANY, "M50NiL Data sheet," data sheet, 2007.

[2] A. ROSE, "Residual stresses as a result of heat treatments and transformations characteristics," *HARTEREI TECH MITT*, vol. 21, no. 1, pp. 1–6, 1966.

[3] H. BÜHLER and A. ROSE, "Darstellung des Entstehens von Eigenspannungen in Werkstücken aus Stahl in ihren Umwandlungsschaubildern," *Arch. Für Eisenhüttenwes.*, vol. 40, no. 5, pp. 411–423, 1969, doi: 10.1002/srin.196904335.

[4] G. DELBART and R. POTASZKIN, "Contribution à l'étude des aciers à structures intermédiaires obtenues par trempe étagée," *Rev. Métallurgie*, vol. 43, no. 3–4, Art. no. 3–4, Mar. 1946, doi: 10.1051/metal/194643030084.

[5] J. PLATEAU, J. DUFLOT, and C. CRUSSARD, "Contribution à la stabilisation de l'austénite," *Rev. Métallurgie*, vol. 49, no. 11, Art. no. 11, Nov. 1952, doi: 10.1051/metal/195249110815.

[6] G. DELBART and F. MARATRAY, "Étude de quelques aciers à faible teneur en éléments d'alliage, traités par trempe classique et par trempe étagée," *Rev. Métallurgie*, vol. 50, no. 11, Art. no. 11, Nov. 1953, doi: 10.1051/metal/195350110781.

[7] Y. MEYZAUD and C. SAUZAY, "Prevision des pourcentages d'austenite residuelle a l'aide des courbes de transformation en refroidissement continu des aciers," *Previs. Pourcentages Austenite Residuelle Aide Courbes Transform. En Refroidissement Continu Aciers*, 1972.

[8] C. AUBRY, "Modélisation et étude expérimentale des cinétiques de revenu/autorevenu d'aciers trempés. Prévision des contraintes résiduelles de trempe dans un acier cémenté en incluant l'autorevenu," phdthesis, Institut National Polytechnique de Lorraine, 1998. Accessed: Dec. 22, 2023. [Online]. Available: https://hal.univ-lorraine.fr/tel-01750727

[9] A. NAVARRO-LOPEZ, J. HIDALGO, J. SIETSMA, and M. J. SANTOFIMIA, "Characterization of bainitic/martensitic structures formed in isothermal treatments below the Ms temperature," *Mater. Charact.*, vol. 128, pp. 248–256, Jun. 2017, doi: 10.1016/j.matchar.2017.04.007.

[10] S. M. C. VAN BOHEMEN, "Bainite growth retardation due to mechanical stabilisation of austenite," *Materialia*, vol. 7, p. 100384, Sep. 2019, doi: 10.1016/j.mtla.2019.100384.

[11] J.-C. LOUIN, "Effets d'hétérogénéités de teneur en carbone sur les cinétiques de transformations de phase et sur la genèse des contraintes internes lors du refroidissement d'aciers," phdthesis, Institut National Polytechnique de Lorraine, 2003. Accessed: Jan. 09, 2024. [Online]. Available: https://hal.univ-lorraine.fr/tel-01750167

[12] S. CATTEAU, "Effets du carbone et de l'azote sur les cinétiques de décomposition de l'austénite dans un acier faiblement allié : étude expérimentale et modélisation," These de doctorat, Université de Lorraine, 2017. Accessed: Dec. 18, 2023. [Online]. Available: https://www.theses.fr/2017LORR0390

[13] G. ROUX, "Prévision des contraintes résiduelles induites par le soudage TIG d'un acier martensitique (X10CrMoVNb9-1)," These de doctorat, Paris 6, 2007.

[14] E. KRÖNER, "Zur plastischen verformung des vielkristalls," *Acta Metall.*, vol. 9, no. 2, pp. 155–161, Feb. 1961, doi: 10.1016/0001-6160(61)90060-8.

[15] J. D. ESHELBY, "The determination of the elastic field of an ellipsoidal inclusion, and related problems," *Proc. R. Soc. Lond. Ser. Math. Phys. Sci.*, vol. 241, no. 1226, pp. 376–396, Aug. 1957, doi: 10.1098/rspa.1957.0133.

[16] D. P. KOISTINEN and R. E. MARBURGER, "A general equation prescribing the extent of the austenite-martensite transformation in pure iron-carbon alloys and plain carbon steels," *Acta Metall.*, vol. 7, no. 1, Art. no. 1, Jan. 1959, doi: 10.1016/0001-6160(59)90170-1.

[17] A. G. HAYNES and W. STEVEN, "The temperature of formation of martensite and bainite in low-alloy steel," *J Iron Steel Inst*, vol. 183, pp. 349–359, 1956.

QDE 2025: Proceedings of the 3rd International Conference on
Quenching and Distortion Engineering
May 6–7, 2025, Vancouver, Canada
https://doi.org/10.31399/asm.cp.qde2025p0083

Rapid Quenching of Steels Using Ultrasound

Noah J. Kantor, Ellen Troyanosky, Qianchen Zeng, Thomas L. Christiansen
Worcester Polytechnic Institute, 100 Institute Road, Worcester, MA 01609

Abstract

This work serves to expand upon the fundamental idea of high convection quenching (HCQ), examining alternative methods to destabilize vapor barrier formation in liquid quenchants. Namely, ultrasound in combination with a brine solution is applied to realize fast yet controllable quenching conditions. With the use of ultrasound in aqueous quenchant solutions, conditions observed in so-called "intensive quenching" and other high convection quenching methods can be replicated in a controlled laboratory environment. This enables systematic studies on the effect of ultra-fast quenching procedures, i.e. under conditions where the vapor film barrier is destabilized, and initial heat transfer is maximized. These rapid quenching concepts are displayed on low-carbon AISI 1018 and high-carbon AISI 52100 steels. The resulting hardness of 1018 under rapid quenching conditions showed increased surface hardness over the control group of approximately 50 HV. In high-carbon 52100, a significant and deep increase of hardness can be seen (around 100 HV) with the addition of ultrasound during quenching. X-Ray diffraction results indicate rapid quenching conditions yield a reduction in retained austenite for 52100.

Introduction

Quenching in water has been historically considered a bad practice for most steels because of the propensity to develop steep thermal gradients. This often leads to various undesirable consequences like distortion, non-uniform hardening, and sometimes even cracking [1,2]. However, quenching with water can, in specific instances, be effective or even necessary. For example, spray quenching is a common method to continuously harden steel in processes where direct flame hardening or induction are used. Furthermore, in the case of low-hardenability (LH) steel alloys, water can sometimes be the *only* effective quenchant to induce a martensitic transformation.

Counterintuitively, rapid quenching in highly agitated water (convection) or brine solution is claimed to reduce the probability of quench cracking that is typically associated with, say, an immersion water quench. This process is known as intensive quenching [3,4]. Curiously, intensive quenching has the capability to produce material properties that can exceed less aggressive (slower cooling) liquid quenching media such as standard oil or aqueous polymer immersion quenching. Other notably reported benefits resulting from intensive quenching include increased surface hardness, minimal distortion, high compressive stresses on the surface, and substantially increased fatigue performance [4,5]. Intensive quenching has also been considered an alternative to carburizing, demonstrating hardness comparable to carburizing processes along with a favorable residual stress profile in low carbon steels. [4,5]. While the enhanced performance yielded from intensive quenching is proven, the mechanisms responsible for the enhanced properties remain largely uninvestigated, or at least debatable.

Current literature on intensive quenching claims that enhanced properties, notably the high surface compressive residual stresses, are brought on by near instantaneous transformation of the surface into martensite. The cooling and transformation mechanism involves so called "pre-transformation shrinkage" of the austenitic core during cooling. This shrinkage is said to offset the expansion associated with the subsequent martensitic transformation within the core of the quenched part. Increased hardness from intensive quenching has been attributed to "finer grain structure" and "super-strengthened" martensite in past literature [4,5]. Hitherto, no self-contained or consistent explanation has been provided in the literature.

What remains unaddressed is the mechanism of martensite formation. Revisiting the seminal, but largely overlooked, work from Zhao and Notis [6] can perhaps provide a more consistent and physically meaningful explanation to HCQ. In 1995, these authors reviewed older data on martensite formation and put forward a new interpretation on how martensite transforms which can be applied to help explain "anomalies" observed in literature. This article explains how *all* steels and even plain iron can form both athermal martensite as well as isothermal martensite. In TTT diagrams martensite formation should be conceived as C-curves, similar to what is typically only seen with diffusional transformations. Moreover, martensite formation can be suppressed with sufficient cooling rates [6-8].

Typical liquid immersion quenching follows a classic quench curve. Initial submersion of a given austenitized part in a liquid quenchant (water, oil, polymer) causes instant boiling and thus creates a surrounding vapor barrier. During the rapid formation of the vapor barrier, uneven cooling is brought on by non-uniform quenchant boiling. Characteristically, low initial cooling rates occur due to the formation of the insulating vapor barrier. Upon further cooling, the vapor barrier is destabilized, leading to nucleate boiling, where heat flux is at a maximum, and the cooling rate is highest [2,5,9]. Applying sufficient convection can destabilize and even eliminate this vapor barrier, achieving rapid initial cooling rates. Various methods are used to disrupt and eliminate vapor barrier formation. Most of these methods include some form of convective agitation through direct movement of the quenchant. Intensive quenching achieves this through agitation, and alteration of the Leidenfrost temperature using dilute salt solutions within water. One particularly niche method of disruption uses sonic waves, particularly ultrasound, to create localized agitation. Patents going back to the 1950's show this technology being used and effective in the quenching of hardenable steels [10,11]. The present contribution seeks to replicate rapid quenching conditions akin to HCQ for a low and high carbon steel under highly controlled laboratory conditions. To this end an ultrasonic quenching bath is applied to achieve these quench rates.

Experimental Methods

Billet Material

Each of these materials, the low carbon AISI 1018, as well as AISI 52100 were purchased in soft annealed and spheroidized condition from McMaster-Carr. The respective compositions can be seen in Table 1.

Table 1: Average compositions of materials used in experiments, collected using Spark OES

AISI	Fe	C	Si	Mn	Cr	Cu	P	S
1018	Bal.	0.211 ± 0.011	0.244 ± 0.010	0.814 ± 0.003	0.161 ± 0.001	0.191 ± 0.001	0.009 ± 0.001	0.016 ± 0.001
52100	Bal.	0.976 ± 0.037	0.296 ± 0.021	0.326 ± 0.006	1.456 ± 0.026	0.217 ± 0.001	0.009 ± 0.001	-

Fixturing

Steel samples were portioned into billets measuring 25 mm in diameter by 25 mm in length. Samples were fixtured by wrapping in stainless steel wire, such that they could be held without contact on the sample. One half of the face (half the circumference of the billet) was quenched directly. Cross sections were taken halfway across the billet's length.

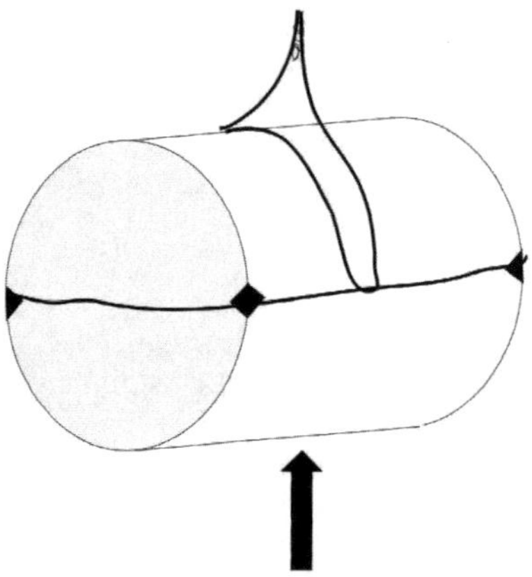

Figure 1: Drawing of fixtured billet with the specified direction of quench

Procedure

Austenitization treatments were performed in a KSL-1500X-S MTI furnace with an inert nitrogen gas flush unit installed. To minimize oxidation and decarburization during heat treatment, samples were "submerged" in steel powders within a small ceramic crucible. The samples of 1018 were covered with H13 powder with particle sizes ranging from 65-150μm. Similarly, samples of 52100 samples were covered with 440C powder with the same approximate particle size.

Austenitization occurred at 925°C for a duration of 30 minutes with all samples. A nitrogen purge was used immediately after placing the fixtured billets in the furnace. A steady inflow of nitrogen was present throughout the duration of the heat and soak. Following the 30-minute austenitization heat and soak, each billet was promptly removed from the muffle furnace and taken out of the powder before exposure to the quenchants. The handling of hot billets was done with the fixtured stainless steel wire to mitigate any direct contact with the sample.

The liquid quenchants used were all composed of distilled water. Three separate quenchant conditions were used. The first condition was immersion in plain distilled water. The second condition used water immersion, but with the addition of ultrasound (abbreviated US). Condition three used a brine solution composed of 10% salt by mass along with the addition of ultrasound (abbreviated BUS). The salt additive used in the brine was non-iodized NaCl. All quenching tests were performed at a bath temperature of 24°C.

The ultrasound was created using the Crest Ultrasonics P1100D – 45 ultrasonic benchtop tank, operated at a frequency of 45 kHz, with an average power output of 180 W. The bath had an internal volume of 12.30 liters (3.25 gallons), which was kept full during each test. For the ultrasonic trials, all samples were held 1 cm away from one of the four ultrasonic transducers at the bottom of the tank. A billet of 1018 was also quenched in an ambient temperature fluidized bed with 120 mesh alumina media, fluidized with compressed air flowing at 4 SCFM. This was done to establish a baseline quench at a relatively low cooling rate, comparable to that of a medium pressure gas quench.

Metallography

Each sample was examined in the as-quenched condition, with no temper. Each billet was precisely cut such that the cross-section would expose a top-down view of the most aggressively quenched location in the sample. Cuts were performed using a Buehler IsoMet High Speed Pro precision saw to obtain cross-sections for examination. Blades were chosen to fit the specific material being cut. Each sample was mounted in a Buehler Simplimet 4000 mounting system, using Buehler PhenoCure compression mounting compound. All mounted samples were polished on the Buehler AutoMet. The grinding procedure consisted of 240, 320, 500, 1000, and 4000 grit SiC abrasive papers. The finishing polishing steps included a 3 and 1 μm monocrystalline diamond suspension made by Buehler. Samples were then etched using Nital reagent. The 1018 samples were swabbed using cotton balls soaked in 3% Nital five times, terminated with deionized water, and cleaned using ethanol. 52100 samples were swiped ten times with 5% Nital-soaked cotton balls. Micrographs were obtained using a Nikon Epiphot 200 Microscope. Vickers hardness tests were completed using a Wilson VH3300 automated hardness tester with a load of 100 g, and a 10 s dwell time. A minimum of 3 hardness profiles were recorded for all quenching conditions on each material to determine standard deviation at each given depth.

X-ray Diffraction

XRD analysis was performed with a PANalytical Empryrean X-ray diffractometer using Cu K-α radiation. A 2θ range of 30-105° was scanned with a step size of 0.03 degrees and step time of 30 seconds. Samples that were previously cross-sectioned and prepared for metallography were used. These cross-sections were in the polished and etched state prior to XRD analysis. The surface region of the cross-section was analyzed for both materials in the immersion quench and the BUS conditions. Results were processed using Origin software and the recorded line profiles smoothed by using Savitzky-Golay method.

Results and Interpretation

AISI 1018

Hardness profiles of the quenched 1018 are shown in Figure 2. The hardness profile of fluidized bed quenched 1018 (Figure 2a) is essentially a plateau, with a hardness level corresponding to a mixed pearlitic and ferritic structure. All water quenched 1018 samples experienced near-peak hardness achievable for 1018 in the surface region, as seen in Figure 2b. The 1018 ultrasonically quenched samples (US and BUS) both demonstrated higher peak hardness near the surface layer when compared with the immersion quenched sample. In particular, the billet with brine quenchant and ultrasound (BUS) showed the highest peak hardness at the surface (outermost 1 mm), nearing 475 HV, although it experienced the fastest drop in hardness when looking across the profile depth. The ultrasonic sample *without* brine (US) demonstrated consistently higher hardness than the immersion quenched sample across the entirety of the depth profile.

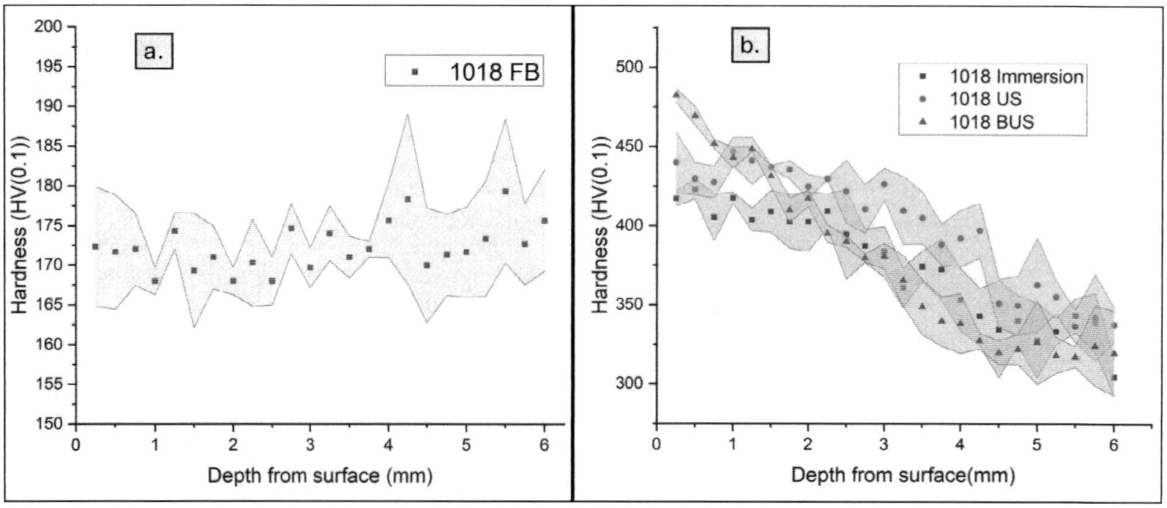

Figure 2: Hardness results of quenched 1018. a. fluidized bed (FB), and b. immersion, ultrasonic (US), and brine with ultrasonic (BUS)

Cross-sectional micrographs were taken for all quenched samples to examine the microstructure at various radial depths into each billet. Figure 3 shows the surface region of all 1018 samples including the fluidized bed quenched sample. The microstructure in Figure 3a is pearlitic (hypo-eutectoid) even in the surface region. This is fully consistent with the hardness depth profile in Figure 2a. It is evident that this less aggressive quenching method is insufficient to develop martensite, i.e. the material has a low-hardenability.

All water-quenched samples show a primarily martensitic structure at the surface. This composition is normally expected to yield lath martensite upon sufficently fast quenching. Figures 3c and 3d show a differing microstructure than the immersion quench seen in Figure 3b. This difference between immersion, and US and BUS conditons may be attributed to a mixed martensite morphology, consisting of both lath and twinned martensite in the rapidly quenched samples. However, further characterization would be required to unravell the exact morphology of the formed martensite for the rapidly quenched samples. The core microstructures of immersion and BUS samples, Figure 3e and Figure 3f, respectively, are both predominantly martensitic, but also contain Widmanstatten ferrite at the prior austenite grain boundaries. The morphology of the Widmanstattem has the characteristic "fish-bone" structure; this feature was absent in the surface region. This type of microstructure is consistent with a slower cooling rate in the core of the quenched samples.

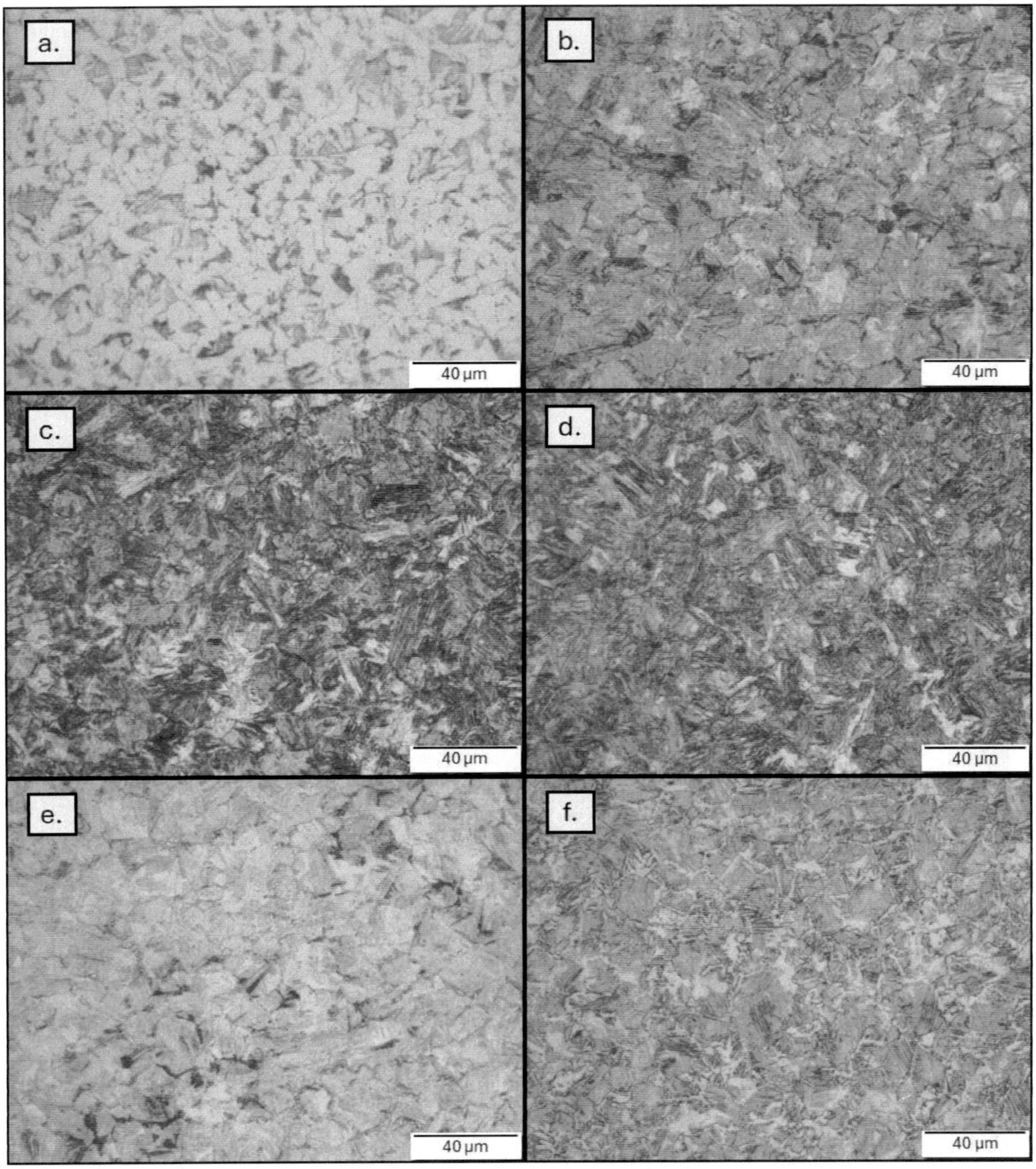

Figure 3: Micrographs of 1018, measured from the sample's cross-section for: a. Fluidized bed quench surface, b. Immersion quench surface, c. Ultrasonic, d. Brine solution with ultrasound surface, e. Immersion quench core, f. Brine solution with ultrasound core

X-Ray diffractograms, seen in Figure 4, were used to compare the least aggressive (water) quenching method, immersion, with the most aggressive method, brine with ultrasound (BUS). There is no discernable difference in terms of peak shape and the only phase present is martensite with a low tetragonality (peak positions are essentially the same as ferrite). However, there is a distinct peak shift between the two conditions; the BUS condition is shifted to lower diffraction angles (see Figure 4B). This could be indicative of a change in stress state when compared with the immersion sample.

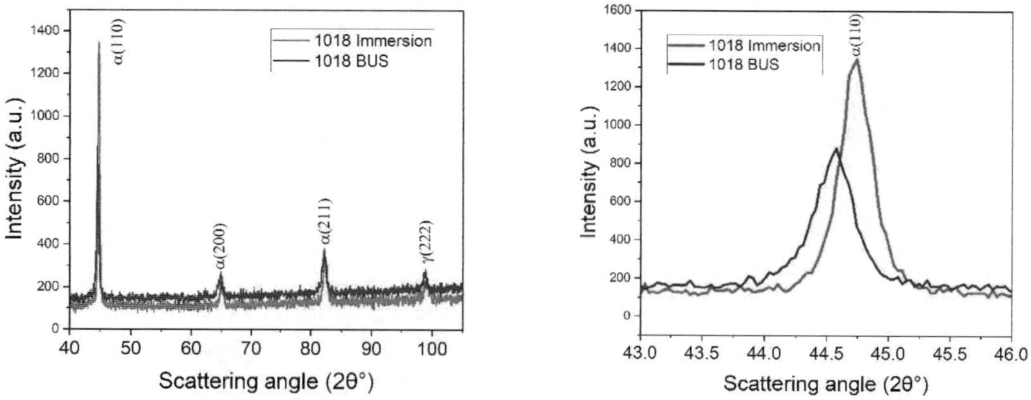

Figure 4: X-Ray diffractograms of 1018 samples, taken from the surface of the cross-section for immersion, and brine and ultrasound (BUS) quenched billets

AISI 52100

In an identical manner to the 1018 billets, the 52100 samples were cross sectioned to reveal their respective microstructures. Figure 5 indicates that all samples are mostly composed of martensite. The martensite morphology appears similar between the various quench conditions. Prior austenite grains can be seen along with undissolved carbides, likely Chromium carbides, which decorate the boundaries. Dense martensite packets can be seen across all the samples. Due to the high carbon content, twinned martensite has formed.

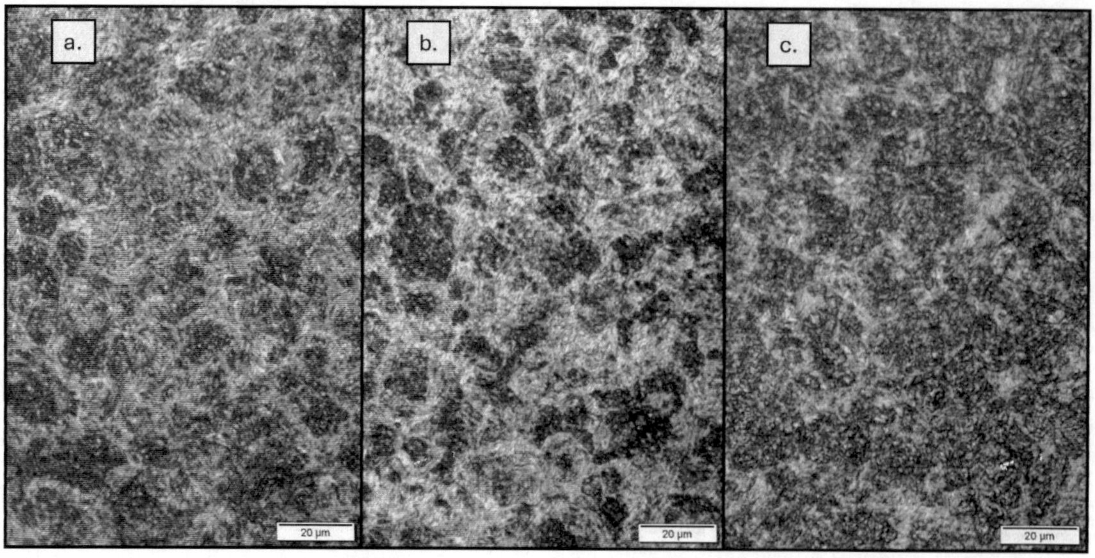

Figure 5: Micrographs of 52100 quenched billet sample cores: a. Immersion quenched, b. Ultrasonic quenched, and c. Brine and Ultrasonic quenched

Hardness results for the 52100 samples in Figure 6 indicate near-complete through hardening across the sample depths, irrespective of quench method. This is consistent with the higher hardeneability of 52100 compared to 1018. The immersion quenching leads to a peak hardness of approximately 850 HV near the surface, and a gradual decrease into the core. The ultrasonically quenched samples (both US and BUS) exhibit a significantly greater hardness, consistently 75-100 HV higher over the full depth. Despite this increase with both ultrasound conditions (US and BUS), the brine sample (BUS) did not appear to have any significant hardness increase over the US sample.

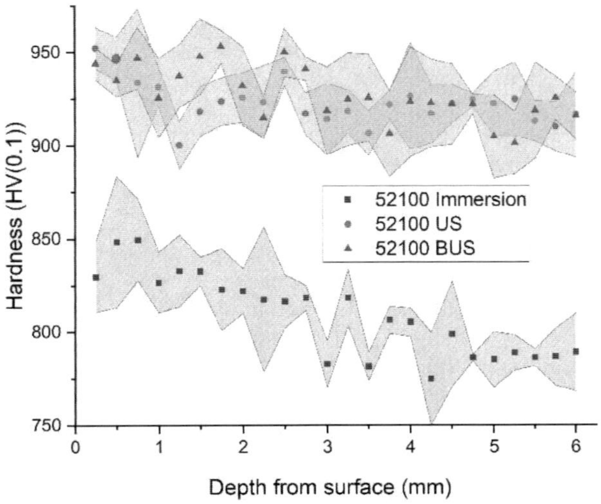

Figure 6: Hardness results of 52100 quenched billet samples: Immersion, ultrasound (US), brine and ultrasound (BUS)

XRD analysis of the quenched 52100 is shown in Figure 7. The immersion quenched condition contains a significant amount of retained austenite whereas for the brine with ultrasound (BUS) condition, retained austenite is virtually absent. The broad and assymmetric martensite peaks reflect tetragonally distorted martensite, consistent with the high carbon content in this alloy (the individual components of the BCT splitting is not indicated; only the main BCC peak in Fig. 7). It is difficult to conclude anything on peak shift of the the different phases.

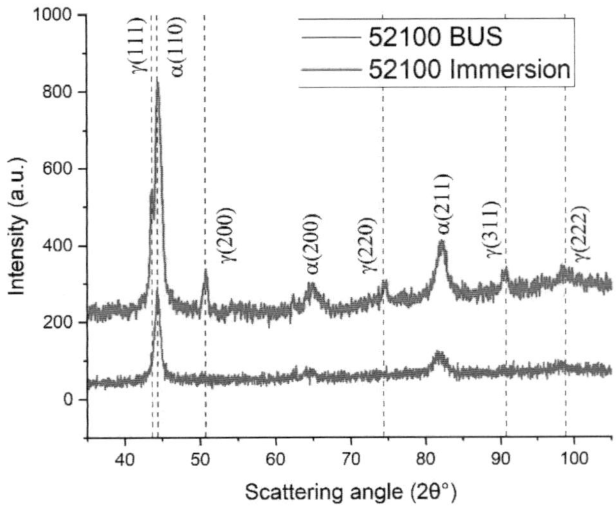

Figure 7: X-Ray diffractograms of 52100 samples, taken from the surface of the cross-section for immersion, and brine and ultrasound (BUS) quenched billets

Discussion

The interpretation of martensite formation as provided by Zhao and Notis [6] could shed new light on the mechanism behind fast quenching. Due to the short timescale of cooling and the very high local cooling rates achieved, it could be inferred that upon extreme initial cooling conditions, the surface layer of a given quenched part could experience suppressed or delay in martensite formation. Hence, the extreme local cooling rate in the surface region would give way to a lowering of the M_s temperature and a concomitant change of martensite morphology from lath to twinned in low carbon steels. Moreover, the isothermal component, which is considered to be fast, should be taken into account. Adopting this concept for HCQ means that if the surface were to transform *after* the subsurface (the most extreme situation), transformation stresses would lead to compressive stress in the surface region. Normally, only the thermal stresses upon quenching will lead to the development of compression within the surface. In general, any lowering of M_s or delay in formation of martensite in the surface region would favor compression in the surface. Martensite morphology has been reported to have a direct connection to carbon content where lath is most frequently seen in low carbon steels and twinned morphology is normally seen in higher carbon steels. For low carbon steels, extremely rapid cooling can lead to a change of morphology from lath to twinned martensite. This is expected to result in increased hardness as twinned martensite should be the hardest of the two morphologies.

Turning to the present result, the hardening response of the low carbon 1018 ostensibly depends on the severety of the quench applied. According to hardness data in Figure 2b, the BUS sample appears to have a stark increase in peak hardness at the surface, where cooling rate is highest. This result could be corelated with the observed change of martensite morphology with cooling rate. Here, it could be postulated that the ultrasonic conditions lead to a mixed type of martensite (lath & twinned). This would be consistent with the behavior described above (obervations reported in Zhao and Notis), i.e. supression of lath martensite at high cooling rates for low carbon steels [6].

XRD results for 1018 indicate a different stress state depending on the cooling conditions: possibly compression in the surface of the BUS quenched state, although more in-depth stress analysis would be needed before drawing firm conclusions. Alternatively, an altered stress state and hardness could be a direct result of plastic deformation brought on by extremely steep thermal gradients locally exceeding the yield strength of the material. Curiously, it is also seen that the hardness profile for the 1018 BUS sample quickly drops off throughout the sample's depth, even reaching below that of the immersion quenched sample. It could be plausible that a change in morphology could alter the heat conduction regime though the sample, e.g. by different portions of martensite and austenite at different depths during cooling. Effectively, this could result in a different cooling profile and hence hardness depth profile.

52100 displayed the most prominent increase in hardness between the two materials. This vast change brought on by the use of ultrasound indicates that there is in fact some alternate hardening regime occuring. Due to its carbon content, it is likely not a change in martensite morphology (twinned martensite will form irrespective of cooling rate in high carbon steels) [6,7]. However, XRD data from the surface of the 52100 indicates that there is very little retained austenite in the BUS sample as compared to the immersion sample. Certainly, a higher martensite fraction would result in an increase in overall hardenss. The reasoning behind how this austenite fraction changed has yet to be determined. It should be noted that a high fraction of martensite is observed over the full depth of the sample and not just in the surface region.

Conclusions

This study investigated the response of two materials, low carbon AISI 1018 and high carbon AISI 52100, to quenching under extreme cooling conditions using ultrasound and brine. Three quenching conditions were evaluated: immersion quenching in still water, quenching in water with the use of ultrasound (US), and quenching in a 10% brine solution with ultrasound (BUS). Each material demonstrated a unique response to the addition of ultrasound, as seen through the difference in hardening, microstructure, and with the collected diffractograms. 1018 demonstrated a clear increase in surface hardness (~50 HV) under the BUS condition when compared with the standard immersion quench in water. The 1018 US condition demonstrated consistently higher hardness than the immersion sample across the entire depth. XRD results for 1018 indicated a peak shift, indicative of a change in stress state. 52100 samples experienced significant increase in hardness (~75-100 HV) under the US and BUS conditions. Diffractograms indicate that these quenching conditions yielded little retained austenite when compared with the immersion quench. Regardless of the cooling and transformation mechanism, which is not yet fully understood, it was observed that rapid quenching conditions are possible to obtain in a laboratory setting.

Acknowledgments

Support from Center for Heat Treating Excellence (CHTE) is gratefully acknowledged.

References

[1] Totten, G.E., et al., *Handbook of Quenchants and Quenching Technology*, ASM international (1993), p 444-446.

[2] Cook, W.T., "A Review of Selected Steel-Related Factors Controlling Distortion in Heat Treatable Steels," *Heat Treating: Proceedings of the 18th Conference,* October 1998, p 12-21. doi:10.1361/cp1998ht012

[3] Kobasko, N.I., Prokhorenko, N.I., "Effect of the Quenching Rate on the Formation of Cracks in Steel," *Metal Science and Heat Treatment*, Vol. 6, No. 45 (1964), p 104-105. doi:10.1007/BF00655387

[4] Aronov, M.A., et al., "Application of Intensive Quenching Methods for Steel Parts," *Proceedings of the 21st ASM Heat Treating Society Conference,* Indianapolis, Indiana, November 2001, p 12-21. doi:10.1361/cp1998ht012

[5] Aronov, M.A., Kobasko., N.I., Powell, J.A., "Basic Principles, Properties and Metallurgy of Intensive Quenching," *SAE Transactions,* Vol. 111, Section 2: Journal of Commercial Vehicles (2002), p 37-44. doi:10.4271/2002-01-1338

[6] Zhao, J., Notis, M.R., "Continuous Cooling Transformation Kinetics Versus Isothermal Transformation Kinetics of Steels: A Phenomenological rationalization of Experimental Observations," *Materials Science and Engineering: R: Reports,* Vol. 15, Issues 4-5 (1995), p 135-207. doi:10.1016/0927-796X(95)00183-2

[7] Roitburd A.L., Kurdjumov, G.V. "The Nature of Martensitic Transformations," *Materials Science and Engineering,* Vol. 39, Issue 2 (1979), p 141-167. doi:10.1016/0025-5416(79)90055-7

[8] Villa, M., Somers, M.A.J., "Thermally Activated Martensite Formation in Ferrous Alloys," *Scripta Materialia,* Vol. 142 (2018), p 46-49. doi:10.1016/j.scriptamat.2017.08.024

[9] Totten, G.E., et al., *Handbook of Quenchants and Quenching Technology*, ASM international (1993), p 70-71.

[10] T. J. Bulat, U.S. Patent 2,920,988

[11] R.F. Harvey, U.S. Patent 2,798,832

[12] Moskaleno, A.A., et al., "Quenchants Characterization by Acoustical Noise Analysis of Cooling Properties of Aqueous Poly (Alkylene Glycol) Polymer Quenchants," *Proceedings of the Second International Conference on Quenching and the Control of Distortion,* Cleveland, Ohio, November 1996, p 117-122.

[13] Redman, R., Kessler, O., "Ultrasonic Assisted Water Quenching of Aluminum and Steel Cylinders," *International Heat Treatment and Surface Engineering,* Vol. 6, Issue 3 (2012), p 115-121. doi:10.1179/1749514812Z.00000000021

[14] Kobasko, N.I., "Investigation of Batch Intensive Quenching Processes When Using Hydrodynamic Emitters in Quench Tanks," *Eureka: Physics and Engineering,* No. 6 (2016) p 29-36. doi:10.21303/2461-4262.2016.00212

QDE 2025: Proceedings of the 3rd International Conference on
Quenching and Distortion Engineering
May 6–7, 2025, Vancouver, Canada
https://doi.org/10.31399/asm.cp.qde2025p0092

Shaft distortion as a result of local flow inhomogeneities during oil quenching in an industrial quench tank

Gabriel Ebner, Dr.-Ing. Thomas Lübben, Dr.-Ing. Holger Surm
Leibniz-Institut für Werkstofforientierte Technologien (IWT), Bremen, Germany
ebner@iwt-bremen.de, +49 421 218-51407

Abstract

The quenching process during the heat treatment of components can contribute to the resulting extent of distortion. Location dependencies of the oil flow in the quench tank lead to gradients in heat transfer coefficient, temperature, elongation and microstructure transformation. This ultimately results in distortion. To gain a deeper understanding of the associated mechanisms impacting distortion the aim of this work is to work out the relationships between flow inhomogeneities in an industrial quench tank and the distortion of gear shafts. For this purpose, oil flow-modifying measures are used to induce specific shape-changes on case-hardened gear shafts from the commercial vehicle sector, which are quantified by runout and coordinate measurements. Extensive flow measurements using impeller anemometry provide information about the existing quenching conditions resulting from the directed flow in the oil bath. In addition, shafts are quenched in a laboratory quench tank with defined inhomogeneities in the flow velocity. Recordings of the quenching process using a video camera in combination with coordinate measurements on the quenched shafts provide additional information for understanding the development of shaft distortion. The results of the different experiment show that flow components that are perpendicular to the longitudinal axis of a shaft have a big impact on the quenching process and the distortion. These asymmetric conditions lead to earlier rewetting at the affected positions and therefore to inhomogeneous cooling of the part. Furthermore the increase of the flow velocity leads to higher runout deviations.

Introduction

In this work, the relationships between oil flows in the quench tank and the development of runout on gear shafts in the context of case hardening were investigated. One of the questions to be investigated was the role of rewetting during the quenching process in the development of these form changes. According to the working hypothesis on which this investigation was based, inhomogeneities in the flow lead to asymmetric rewetting on the component, which in turn leads to asymmetric heat transfer. The resulting temporal gradients during the transformation of the microstructure generate stresses that can ultimately lead to distortion. The following points were addressed in this context:

- Determination of the influence of the circulation intensity and flow-influencing installations in an industrial quench tank on the runout deviation of shafts
- Determination of flow distributions in an industrial quench tanks that lead to particularly large runout deviations
- Evaluation of the role of rewetting in the development of runout deviations using defined flows in laboratory tests
- Identification of distortion-critical flows in the oil quenching of shafts
- Derivation of instructions for the design of quench tanks, charging racks charging quenching process management

Experimental Methods

In order to achieve the research objective, a series of tests were carried out in an industrial quench tank of a multi-purpose chamber furnace and in a laboratory quench tank. A rig to measure the runout deviations of gear shafts was set up in which the runout deviations and the curvature directions of the shafts were measured before and after heat treatment using dial gauges and an angle scale. For the flow measurements, a vane wheel anemometer was used. During the experiments in the industrial quench tank, several batches of shafts from the passenger car and commercial vehicle sector were case-hardened. In addition to the reference batches under normal conditions, quenching was also carried out in which flow-influencing measures were taken in the quench tank in order to specifically generate extreme effects in the distortion of the shafts. By successively introducing batch racks, flow influencing measures and entire batches into the quench tank, the resulting flow distributions around a cold batch were measured in detail with the vane wheel anemometer. This was necessary in order to be able to investigate correlations between the observed distortion phenomena and the underlying flow conditions at a later stage. In the laboratory quench tank experiments, cylinder samples were quenched from a tube furnace into a quench tank with two windows for observing the quenching process. In this way, the influence of a defined asymmetrical flow during quenching on the rewetting of the samples was investigated. In the first step, the quench tank was modified to generate defined flows. For this purpose, a pump and corresponding piping were installed, which enabled the samples to be quenched in a free jet from below and laterally at 45°. The flow velocity was varied via the pump and the generated flows were evaluated using further flow measurements. Coordinate measurements before and after heat treatment were used to determine the amount and direction of runout deviations as a result of asymmetric quenching on the samples. The rewetting processes at different quenching parameters were documented using video recordings. Finally, the results of both series of tests were combined in order to identify critical flows that have a particular impact on the development of runout deviations on shafts.

Results

Distortion caused by flow influencing measures

In order to create extreme effects during shaft distortion, a plate was inserted under one half of the batch carrier to block the flow coming from below for this part of the batch. Figure 1 shows the batch carrier with the commercial vehicle shafts and the blocking by the inserted plate under Half 1. Batches were case-hardened in this configuration and the resulting distortion of the shafts was then measured. In addition, batches without flow influence were also case-hardened to define a reference. It was found that the shafts in Row 2 and Row 3 in particular showed increased runout deviations compared to the reference batches. In addition, a pronounced direction of curvature was measured on these shafts, which could be reproduced over several repeat tests and at different quenching speeds.

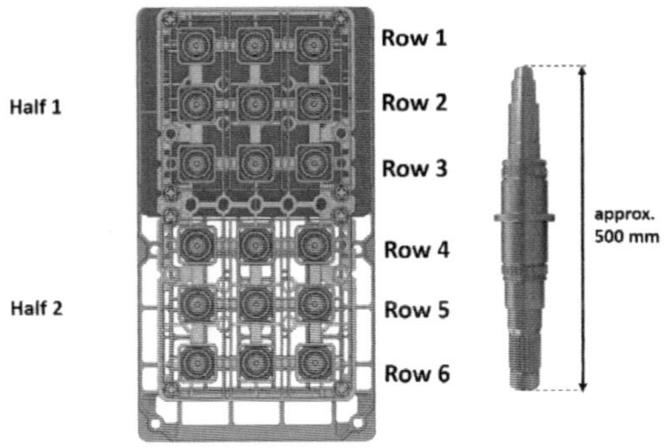

Figure 1: Batch carrier with blockage under Half 1

Flow measurements in the industrial quench tank

In the industrial quench tank the local flow velocities were recorded at different speeds of the circulation system and at various points inside the quench tank. The velocities of the intermediate positions were subsequently interpolated from the measuring grid. Loaded batch carriers were also inserted into the quench tank, including the blocking of a half batch described above, and the resulting flow distributions were measured. For this purpose, the vertical component of the flow was measured using horizontal measuring grids above the batch carrier. For the horizontal flow component, on the other hand, vertically oriented planes were recorded between the rows of shafts in the batch carrier. Due to the blocking of a half batch from below, only a very low flow velocity was measured in the affected area in the vertical direction, as expected, as can be seen on the left in Fig. 2. However, as described above, the highest runout deviations and a pronounced direction of curvature were measured on the shafts in this area. Due to the reduced cross-section, the intensity of the flow in the non-blocked part of the batch was significantly higher than in the reference batch. Only by measuring the horizontal component did a more complete picture emerge of the flow conditions prevailing in the quench tank as a result of the flow influencing measures. It was found that the half-sided blocking of the batch carrier caused a deflection of the oil flow. The blockage thus led to the formation of a strong horizontal component, whereby the affected shafts were subjected to a lateral flow (Fig. 2, right). Due to the predominant absence of a further flow component in the blocked part of the batch, the lateral flow must have been therefore responsible for the development of the conspicuous effect during warpage and was therefore investigated further in the following.

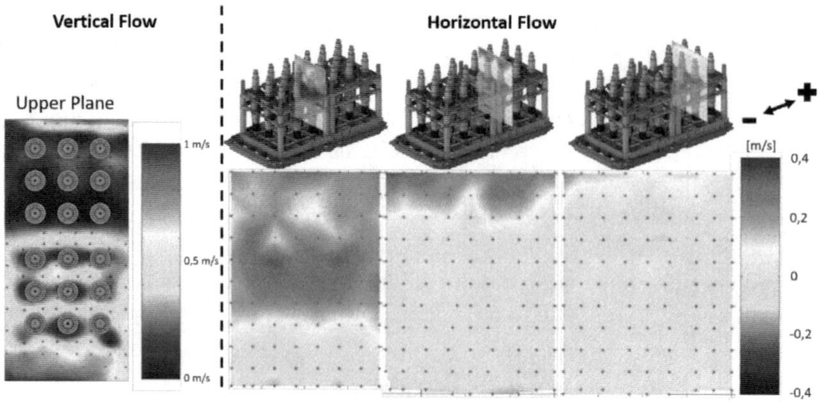

Figure 2: Flow distribution in the batch carrier between the shaft rows in horizontal direction with blocked half batch

Quenching tests in the laboratory quench tank

In addition to the tests on the serial shafts, quenching tests were carried out on cylinder samples made of 20MnCr5 (AISI-5120H) in a laboratory quench tank to observe quenching processes. The flow conditions from the industrial quench tank were reproduced on a smaller scale. The aim was to investigate the effects of the previously shown asymmetric quenching on the rewetting processes. For this purpose, the curvature of the samples as a result of this quenching was evaluated. The cylinder samples were first heated in a vertical tube furnace and then quenched in the quench tank below. The quench tank had two windows for recording the quenching with a camera. The samples were subjected to a free jet, which was directed at 45° from below towards the center of the sample. This configuration simulated the situation from the industrial quench tank, which was present in the transition area between the half batches. The flow velocity could be varied using a pump. The local velocity of the flow along the samples was measured using a vane wheel anemometer in the immediate vicinity of the sample surface. It was found that the highest velocities were measured in the center area of the sample that was exposed to the flow. The maximum speed increased linearly to up to 1.7 m/s when the pump output was increased. Coordinate measurements also made it possible to evaluate the dimensional and shape changes of the cylinders as a result of this quenching. The effects on the rewetting of the samples during the quenching process can be seen in Fig. 3. It shows images taken at different times after the samples were immersed in the oil and at different pump outputs. It can be seen that the asymmetrical flow led to accelerated rewetting in the affected area. The increase in the flow velocity led to an acceleration of this process, so that rewetting took place earlier. The collapse of the vapor skin thus generally led to faster cooling in the rewetted area in contrast to the side of the sample facing away from the flow.

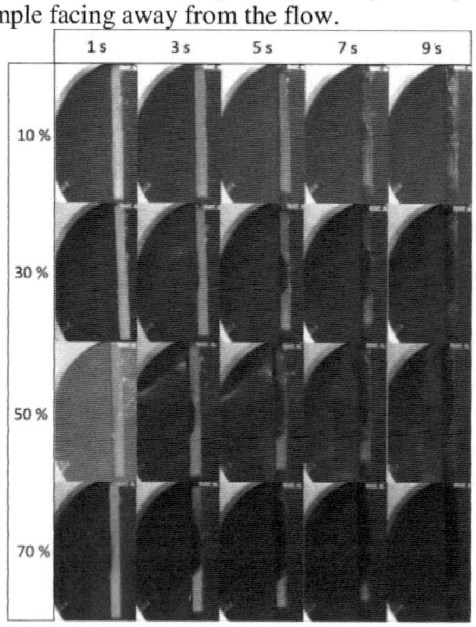

Figure 3: Development of wetting during quenching processes under asymmetric flow, different relative pump capacities and after different times

The runout deviations of the cylinder samples quenched under these conditions were then evaluated. Figure 4 shows the correlation with the set pump power. Two directions of curvature occurred during the tests, both of which were oriented along the direction of flow. In the positively curved samples, there was initially an increase in runout deviations with increasing pump power. As the pump power increased further, the amount of runout decreased again until finally a negative curvature was present. It can be assumed that all quenched samples exhibited a negative curvature at the beginning of the quenching process due to the contraction of the side that cooled first. The specimens quenched at high pump capacities therefore underwent a twofold change in the direction of curvature during the quenching process. The negatively curved samples, on the other hand, showed a different behavior. The negative curvature initially present in these samples due to the one-sided cooling was always increased with increasing pump power and amounted to up to 600 µm. The two directions of curvature could predominantly be assigned to two production batches with slightly different compositions of the melt. The development of the two directions occurred under otherwise constant quenching conditions. For this reason, it is conceivable that the chemical composition caused temporal differences in the microstructure transformations, each of which resulted in the development of a specific direction of curvature. It should be noted that in both cases, at flow velocities usual for industrial quench tanks, a strong increase in runout deviations occurred as a result of the asymmetrical quenching process caused by the lateral flow.

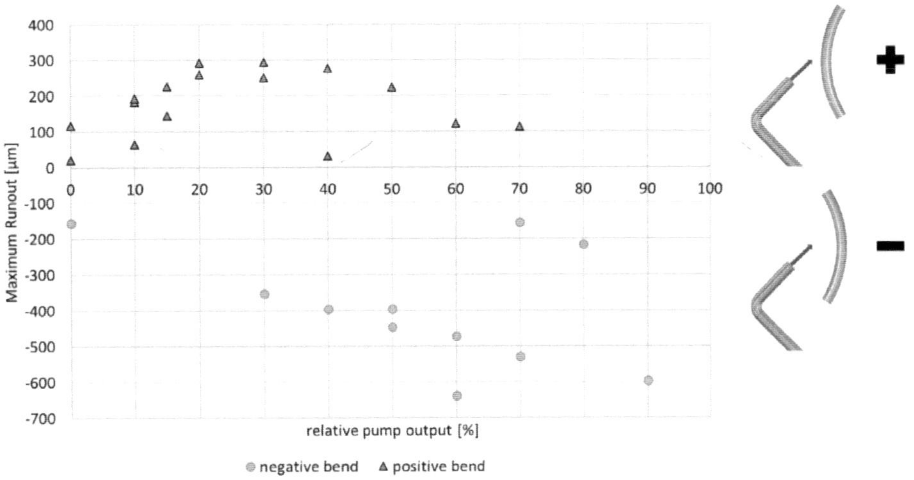

Figure 4: Runout deviations as a result of the lateral flow at different relative pump capacities

Conclusions

Case hardening tests were carried out on gear shafts in the quench tank of a multi-purpose chamber furnace. The quenching in oil was partly carried out by influencing the flow in order to cause extreme distrortion effects. Flow measurements in batch setups showed a horizontal component in the quench tank, which led to asymmetrical quenching of the shafts in the affected area. It was precisely on these shafts that runout measurements showed a conspicuous increase in the maximum runout deviations as well as a pronounced direction of the curvatures along the direction of the horizontal component of the flow. In addition, tests were carried out in a laboratory quench tank in order to investigate rewetting processes during quenching in more detail. For this purpose, the described flow situation from the industrial quench tank was reproduced in the form of a lateral free jet for quenching cylindrical samples. Flow measurements along such a cylinder sample revealed the local flow velocities. Video recordings of the quenching processes showed how the rewetting of the sample was influenced by the type and intensity of the prevailing flow. The result of the asymmetric quenching was also strongly increased runout deviations and an alignment of the curvatures along the flow direction.

The investigations in this project have shown that flow components transverse to the longitudinal axis of shafts and elongated components can have a highly detrimental effect on the occurrence of runout deviations during oil quenching. This should be taken into account when quenching such parts and when designing quench tanks to reduce distortion. In addition, the circulation intensity should only be as high as necessary, as it has also led to a significant increase.

QDE 2025: Proceedings of the 3rd International Conference on
Quenching and Distortion Engineering
May 6–7, 2025, Vancouver, Canada
https://doi.org/10.31399/asm.cp.qde2025p0096

Influence of Quenchant Temperature on Microstructure and Hardness in Jominy End Quench Test for AISI 4135 Steel

J. S. Alabi[1], E. Heidari[2], M.F. Buchely[2], K. Chandrashekhara[1], S. N. Lekakh[2], R. J. O'Malley[2]
[1]Department of Mechanical and Aerospace Engineering
[2]Department of Materials Science and Engineering
Peaslee Steel Manufacturing Research Center
Missouri University Science and Technology, Rolla, MO 65409, USA

V.Athavale[3], and A. Kumar[3]
[3]Nucor Steel Memphis, Inc.
3601 Paul R. Lowry Road, Memphis, TN 38109, USA

Abstract

The temperature of quenchant and the severity of quench can significantly influence the mechanical properties, phase transformation, and hardness of the steel. This study examines the influence of quenchant temperature through experimental and numerical simulations to predict temperature profiles and phase fractions distributions for AISI 4135 steel. The Jominy test specimen was modeled using the SIMheat module of the FORGE software for estimating cooling rates, temperature profiles, and microstructure evolution considering thermophysical properties obtained from the JMATPro database. In experiments, the water quenchant temperature varied from cold (4°C) to room temperature (25°C) to hot water (60°C). The experimental and simulated results were compared, estimating the hardness profile for different quenchant temperatures on AISI 4135 steel. Some specific details obtained from this comparison are discussed.

1.0 Introduction

Heat treatment of steels, particularly the quenching process, aims at controlling the transformation from a parent phase (austenite) into other phases (martensite, bainite, ferrite, to achieve targeted mechanical properties [1]. AISI 4135 is a high-strength low alloy steel used in areas such as aerospace and heavy machinery, with controlled cooling and optimized mechanical properties such as hardness, toughness, strength, and residual stress profiles. Quenchants in the quenching process facilitate heat transfer from the steel sample to the surroundings. Considering the temperature of the quenchant, this seemingly simple parameter may profoundly influence cooling rates and microstructural evolution, making it a key focus of this study [2].

Quenchant temperature influences the rate of cooling during quenching; this influences the mechanical properties. Low quenchant temperatures have the potential to increase the temperature gradient during quenching, increasing the cooling rate and influencing the transformation temperature [3], [4]. Changes in the microstructure, such as the formation of martensite, occur upon quenching and drastically influence mechanical properties. Andreansyah et al. indicated that treatment parameters need to be selected to achieve the properties required for specific applications [5][6]. There have been a number of empirical research studies conducted using the Jominy test. Multiphysics models integrating phase transformation kinetics and mechanical properties prediction have been developed to accurately predict phase transformation and mechanical properties, such as in research involving high-strength steels like AISI 4135.

In the Jominy end-quench test, a steel specimen is quenched at one end to measure the hardenability of an alloy. Numerical models have been developed to simulate the cooling curves and predict the microstructural formation of the phases present, and the hardness profile obtained during the test. M Models have also been developed over time to estimate the hardenability of more complicated shapes of steel components using numerical models [7]. Many models have been developed, with methodology focusing on predicting the microstructure and mechanical properties under differing imposed cooling conditions. Homberg et al. leverage Scheil's additivity rule and the Johnson–Mehl equation to model the austenite-to-pearlite transformation in a study simulating the Jominy end-quench test in AISI 1080 steel [8]. Le Masson et al. developed a two-dimensional numerical method to estimate convection heat transfer coefficients, a vital variable in metallurgical heat treatment simulations. Their work emphasized the importance of precise heat transfer modeling for accurate predictions of cooling rates and subsequent transformations [9].

The Finite Element Method (FEM) provides a numerical method to discretize a continuous domain into more minor elements, forming a mesh structure [10]. Homberg describes a numerical algorithm that allows the simulation of the Jominy end-quench test and the derivation of a continuous cooling curve for steel. There are many commercial solvers available, and for this study, FORGE was used because the solver also allows users to incorporate specific material models, such as JMAK and the Koistinen-Marburger equations [11], [12].

This study employs the Jominy end quench model, including the effect of quenchant temperatures on the mechanical properties of AISI 4135 steel, given its critical role in demanding applications. Developing models in FORGE to predict the evolution of microstructure and the distribution of hardness, the proposed approach introduces a predictive model for simulating microstructural changes and hardness distribution during the Jominy end-quench test. This Multiphysics model for the solution of phase transformations in AISI 4135 steel will give better accuracy with less experimental efforts in industrial heat treatment operations.

2.0 Materials and Methodology

2.1 Experiment

In this study, the AISI 4135 steel grade that was commercially procured was investigated using optical emission spectroscopy (OES) and the LECO combustion method to obtain its composition. The average chemical composition is detailed in Table 1. Before conducting the Jominy end-quench test, the machined cylindrical bars were normalized at 925°C for 1 hour in a steel box with graphite added to mitigate decarburization and then air-cooled to eliminate any microstructural variations introduced by preceding hot working processes, according to recommendations of the ASTM A255 standard [13].

Fig. 1a and Fig. 1b illustrate the microstructural evolution of the steel specimens from their as-received hot rolled condition to the normalized state. The steel variants exhibited pronounced microstructural banding and segregation in the as-received condition, particularly in the mid-thickness region [14]. Although normalization reduced the severity of segregation and resulted in a more uniform microstructure, traces of structural heterogeneity remained. The normalized material was deemed suitable for the Jominy end-quench test.

(a) (b)

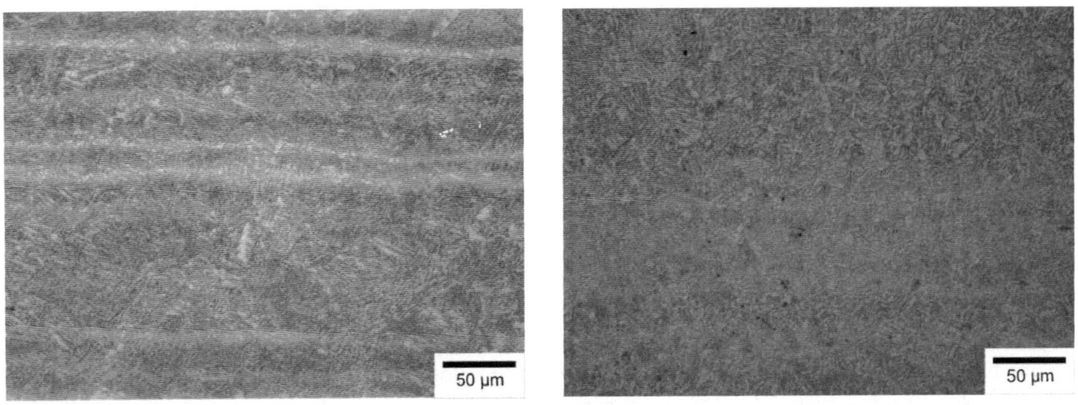

As-Received After Normalizing

Fig. 1. Microstructural of AISI 4135 steel as-received condition and after normalizing.

Table 1: Chemical composition of materials, wt.%. Fe in balance.

AISI/SAE	C	Mn	P (max)	S	Si	Ni	Cr	Mo	Cu	Cb
4135	0.33–0.38	0.70–0.90	0.009-0.015	0.002-0.003	0.15-0.35	1.65-2.00	0.80-1.10	0.15–0.25	0.17 - 0.21	0.02-0.03

The samples were divided into three groups, and each group was treated with tap water in the Missouri science and technology Jominy set up, at different temperatures: 4°C, 25°C, and 60°C, respectively. The study includes estimating the cooling rate's effect on the hardness profile and prediction of the microstructural evolution. For the 4°C condition, ice was added to the bath to maintain the water temperature and it was monitored constantly during quenching. The 25°C condition represented room temperature, with no further modification applied to the water temperature, which reflected a moderate cooling rate suggested by the ASTM A255 standard. For the 60°C condition, a heating element was used to maintain the water temperature.

All the specimens were austenitized at 925°C for 1 hour in a graphite containing steel box to minimize the scale formation and decarburization of the Jominy bars. Then, the specimen was rapidly transferred to the quenching fixture, where a controlled flow of water was applied to one end of the bar, rapidly cooling the steel. After Jominy testing, specimens were machined along the vertical axis to a depth of 0.38 mm to measure the hardness in HRC scale, as shown in Fig 2.b.

In all cases, temperature control and monitoring were done with precision to ensure the reproducibility and reliability of the experimental conditions. This methodology allowed for an effective study of the influence of the water temperature variation on hardness and microstructure in the steel samples.

In this setup (Fig. 2a), a pyrometer positioned 1.5 inches from the quenching end captured the temperature at the surface of the Jominy samples (approximately Jominy Position 24).

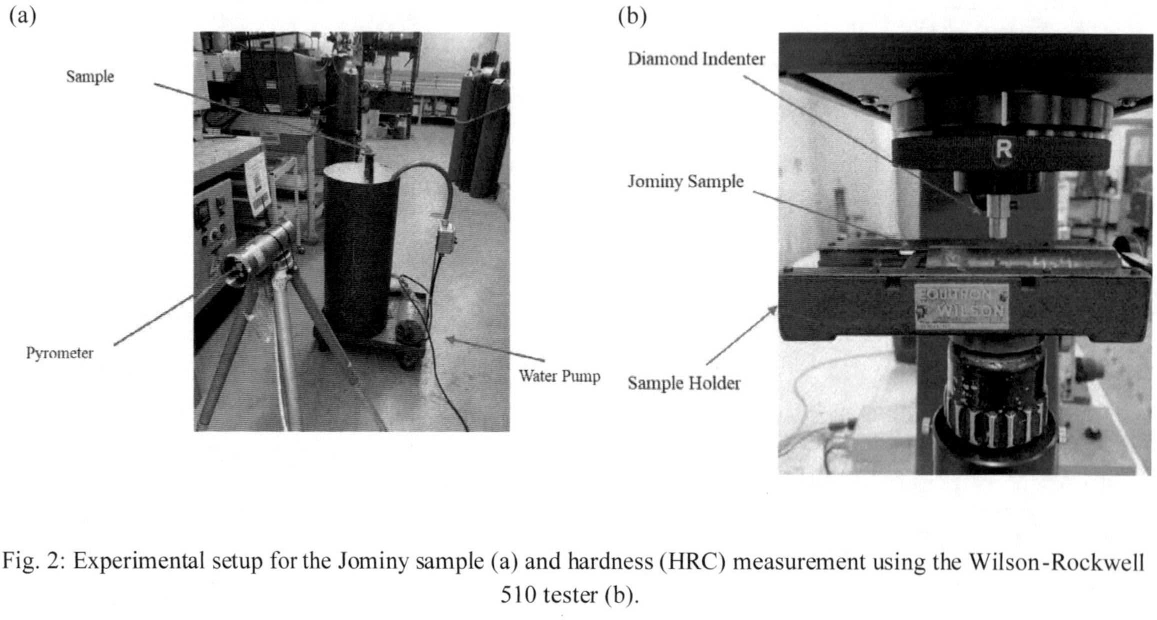

Fig. 2: Experimental setup for the Jominy sample (a) and hardness (HRC) measurement using the Wilson-Rockwell 510 tester (b).

Experiment Validation: In the experimental setup, a pyrometer located 1.5 inches (38.1mm) from the quenching end of the Jominy bar (about Position 24) recorded the specimen surface temperature and cooling curves were collected at a 50 Hz rate using a data acquisition system. Hardness tests were conducted at 1/16-inch (1.5875mm) intervals along the bar using a Rockwell C hardness tester according to ASTM E18 standards. The microstructure was examined using optical microscopy afteretching the samples with a 4% Nital solution to reveal microstructural features, such as prior austenite grains, martensite, bainite, and ferrite. Microstructure quantification was achieved through MIPAR image analysis, which enabled the segmentation and calculation of the proportions of martensite, bainite, and ferrite.

2.2 Modeling Jominy End Quench

Modeling of the Jominy end-quench test was performed using a cylindrical geometry measuring 25 mm in diameter and 100 mm in length, similar to the experimental Jominy bars. A Multiphysics model was employed for the quenching process to account for heat transfer, phase transformations, and strain/stress evolution phenomena involved in the heat transfer process. This study used a two-dimensional axisymmetric model to estimate the heat transfer in the end quenching of Jominy bars using FORGE software for simulation. Material properties and thermal properties such as density, conductivity, and enthalpy are temperature-dependent and defined using the JMatPro model [14]. The heat transfer coefficient estimation conducted by Sugianto et al. [15] was integrated into this model using their experimental data as shown in Fig. 4.

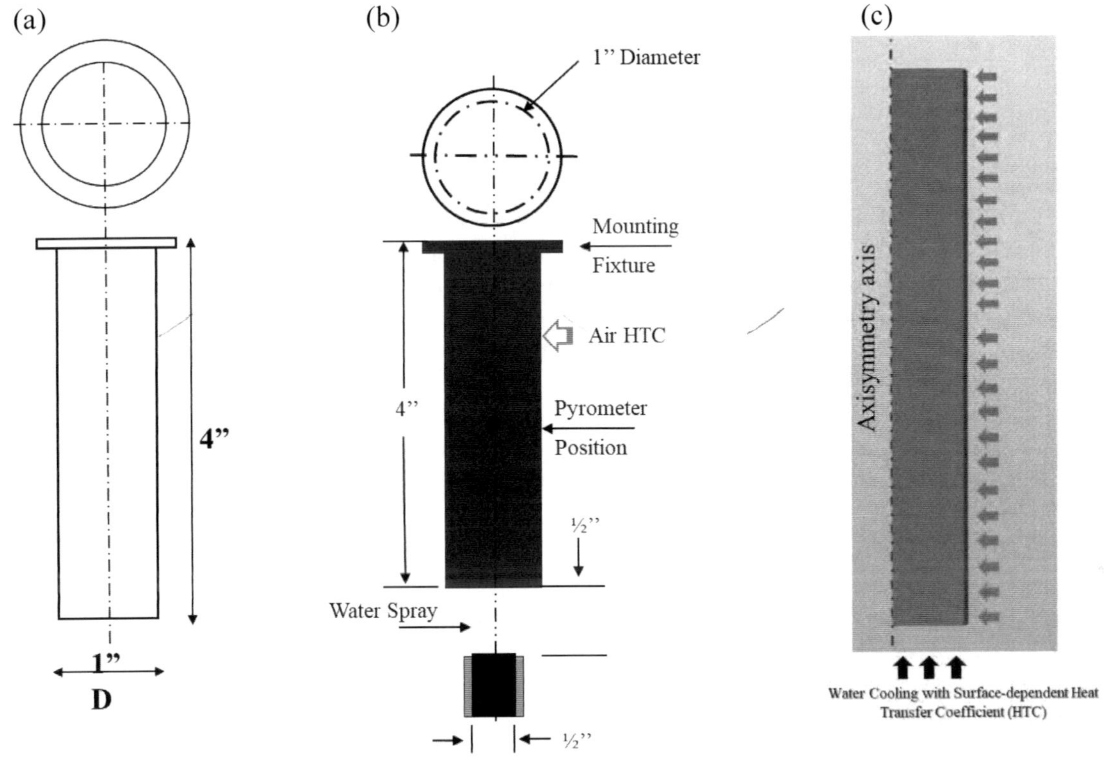

Fig. 3: (a) Geometry of the Jominy sample, (b) application of boundary conditions with an axisymmetric approach, and (c) Simplified model with boundary conditions.

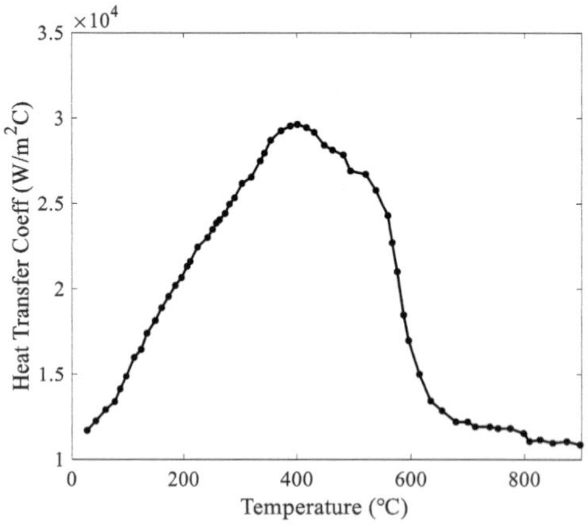

Fig. 4. Temperature-dependent HTC with water [15]

Thermal Transfer Modeling. Modeling of transient thermal behavior during quenchingincludes conduction, convection, and radiation, with an added consideration of latent heat that may be released or absorbed during phase transformations. The modeling considers thermophysical properties, such temperature dependent specific heat and thermal conductivity, which can also vary significantly among different phases (for instance, austenite and martensite). Coupled thermal-mechanical models are needed for accurate temperature evolution, microstructure, and stress development predictions. Transient heat transfer equations govern the quenching process's thermal dynamics modeled using the SIMHeat module in FORGE, are shown below:

Transient heat transfer equations govern the quenching process's thermal dynamics, modeled using the SIMHeat module in FORGE. To account for temperature-dependent material properties, the governing equation is expressed as:

$$\frac{\partial T}{\partial t} = \frac{1}{\rho(T)c_p(T)}[\nabla \cdot (k(T)\nabla T) + Q(T)]$$

where: T is the temperature, $\rho(T)$ is the temperature-dependent density, $c_p(T)$ is the temperature-dependent specific heat, $k(T)$ is the temperature-dependent thermal conductivity, $Q(T)$ accounts for latent heat from phase transformations—modeling of heat transfer by dynamically adjusting thermophysical properties based on local temperature conditions.

***Phase Transformation Modeling*:** The kinetics of phase transformations occurring during quenching, such as martensitic and bainitic transformations, are described using Time-Temperature-Transformation (TTT) and Continuous Cooling Transformation (CCT) diagrams. Nucleation and growth phenomena are modeled using the Johnson-Mehl-Avrami-Kolmogorov (JMAK) formalism, while martensitic transformations without diffusion and influenced by temperature changes are described by the Koistinen-Marburger equation.[12].

Diffusional Transformations (Pearlite, Bainite): Governed by nucleation and long-range atomic diffusion, phase transformations during quenching are described by kinetics models like the Johnson-Mehl-Avrami-Kolmogorov (JMAK) equation.

$$y(t) = 1 - exp\left[-b(T)t^{n(T)}\right]$$

where y(t) is the transformed volume fraction, b(T) and n(T) are temperature-dependent nucleation and growth parameters. Diffusionless Transformation (Martensite): Involves rapid shear transformation without atomic diffusion is modeled using the Koistinen-Marburger equation.

$$y_{mart}(T) = 1 - exp\,[-A_M(M_s - T)]$$

where: y_{mart} (T) is the martensite fraction, A_M is a material constant, and M_s is the martensitic start temperature. Stress effects can influence transformation kinetics and critical temperatures like M_s' which is given as:

$$M_s' = M_s + A\sigma + Bp$$

where: σ is the equivalent stress, p is the hydrostatic pressure, A and B are material-specific constants.

Continuous Cooling Transformation (CCT) and Time-Temperature Transformation (TTT): Transformations under isothermal and non-isothermal conditions are mapped with the TTT and CCT diagrams. TTT diagrams map transformation starts and finishes at constant temperatures, enabling predictions for slow cooling. CCT diagrams represent transformations under continuous cooling, which is crucial for industrial quenching processes. These diagrams are generated using Scheil's additivity rule.

$$\int_{t_0}^{t} \frac{dt}{t(X,T)} = 1$$

where $t(X,T)$ is the transformation time at a given temperature.

Hardness Prediction Formula. The hardness at any location in a Jominy bar is calculated using the rule of mixtures:

$$H_v = P_M H_v^M + P_B H_v^B + (P_F + P_P) H_v^{F+P}$$

where H_v is the Vickers hardness, P_M, P_B, P_F, P_P are the phase volume fractions of martensite, bainite, ferrite, and pearlite, respectively, *and* H_v^M, H_v^B, H_v^{F+P} are the hardness contributions of martensite, bainite, and the combined ferrite-pearlite microstructure.

3.0 Results and Discussion

3.1 Temperature Cooling Profile

Fig. 5 Compares the cooling curves and temperature profiles derived from experiments and simulations. The correlation between the simulation and experiment, despite the slight initial temperature difference due to heat loss during sample transfer, shows the correlation between the experimental setup and the accuracy of simulations. Cooling profiles for quenchants at 4°C, 25°C, and 60°C were analyzed through experiments and simulations. A small adjustment was made to correct the initial temperature differences, further improving the match. With these studies, the role of quenchant temperature in accurately predicting cooling behavior and hardness outcomes.

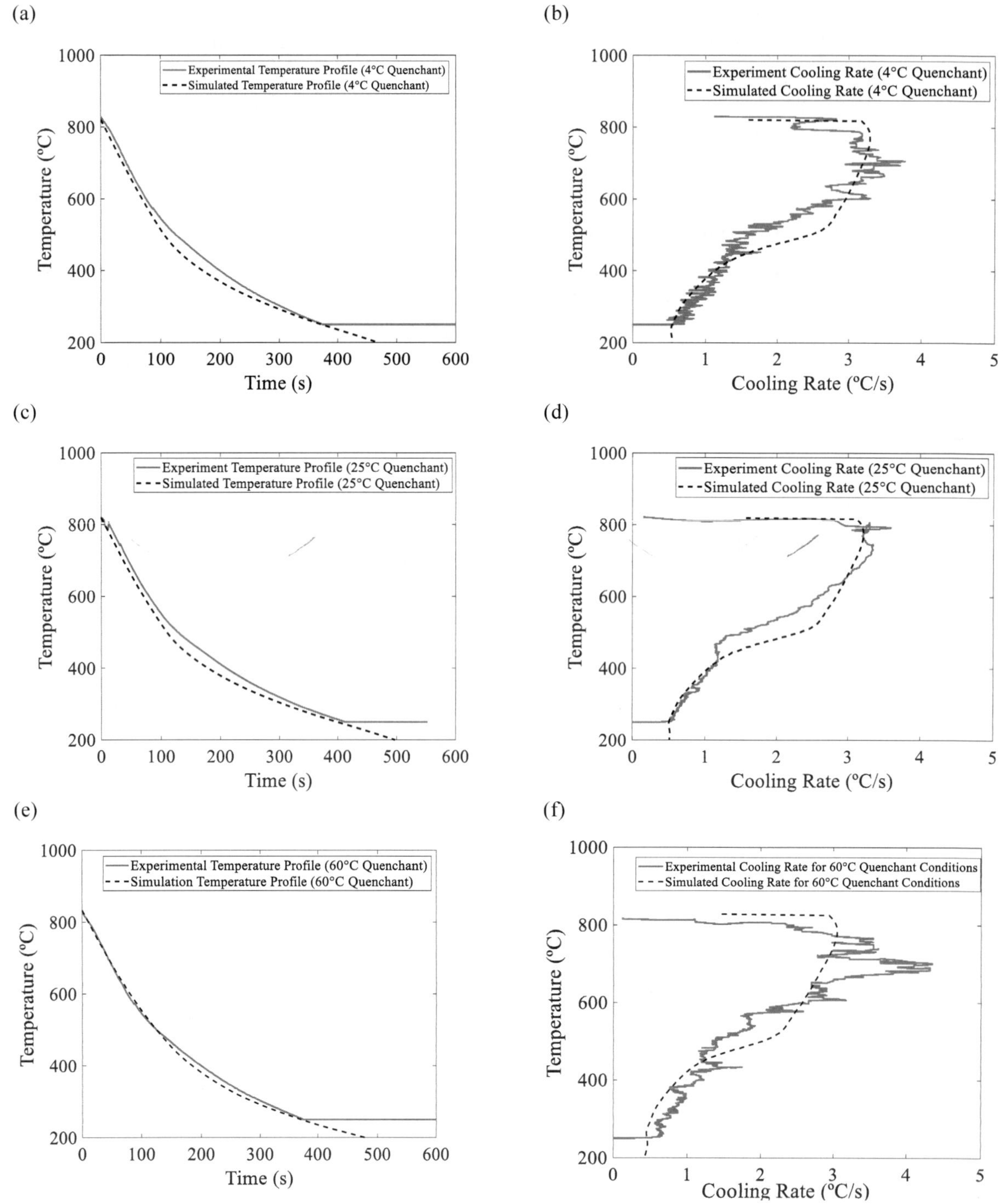

Fig. 5: Temperature cooling profiles and cooling rates for different quenchant temperatures: (a, d) 4°C, (b, e) 25°C, and (c, f) 60°C. Chart (a–c) shows the temperature profiles, while Chart (d–f) depicts the corresponding cooling rates.

Fig. 5 (b, d, f) illustrates the first derivative of the cooling profile for three quenchant temperatures: 4°C, 25°C, and 60°C. The cooling profile was obtained at Jominy position 24 using a pyrometer to measure the surface temperature, then calculating the first derivative to determine the cooling rate. Calculating the average cooling rate between 800°C and 400°C to estimate the average cooling rate in all water temperature condition for both the experiments and simulations. Fig. 5 (a, c, e) also shows the temperature cooling profiles for the three conditions under study, defined by the quenchant temperatures of 4°C, 25°C, and 60°C. The experimental pyrometer measurements at Jominy position 24 recorded average cooling rates of 2.18°C/s for the 4°C quenchant, 2.10°C/s for the 25°C quenchant, and 2.05°C/s for the 60°C quenchant.

In comparison, the simulation results estimated cooling rates for the same quenchants at 2.67°C/s for 4°C, 2.53°C/s for 25°C, and 2.34°C/s for 60°C. These cooling rates highlight the correlation between experimental and simulation cooling rates. However, both the experimental and simulation methods demonstrated that a higher quenchant temperature decreases the cooling rate due to smaller thermal gradients. Possible factors for such discrepancy include idealized boundary conditions and neglected surface effects between the experimentally measured and simulated cooling rates. Notwithstanding, the model has adequately predicted the cooling trends.

3.2 Microstructure Prediction

The calculated phase transformations by the model using the JMAK and Koistinen-Marburger equation reveal significant variations with cooling rate. The distinct cooling profiles for different quenchant temperatures and the time taken to reach the start temperatures for phase formation are important findings in understanding phase transformations in metallurgy.

Fig. 6 shows the phase transformations, indicating that martensite formation and its fractional percentage are highest for the quenchant at 4°C. This is because the more rapid cooling at this temperature suppresses the formation of bainite and, hence, the greater proportion of martensite. A similar trend occurs for the formation of pearlite and ferrite since these phases will not easily form under such high cooling rates. While it predicts the formation of martensite, there is a drop in percentage along the Jominy positions, thus forming other phases, with bainite phases dominating throughout. With the quenchant temperature of 4°C, retained austenite was clearly lower for a 4°C quenchant temperature, resulting in nearly complete martensitic transformation. These results outline the vital importance of the cooling rate or quenchant temperature on phase changes and the resultant microstructure.

Fig. 7, 9, and 11 display the microstructures captured from different positions along the Jominy sample, highlighting variations in microstructure at different locations. In this study, the martensite fraction at these positions was measured using the MIPAR tool.

Fig. 8, 10, and 12 illustrate the martensite fractions at various positions and compare them with the model's predictions to assess the impact of quenchant temperature. As shown in Fig. 12a, the experimental results indicate only slight variations in the martensite percentage, a trend that is also reflected in the predicted results (Fig. 12b). To ensure accuracy, the same image processing method was applied at each position, averaging results from six images per location.

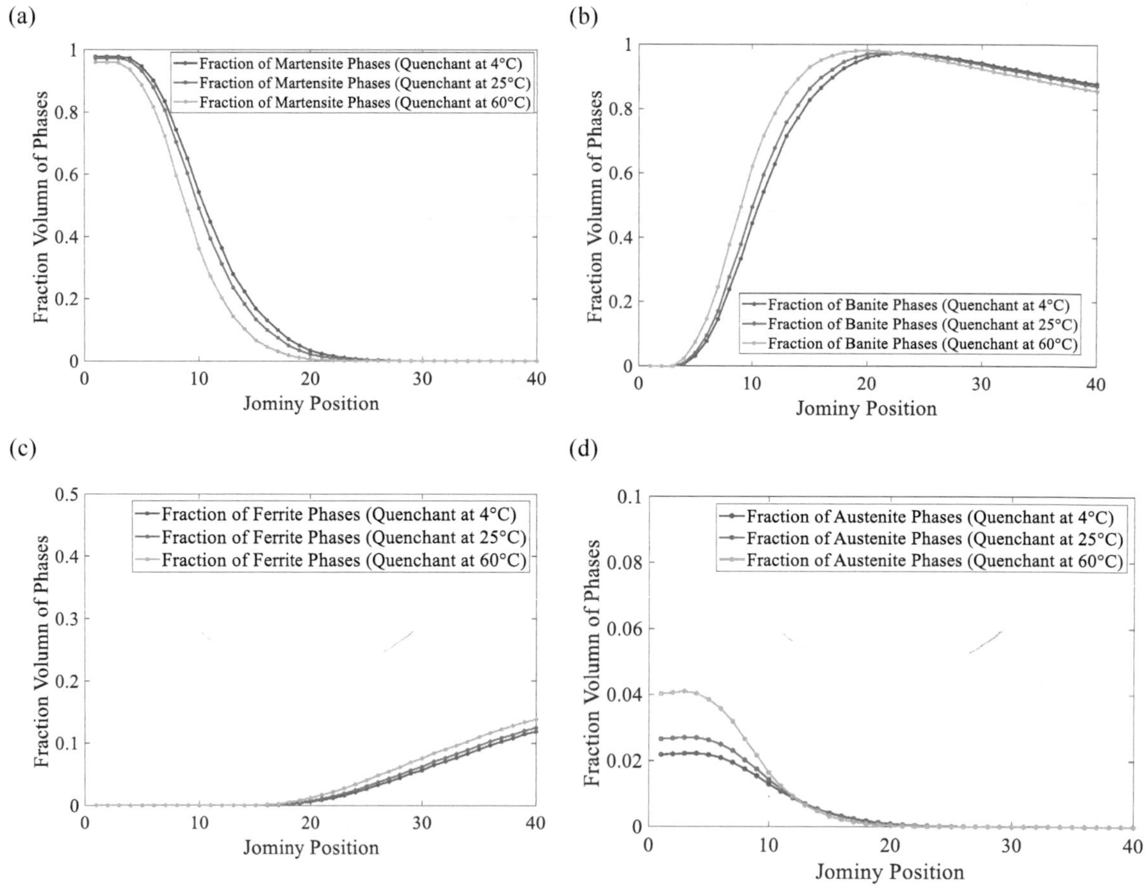

Fig. 6: Phase transformations at quenchant temperatures of 4°C, 25°C, and 60°C: (a) martensite, (b) bainite, (c) ferrite, and (d) austenite formation along the Jominy position.

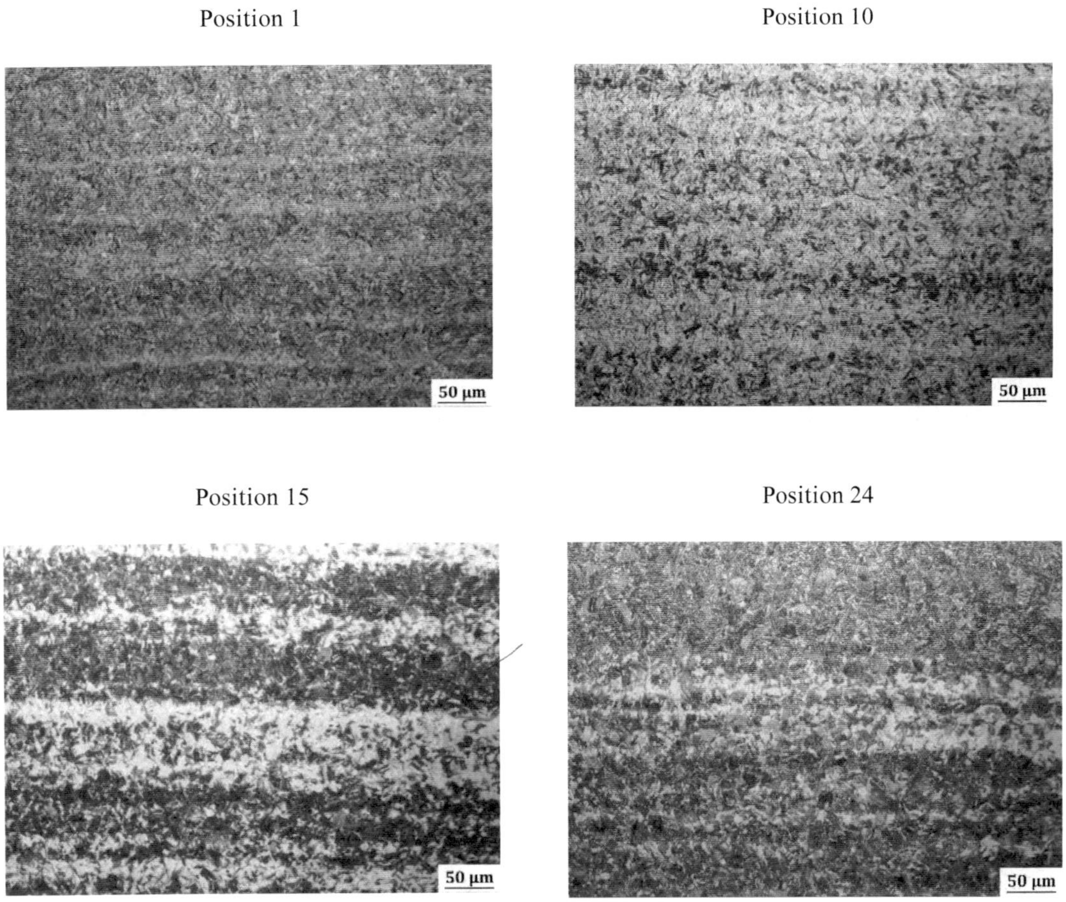

Fig. 7: Microstructure of the Jominy end-quench test at positions 1, 10, 15, and 24 for a quenchant temperature of 4°C.

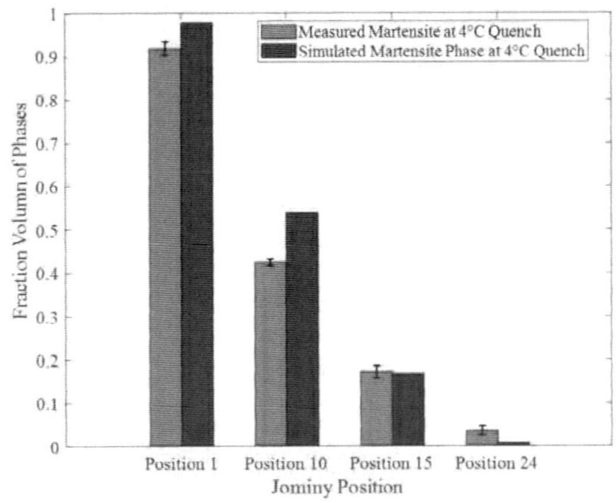

Fig. 8: Comparison between the experimentally measured and predicted microstructures of the Jominy end-quench test at positions 1, 10, 15, and 24 for a quenchant temperature of 4°C.

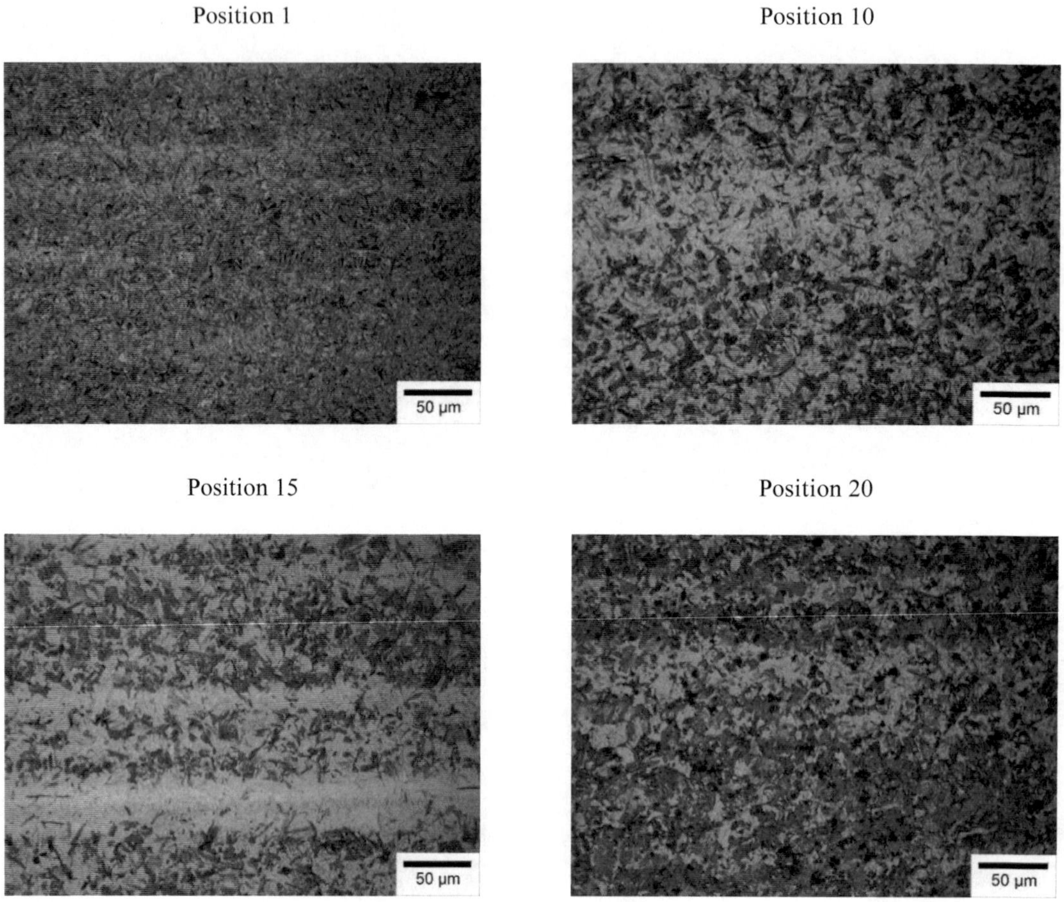

Fig. 9: Microstructure of the Jominy end-quench test at positions 1, 10, 15, and 24 for a quenchant temperature of 25°C

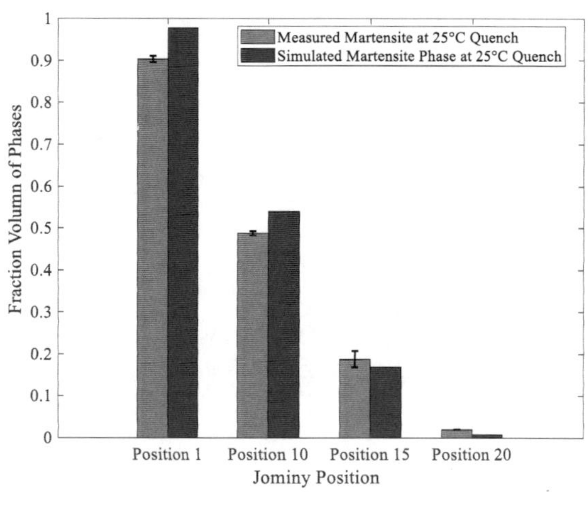

Fig. 10: Comparison between the experimentally measured and predicted microstructures of the Jominy end-quench test at positions 1, 10, 15, and 24 for a quenchant temperature of 25°C.

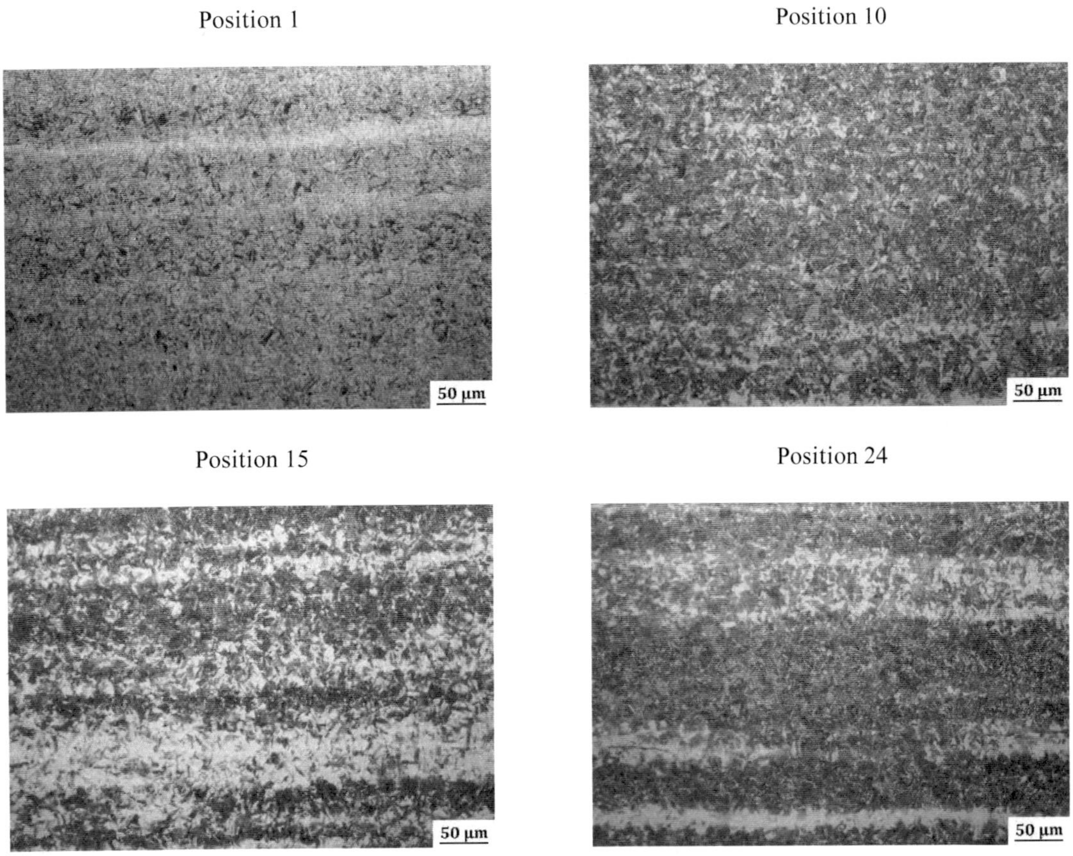

Fig. 11: Microstructure of the Jominy end-quench test at positions 1, 10, 15, and 24 for a quenchant temperature of 60°C

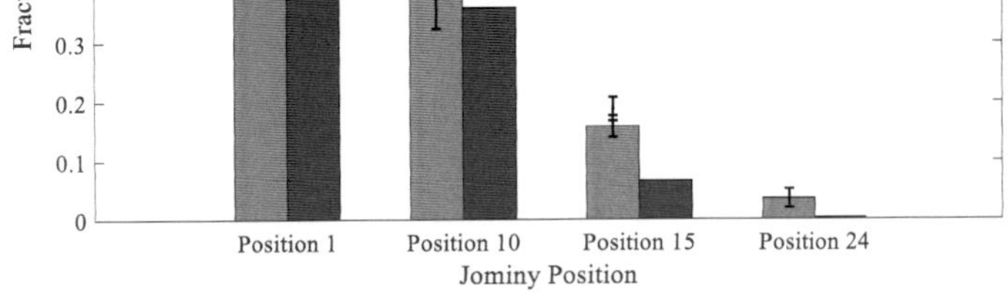

Fig. 12: Comparison between the experimentally measured and predicted microstructures of the Jominy end-quench test at positions 1, 10, 15, and 24 for a quenchant temperature of 60°C.

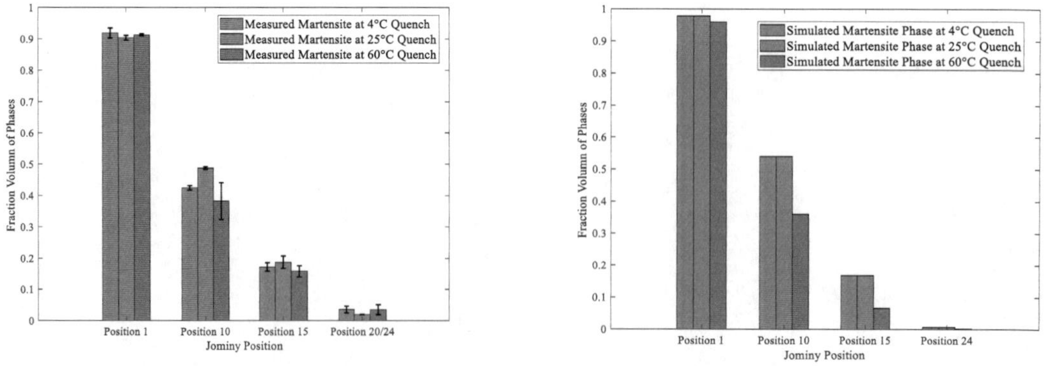

Fig. 13: Predicted and measured microstructure of the Jominy end-quench test at positions 1, 10, 15, and 24 for a quenchant temperature of 60°C

3.3 Hardness Profile

The simulated and experimental hardness results demonstrate the same tendency as the increase of hardenability, this is observed in reducing the temperature of the quenchant down to 4°. However, in general, its expected as a higher cooling rate due to increased thermal gradient promotes phase transformation to martensite.

Fig. 14 and 15 a gives the experimental hardness profile, which starts with the hardness values ranging between 52 HRC and 60 HRC at ~ Jominy position 1. The overall hardenability curve shows a reasonable correlation between experiment and prediction. With a rise in the temperature of the quenchant, there is a noticeable shift resulting in changes in hardness. Experimentally, the hardness values drop from 30 HRC to 26 HRC as quenchant temperatures increase from 4°C to 60°C.

In Fig. 15 (b) simulation results follow the same trend, with small differences in the hardness profiles for different quenchant temperatures. Simulations also reveal slightly higher hardness values under the 4°C quenchant condition. In general, the model is in good agreement with the experimental findings and confirms once more that the quenchant temperature has an observable effect on hardenability. However, in the modelling, the heat transfer coefficient was kept constant for the different conditions, and only the temperature of the fluid, T_∞ , was modified. This assumption is not necessarily valid, because at temperatures below the Leidenfrost temperature, nucleate boiling can occur leading to higher than predicted heat extraction that was not included in this paper, but that will be evaluated for future work. This assumption in the physical model may explain the differences in the data of hardness trends presented in Fig. 6.

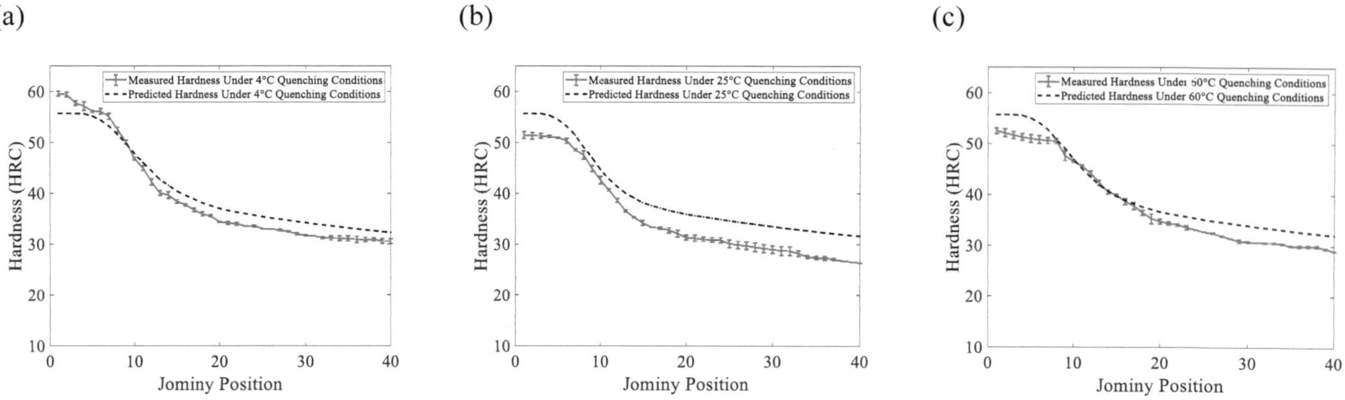

Fig.14: Predicted and experimental hardenability curves for AISI 4135 steel at quenchant temperatures of (a) 4°C, (b) 25°C, and (c) 60°C.

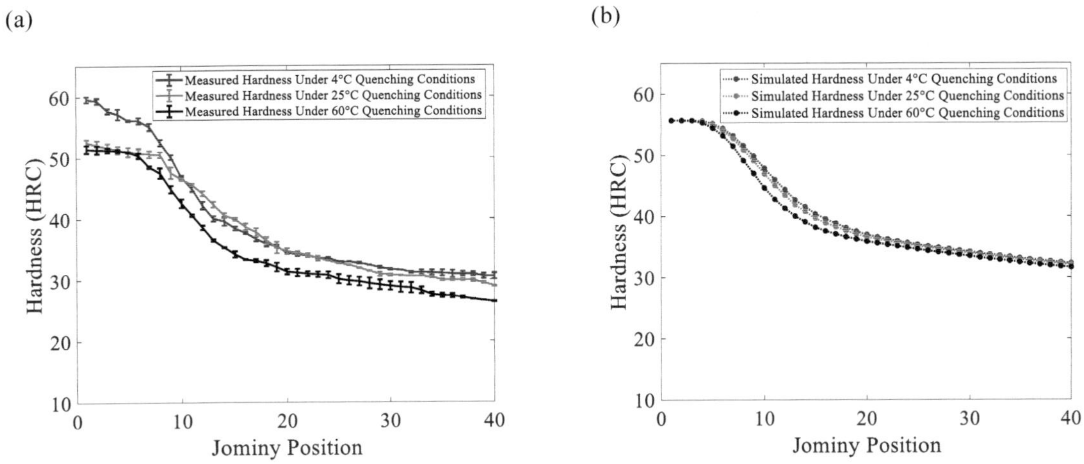

Fig. 15: Hardenability curves for AISI 4135 steel at quenchant temperatures of 4°C, 25°C, and 60°C, showing (a) experimental results and (b) predictions from the model.

4.0 Conclusion

The study was focused on quenchant temperature, and in this work, was found to play a crucial role in cooling behavior, microstructure, and hardness property development in AISI 4135 steel during the Jominy end-quench test. This work demonstrates model's prediction accuracy for cooling rates, phase transformations, and hardness profiles by correlating experimental data with numerical simulations. The correlation of experimental observations with the prediction model also strengthens trends that have been identified in previous experimental studies.

Cooling rates measured during the experimental Jominy end-quench test were 2.18°C/s, 2.10°C/s, and 2.05°C/s for quenchant temperatures of 4°C, 25°C, and 60°C, respectively; the simulations demonstrated consistent predictions with values of 2.67°C/s, 2.53°C/s, and 2.34°C/s for the corresponding quenchant. The hardness profiles revealed a shift in the hardness profile to the upper region with peak hardness for a 4°C quenchant temperature. These results indicate that lower quenchant temperatures, such as 4°C, increase cooling rates, promote more martensite formation, and suppress the formation of bainite, ferrite, and pearlite. On the other hand, a higher temperature of the quenchant, such as 60°C, slows the cooling rate and thus reduces the martensitic content and increases the formation of diffusional phases. In addition, the decrease in retained austenite at lower quenchant temperatures emphasizes the effectiveness of rapid cooling to promote almost complete martensitic transformation. However, the observed results match with what has been seen in experiments, highlighting how effectively our model can mimic the intricate changes in microstructures and show not much significant change concerning quenchant temperature as in the study with 4°C, 25°C, and 60°C.

This numerical model, using the JMAK and Koistinen-Marburger equations, predicts accurately corresponding with experimental results, reduces the reliance on extensive experimental trials, validates previous research findings, and offers a scalable solution for predicting microstructural changes and hardness profiles under varying cooling conditions.

Experimental and predicted microstructures at different Jominy positions were comparatively evaluated, correlating the martensite phase fractions. Microstructures quantified from different positions are shown in Figures 7, 9, and 11, while Figures 8, 10, and 12 provide an overall comparison between experimentally derived and simulated martensite fractions. Results show minor differences in martensite content comparing calculated and experimental results, thus proving the validity of the model.

The present work provides valuable information from an industrial point of view. The model's predictive capability may allow the optimization of quenching parameters, allowing better control of microstructure and mechanical properties with a reduction in trial-and-error approaches. This means reduced production costs, improved efficiency, and better consistency in high-strength steel products. These will immensely benefit industries like aerospace, automotive, and heavy machinery, where the performance of the material is of the essence.

5.0 Acknowledgements

The Peaslee Steel Manufacturing Research Center at Missouri University of Science and Technology supported this research. We also thank Nicolas Poulain and John Swanson of Transvalor for supplying the FORGE Solution.

6.0 References

[1] A. A. Lebedev and V. V. Kosarchuk, "Influence of phase transformations on the mechanical properties of austenitic stainless steels," *International Journal of Plasticity*, vol. 16, pp. 749–767, 2000, doi: 10.1016/S0749-6419(99)00085-6.

[2] G. Krauss, *Steels - Processing, Structure, and Performance* (2nd Edition). ASM International, 2015. [Online]. https://app.knovel.com/hotlink/toc/id:kpSPSPE002/steels-processing-structure/steels-processing-structure

[3] H. Isworo and N. Rahman, "Effect of variations in heating temperature and cooling media on the hardness and microstructure of Steel ST 41 hardening method," *Scientific Journal of Mechanical Engineering Kinematika*, vol. 5, no. 1, pp. 37–50, 2020, doi: 10.20527/sjmekinematika.v5i1.136.

[4] S. Sopiyan, B. Basori, and F. B. Susetyo, "The effect of water temperature as quenching media on the characteristics of HV 350 weld deposits," *Sintek Jurnal: Jurnal Ilmiah Teknik Mesin*, vol. 14, no. 2, p. 118, 2020, doi: 10.24853/sintek.14.2.118-122.

[5] M. Andreansyah, R. D. Anjani, and V. Naubnome, "Pengaruh Proses Heat Treatment (Quenching dan Tempering) Terhadap Kekerasan dan Struktur Mikro Baja Karbon Menengah," *Jurnal Serambi Engineering*, vol. 9, no. 1, 2023, doi:10.32672/jse.v9i1.791.

[6] M. Pita and L. Lebea, "Investigating the effect of cooling media on hardness, toughness, coefficient of friction, and wear rate of mild steel heat treated at different temperatures," *Material Design & Processing Communications*, vol. 2022, pp. 1–10, 2022, doi: 10.1155/2022/3564875.

[7] M. Pietrzyk and R. Kuziak, "Computer aided interpretation of results of the Jominy test," *Archives of Civil and Mechanical Engineering*, vol. 11, no. 3, pp. 707–722, 2011, doi: 10.1016/S1644-9665(12)60111-3.

[8] D. Hömberg, "A numerical simulation of the Jominy end-quench test," *Acta Materialia*, vol. 44, no. 11, pp. 4375–4385, 1996, doi: 10.1016/1359-6454(96)00084-5.

[9] P. Le Masson, T. Loulou, E. Artioukhine, P. Rogeon, D. Carron, and J.-J. Quéméner, "A numerical study for the estimation of a convection heat transfer coefficient during a metallurgical 'Jominy end-quench' test," *International Journal of Thermal Sciences*, vol. 41, no. 6, pp. 517–527, 2002, doi:10.1016/ S1290-0729(02)01345-5.

[10] S.-J. Lee, E. J. Pavlina, and C. J. Van Tyne, "Kinetics modeling of austenite decomposition for an end-quenched 1045 steel," *Materials Science and Engineering: A*, vol. 527, no. 13–14, pp. 3186–3194, 2010, doi: 10.1016/j.msea.2010.01.081.

[11] J. W. A., "Reaction kinetics in processes of nucleation and growth," *American Institute of Mining, Metallurgical, and Petroleum Engineers*, vol. 135, pp. 416–458, 1939, 2024. [Online]. https://cir.nii.ac.jp/crid/1570572700016481152.bib?lang=en

[12] D. P. Koistinen and R. E. Marburger, "A general equation prescribing the extent of the austenite-martensite transformation in pure iron-carbon alloys and plain carbon steels," *Acta Metallurgica*, vol. 7, no. 1, pp. 59–60, Jan. 1959, doi: 10.1016/0001-6160(59)90170-1.

[13] *ASTM A255 Standard Test Method for End-Quench Test for Hardenability of Steel*. ASTM Standards, 1990.

[14] N. Saunders, U. K. Z. Guo, X. Li, A. P. Miodownik, and J.-Ph. Schillé, "Using JMatPro to model materials properties and behavior," *JOM*, vol. 55, no. 12, pp. 60–65, Dec. 2003, doi: 10.1007/s11837-003-0013-2.

[15] J.S. Alabi, E. Heidari, M.F. Buchely, K. Chandrashekhara, S.N. Lekakh, R.J. O'Malley, V. Athavale, A. Kumar, "Impact of quenching intensity conditions on using a finite element model to investigate the microstructure and hardenability of low-alloy steel," *Heat Treating Conference,* 2024, pp. 272–280, doi: 10.31399/asm.cp.ifhtse2024p0272.

QDE 2025: Proceedings of the 3rd International Conference on Quenching and Distortion Engineering
May 6–7, 2025, Vancouver, Canada
https://doi.org/10.31399/asm.cp.qde2025p0111

Limitations of Standard Probes for the Measurement of Spray Quenching

Robert C. Goldstein[1], D. Scott Mackenzie[2]
1 - Fluxtrol, Inc. Auburn Hills, USA
2-Quaker Houghton, Inc. Conshohocken, PA
rcgoldstein@fluxtrol.com

Abstract

Spray quenching is an important technique used to quench parts after induction hardening. It has the advantages of rapid cooling of workpieces, and a repeatable process through the control of spray pattern, quenchant velocity, and pressure. There have been numerous efforts to characterize the heat transfer coefficient as a function of surface temperature in the literature. One method has been to use standard probes used to produce cooling curves for immersion quenchants. However, due to the high surface heat transfer, steep gradients can occur, resulting in thermal lag and an underestimation of the heat transfer coefficient. This improper estimation of heat transfer coefficient can have a significant impact on the prediction of the resultant microstructure, component physical dimensions and residual stress distribution. An examination of these thermal gradients as a function of probe diameter and material was accomplished, to determine the proper probe geometry to measure and calculate the resultant heat transfer coefficient was determined.

Introduction

The heat extraction characteristics of a quenchant can be measured using probes instrumented with thermocouples. Various techniques have been used, including cylindrical and spherical probes, as well as other geometries. The probes have been fabricated from a variety of different metals including stainless steel, silver, and nickel alloys. A summary of the various methods is shown in Table 1.

Table 1. International methods used to measure cooling curves of a quenchant [1].

	Method				
	ISO 9950	AFNOR NFT 6077B	JIS K2242	Z8 E 45003	ASTM D6200
Country	International	France	Japan	China	USA
Probe Alloy	Inconel 600	Silver 99.999%	Silver 99.999%	Silver 99.999%	Inconel 600
Probe Dimensions, mm	12.5 x 60	16 x 48	10 x 30	10 x 30	12.5 x 60
Reference Fluid	Various	Various	Dioctyl Phthalate	Dioctyl Phthalate	Various
Vessel Dimensions, mm	115±5 dia.	138 dia. X 99 high	300 ml beaker	300 ml beaker	115±5 mm dia.
Oil Volume, ml	2000	800	250	250	2000
Oil Temperature °C	40 ± 2	50 ± 2	80, 120, 160	80 ± 2	40 ± 2
Probe Temperature, °C	850 ± 5	800 ± 5	810 ± 5	810 ± 5	850 ± 5

One of the most accepted methods of measuring cooling rate data is based on the 12.5 mm diameter probe manufactured from Inconel 600 alloy as specified in ASTM D6200 or ISO 9950 [2] [3]. This method was originally developed by the Wolfson Heat Treatment Center and was recommended for adoption by the International Federation for Heat Treatment (now International Federation for Heat Treatment and Surface Engineering or IFHTSE).

The Wolfson probe now used by ASTM D6200 is cylindrical probe, 12.5 mm in diameter with a length of 60 mm. A sheathed Type K thermocouple is placed at the geometrical center of the probe. It is manufactured from nickel alloy 600 (UNS N06600) purchased to AMS 5665. This probe has a nominal composition of 76% Ni, 15.5% Cr, 8% Fe. The probe is usually welded to a support tube with a length of 200 mm of the same material as the probe. Prior to use, the probe is conditioned in a furnace to provide a uniform surface condition, and then cleaned using 600 grit sandpaper.

The probe is heated in a furnace to the desired temperature (for steels 850°C is used). Once the probe has reached the desired temperature, it is transferred to the quench bath, which consists of 1000 or 2000 ml of quenchant at the desired temperature. This transfer can be done manually or automatically. Typically, 60°C quenchant temperature is used for normal quench oils, 120°C is used for martempering oils, and 40°C is used for polymer quenchants. However, these temperatures can be changed to suit the specific application.

Results obtained by the different test conditions vary depending upon the material, geometry, and surface condition of the probe [4]. Cooling curves produced in this manner illustrate the three phases of cooling and demonstrate the influence of the various factors (agitation, quenchant temperature, contamination and degradation, etc.) have upon quenching performance [5] [6]. In addition to the basic cooling curve data, heat transfer data can be obtained [7] which is useful for modeling.

The primary focus of most cooling curve applications has been immersion quenching. However, some applications using the standard ISO 9950 probe to spray quenching [8] and gas quenching [9] have been investigated. However, most of this work has focused on the characterization of wetting behavior in immersion quenching media. This is extremely important for furnace heat treatments; however, in the majority of in-line induction surface heat treating applications spray quenchants are employed to achieve improved component performance [10].

Spray quenchants are known to be significantly faster than immersion quenchants, leading to more rapid surface cooling and with the proper thermal profile, higher surface compressive residual stresses. Peak heat transfer coefficients for immersion quenchants are typically in the 1000's of W/m^2K whereas for spray quenchants, high velocity flowing fluids or strongly agitated immersion quench systems with spherical bodies, these have been reported to be on the order of 10's of thousands up to even millions of W/m^2K in some instances as the water velocity is much higher and the spraying process/flowing fluid/shape is reported to break up the boiling film [10] [11] [12].

The impact of higher heat transfer coefficients can be extremely significant on component dimensional movement, residual stresses and resultant component loading capabilities. In practice, different concentrations of polymers are often added to water spray quenchants to control excessive dimensional movement and reduce the probability of cracking in induction heat treated parts. Li, et. al published a paper which studied the influence of quench intensity using average heat transfer coefficients between 5,000 and 25,000 W/m^2K for a full float induction scanning heat treated axle using Dante software. Their findings showed that this variation in the heat transfer coefficient resulted in axial maximum transient growth of the axle of close to 0 for the low value, but closer 3 mm for highest heat transfer coefficients. The resultant surface compressive residual stresses prior to tempering varied by a factor of almost 2 to 1 [13].

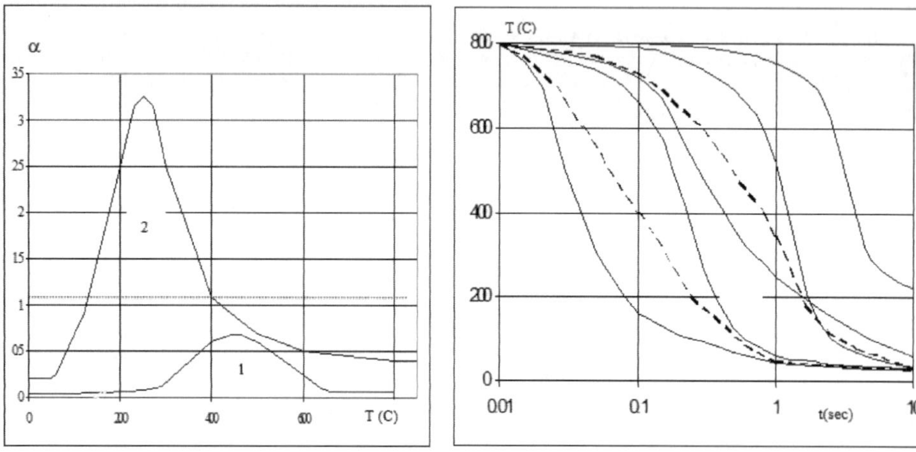

*Figure 1 Non-linear heat transfer coefficients measurement with 8 mm silver probe and comparison with calculations using a mean heat transfer coefficient1- Fast oil unagitated [11], where 2-Water unagitated; 3-Oil shower (0.7 $m^3/(s{\bullet}m^2)$); 4-Water shower (0.28 $m^3/(s{\bullet}m^2)$); 5-Water shower (2 $m^3/(s{\bullet}m^2)$). ** Dashed lines represent simulation with mean value heat transfer coefficients for cases 2 and 4*

This substantive change in residual stresses and dimensional movement occurred despite using relatively moderate values compared to those reported in some of the literature. It is important to keep in mind that the values used in this study were average values; whereas the much higher values reported in the literature are peak values. Nemkov, et al [11] studied the difference between using non-linear heat transfer coefficients from the ELTA (www.nsgsoft.com) database, mean transfer coefficients those measured using a 16 mm diameter silver probe from IVF using ELTA software. Their findings showed significant differences in the shape of the cooling curves along with the time until first martensitic formation on the surface, which is one of the critical driving forces behind residual stress formation, dimensional movement and crack probability (Figure 1).

Limited information is available regarding the non-linear heat transfer coefficients used in the ELTA database other than that they were created during "original experiments conducted in St. Petersburg, Russia". However, these values were compared to experimentally measured cooling surface temperatures for a spray cooled test sample on a Gleeble test stand by Whitely, et al [14] (Figure 2). The sample cooling results matched very closely with the 0.8 m/s water spray quench value in the ELTA database with peak heat transfer coefficient value of 320,000 W/m^{2}K [14]. The main deviations observed were once surface temperatures fell below 100 C, which are in the region well below the maximum heat transfer coefficient value.

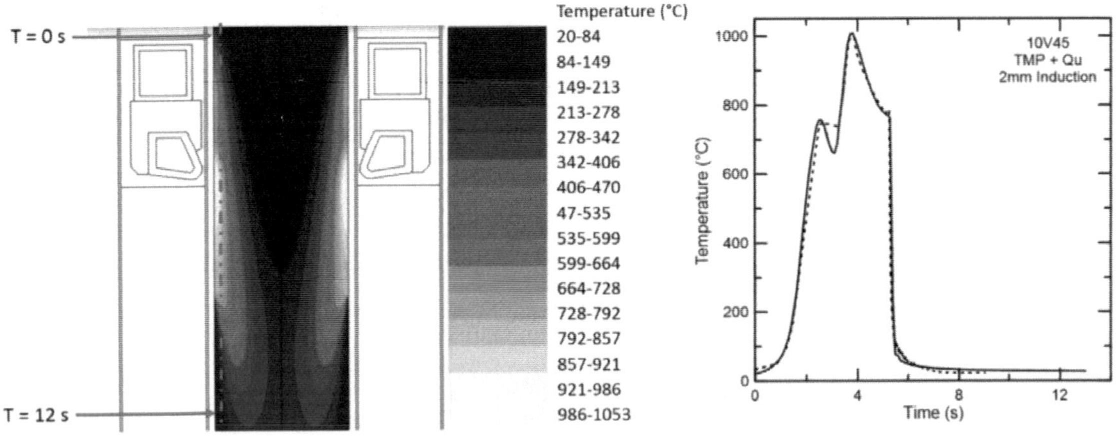

Figure 2 Modeling and experimental comparison between Gleeble simulation results using 0.8 m/s spray quench from ELTA database [14].

This excellent agreement with test results means that these values are likely real; however, to the author's knowledge there has been no published values for measured heat transfer coefficient values this high using standard quench probes available on the market. The cause of the deviation is thermal lag in the test probe as test probes typically measure temperatures at the center of the test probe as opposed t the surface of the probe for durability. The following study is intended to better understand the thermal lag within a quench probe and determine the probe characteristics necessary to measure accurately spray quenchant rates with the characteristics of those in the ELTA database.

Design of the Case Study

Thermal lag is the current hypothesis for the source of the measurement error, so much of the effort will focus on adjusting the thermal lag within the test probe. The standard, Inconel 600 probe will be used to develop the baseline data for comparison. A drawing of the standard quench probe considered is shown in Figure 3.

There are three main variables in determining the level of thermal lag, diameter of the test probe, thermal conductivity of the test probe and location of the temperature measurement point. Since the desire is to have a standardized, multi-use test probe for spray quenchant characterization, we will focus on a central temperature measurement point and limit the variables to test probe diameter and test probe material as there is a risk that non-centric measurement points would increase dimensional movement over multiple cycles if internal or pop off in the case of surface mounted wires. Two materials were chosen for comparison, Inconel 600 (material used for standard probes) along with Silver, which was used previously by IVF for standard quench probes and has a thermal conductivity more than an order of magnitude higher.

The authors reviewed available sheathed thermocouples to determine the theoretical limits for the size of potential quench probes. Based upon this review, it was determined that 1.5 mm could be used as a minimum value for the study and the maximum size was set to that of the current probe (12.5 mm) as a larger probe would only increase the thermal lag in the system. The intermediate diameters considered in the study were 3 and 6 mm. One minor difference in the study is due to the minimum diameter being 1.5 mm and the standard quench probe having a 1.5 mm diameter hole, the 12.5 mm probes modeled will have the 1.5 mm diameter hole while the other quench probes modeled will be solid. Two quenchants were used in this study from the ELTA database, a 12% polymer spray quench (slowest spray quench in the database) and the 0.8 m/s spray quenchant that matched the Whitely results. The heat transfer coefficients versus temperature are displayed for these two conditions in Figure 4a. More importantly, the resulting heat fluxes are shown in Figure 4b. These show cooling rates that are on the order of 10 times higher initially, compressing to approximately 3 times higher in the area just above the Ms for most induction heat treated steels and then expanding again to more than 10 times as we transition through the martensitic transformation range. Also of note is that even though the 12% polymer quenchant is on the order of some of the highest reported heat transfer coefficients in the literature (18,000 W/m^2K), it is important to note that this maximum value occurs at 500 – 600 C as opposed to 250 -300 C, which will result in approximately double the heat flux compared to those previously measured results.

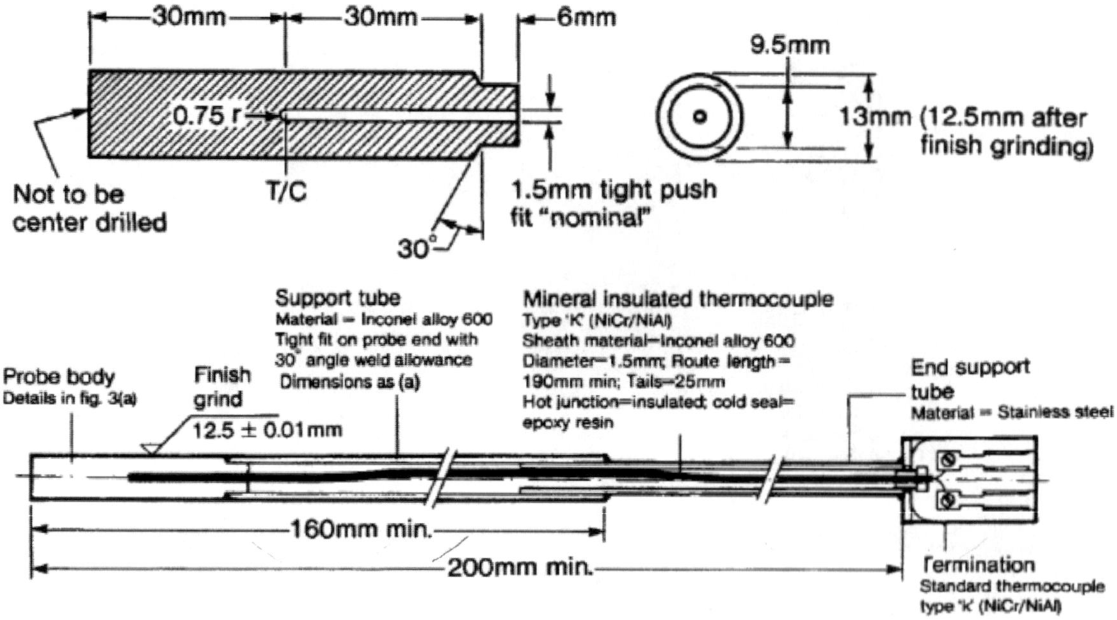

Figure 3. Standard, Inconel 600 probe geometry from the ASTM D6200 Standard

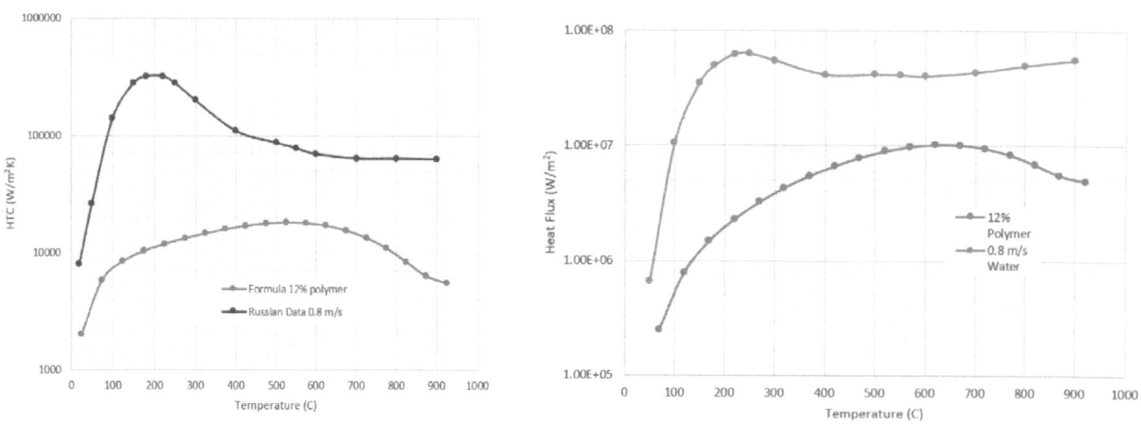

Figure 4. a) Heat transfer coefficient versus surface temperature for 12% polymer spray quench and 0.8 m/s spray quench from ELTA Database and b) resulting heat flux versus surface temperature.

ELTA program will be used for 1D transient thermal analysis. As thermal lag evaluation is the goal, the first 0.5 seconds of the cooling cycle will be evaluated, as with this cooling time, ELTA calculates the data approximately every ms. An initial temperature of 900 C was used for the analysis as induction surface heat treated components are usually heated to higher temperatures than furnace heat treated components due to the reduced time.

Calculation Results

Standard Inconel Quench Probe

Figure 5a shows the plots of surface, internal radius (0.75 mm for the standard probe) and delta T for both the 12% polymer spray quench and the 0.8 m/s water spray quench data. For the polymer spray quench, there is minimal change in the internal radius temperature until approximately 0.35 s, the thermal lag in the measurement. By this point in the process, surface temperature has fallen to approximately 380 C. It will be difficult to accurately discern differences occurring in the curve prior to this point in time. During this period, heat transfer coefficient changes on the order of approximately 3:1 and heat flux varies by approximately 2:1. Averaging of data in this range would have a moderate effect on the modeling results for a real process.

For the 0.8 m/s water spray quench, the situation is much different. Surface temperature reaches close to 100 C in under 0.1 s, while the internal radius temperature doesn't change significantly until approximately 0.25 s. To accurately see the shape of the surface temperature curves and delta T during the initial cooling phases, Figure 5b shows the curves during the first 0.05 s of the process. Over this temperature range, the heat transfer coefficient and heat flux vary by a factor of around 5:1, with the larger variation being in heat flux. Since majority of the 0.25 s time until the internal temperature probe responds the surface temperature is around 100 C, where heat flux is falling rapidly and well below the average value above, this will tend to significantly underestimate peak heat transfer coefficients, and it will be very difficult to discern the shape of the curve. This is important as we will pass through the Ms for nearly all induction heat treated steels and the Mf for most of them in this region of the curve, which will have a significant impact on the resultant residual stresses, dimensional movement and resultant component performance. Clearly, this temperature probe is not appropriate for measuring this type of quenchant.

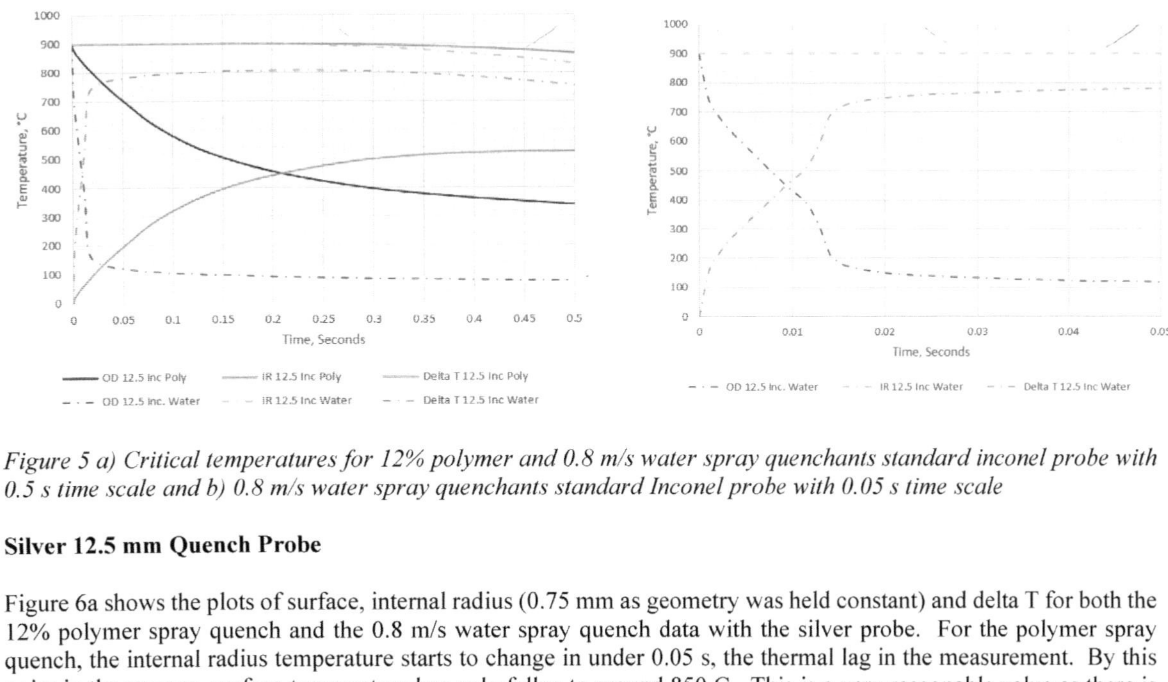

Figure 5 a) Critical temperatures for 12% polymer and 0.8 m/s water spray quenchants standard inconel probe with 0.5 s time scale and b) 0.8 m/s water spray quenchants standard Inconel probe with 0.05 s time scale

Silver 12.5 mm Quench Probe

Figure 6a shows the plots of surface, internal radius (0.75 mm as geometry was held constant) and delta T for both the 12% polymer spray quench and the 0.8 m/s water spray quench data with the silver probe. For the polymer spray quench, the internal radius temperature starts to change in under 0.05 s, the thermal lag in the measurement. By this point in the process, surface temperature has only fallen to around 850 C. This is a very reasonable value as there is only moderate change in this region. As the cooling progress progresses, the internal radius temperature and the surface temperature slopes mirror each other fairly-well with a lag of around 0.05 s. With these curve characteristics, it is possible to make a good approximation of the real heat transfer coefficient.

For the 0.8 m/s water spray quench, the situation is somewhat different. The internal radius temperature begins to change at around 0.02 s. Surface temperature at 0.02 s has already declined to approximately 650 C. To accurately see the shape of the surface temperature curves and delta T during the initial cooling phases, Figure 6b shows the curves during the first 0.16 s of the process. Over this temperature range, the heat transfer coefficient and heat flux vary have relatively small variation, which is good. However, as we track the shape of the surface curve as we pass between 400 and 150 C, these dynamic changes are not necessarily mirrored well on the internal radius where temperature is measured. This temperature range is important as it falls into the region where martensitic transformation is occurring in most induction hardened steels. While the results are substantially better than for the Inconel probe of the same size, this probe is still somewhat too slow to characterize accurately quenchants of this type.

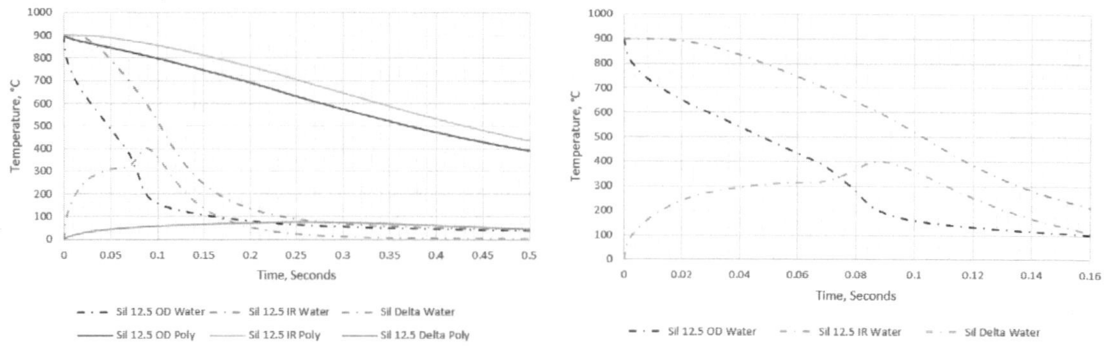

Figure 6. a) Critical temperatures for 12% polymer and 0.8 m/s water spray quenchants silver 12.5 mm probe with 0.5 s time scale and b) 0.8 m/s water spray quenchants silver 12.5 mm probe with 0.16 s time scale

Inconel Quench Probe Size Dependency

Since the main challenge identified above is with the 0.8 m/s water spray quenchants, calculations for the small diameter Inconel quench probes will focus on the requirements for water spray quenchants as opposed to the 12% polymer spray quenchant. Curves for the 1.5, 3 and 6 mm diameter Inconel 600 probes are presented in Figure 7 for the first 0.1 s of cooling to visual the critical areas. There is too much thermal lag in both the 3 and 6 mm probes as the probe temperature is well below 200 C before any response is visualized on the internal surface. For the 1.5 mm probe, the response is much faster; however, the shape changes present on the S-curves of the surface are not reflected in the shape of the internal surface cooling curve. Based upon this, it is the opinion of the authors that Inconel 600 probes are not a realistic option for characterization of water spray quenchants for induction heat treating applications. This is without even considering the practical considerations based upon the size of the probe relative to the size of the individual streams of water typically present in induction spray quenching applications.

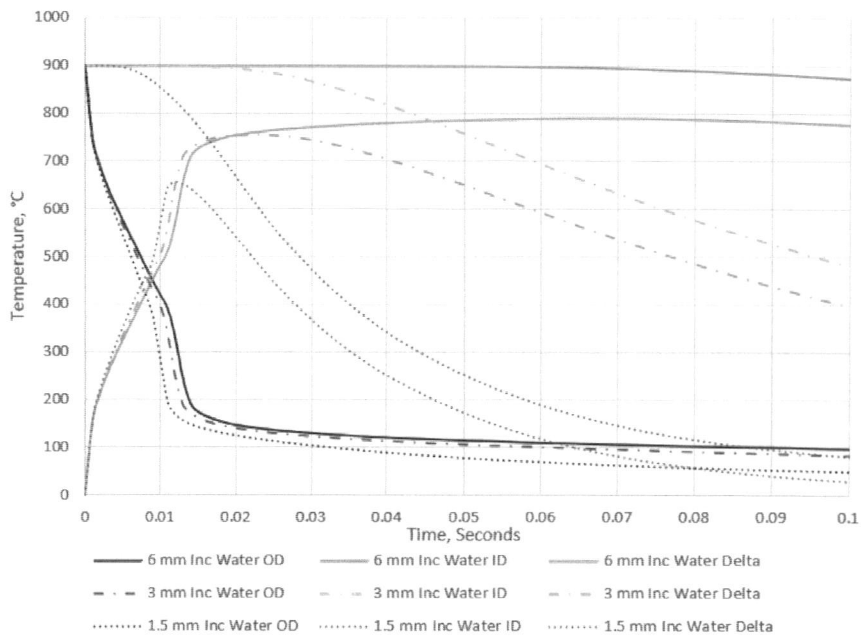

Figure 7 Critical temperatures for 0.8 m/s spray quench from ELTA Database for 1.5, 3 and 6 mm Inconel 600 probes.

Silver Quench Probe Size Dependency

Since the silver quench probe was already considered sufficient for the 12% water spray quenchant, calculations for the small diameter silver quench probes will only focus on the requirements for water spray quenchants as well. Curves for the 1.5, 3 and 6 mm diameter silver probes are presented in Figure 8 for the first 0.1 s of cooling to visual the critical areas. For the silver probes, in all cases the delta T is significantly lower than even the smallest diameter Inconel probe. For the 6 mm the internal temperature begins to respond at about 0.05 s and the surface temperature has only fallen to about 750 C, which is reasonable. As we follow the shape of the curves, inflection points are reflected on the internal surface temperature, although they are not nearly as pronounced as on the surface curves, meaning that there will be some dampening of the signal when doing the calculations. For the 3 and 1.5 mm curves, there are progressively smaller delays to internal response and the shapes of the internal curves match much better to the surface temperature curves. These sizes could both be considered accurate for the measurement of the heat transfer coefficient.

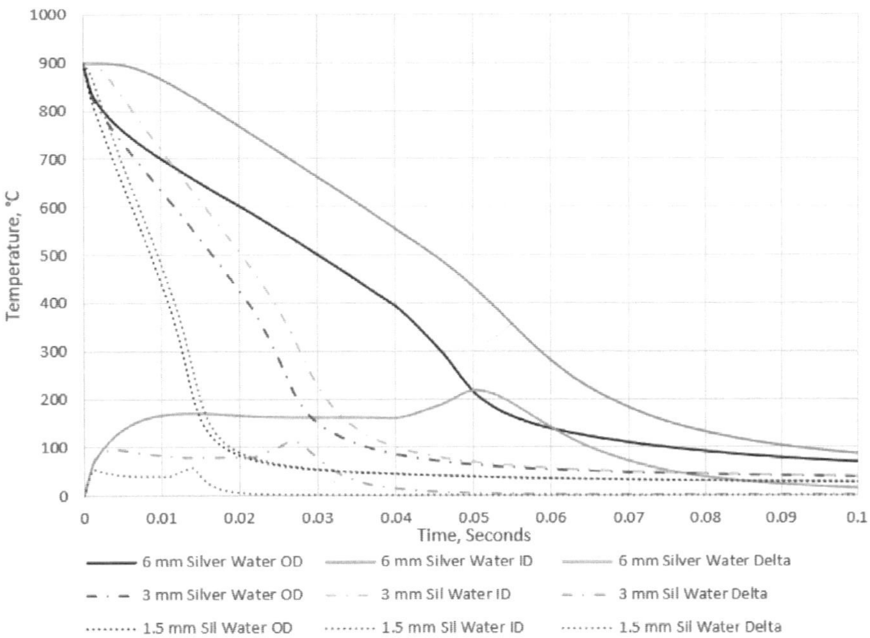

Figure 8. Critical temperatures for 0.8 m/s spray quench from ELTA Database for 1.5, 3 and 6 mm silver probes.

Conclusions

Spray quenching is an important technique used to quench parts after induction hardening. It has the advantages of rapid cooling of workpieces, and a repeatable process through the control of spray pattern, quenchant velocity, and pressure. However, there is very little data available in the literature on spray quenchants commonly used for induction heat treating as it is very difficult to accurately characterize them due to their very high heat transfer coefficient values. This makes it difficult to accurately model the resulting residual stresses, dimensional movement and resulting component properties for induction heat treated components.

A study was conducted using ELTA that varied the diameter of quenching probes made of Inconel 600 (industry standard) and Silver for two spray quenchants from the ELTA database, 12% polymer quench and 0.8 m/s water spray quench. The 12% polymer quench solution, which is on the upper end of concentrations typically used in the field and hence, would tend to be on the "slower" end of induction spray quenchants and the 0.8 m/s water spray quench is believed to be on the "faster" end of induction spray quenchants. The calculations showed that using the standard 12.5 mm OD Inconel 600 probe to characterize the 12% polymer spray quench would likely result in moderate error in the characterization of this quenchant, which would have an influence on the induction heat treating modeling process. Calculations showed that for the 12% polymer spray quench, a 12.5 mm OD silver quench probe would be the preferred geometry for a quench probe.

For 0.8 m/s water spray quench, Inconel 600 quench probes were all too slow to accurately characterize a quench with these properties. Silver probes did show promise for characterizing this level of quench intensity. A probe size of somewhere between 3 and 6 mm would likely be the best compromise between robustness and accuracy. While not studied, this level of accuracy could also be achieved as well with larger diameter probes with wall thicknesses representative of the relative mass of the probes.

Additional work is planned to convert the internal temperatures modeled in this exercise to heat transfer coefficients using the formulae in standard quench characterization devices. Additional studies should also be conducted on quench characterization, as it is known that spray quenchants heat transfer coefficients depend not only on quench medium chemistry and average velocity, but also on local hole size, hole angle, hole depth and other factors.

References

[1] G. Totten, C. Bates and N. Clinton, Eds., Handbook of Quenching and Quenchants, Metals Park, OH: ASM International, 1993.

[2] ASTM International, Standard Test Method for Determination of Cooling Characteristics of Quench Oils by Cooling Curve Analysis, ASTM D 6200, Conshocken: ASTM, 2012.

[3] International Organization for Standardization, "Industrial quenching oils -- Determination of cooling characteristics -- Nickel-alloy probe test method," Geneva, 1995.

[4] C. E. Bates, "Recommended Practice for Cooling Rate Measurement and Quench Factor Calculation," Warrendale, PA, 1987.

[5] D. S. MacKenzie, "Care and Maintenance of Oil Quenchants used for Quenching Automotive Components," in *European Conference on Heat Treatment and Surface Engineering, 11-13 May*, Prague, Czech Republic, 2016.

[6] D. L. Moore and D. A. Guisbert, "Influence of Test Conditions on Cooling Curve Data," in *Proc. 14th ASM HTS Conference & Exposition*, Indianapolis, IN, 1993.

[7] E. Troell, H. Kristofferson, J. Bodin, S. Segerberg and I. Felde, "New Tool for Calculation of Heat Transfer Coefficients, Microstructure and Hardness from Cooling Curve," in *IFHTSE 5th International Conference on Quenching and Control of Distortion*, Berlin, Germany, 2007.

[8] S. Gallegos-Cantú, D. S. MacKenzie and B. Hernández-Morales, "Cooling Curve Analysis of a PAG Polymer Quenchant Using Two Spray Quench Rings Designs," in *IFHTSE 2024,September 30–October 3, 2024* , Cleveland, OH, 2024.

[9] E. Troell and S. Segerberg, "Cold Chamber Gas Cooling for Low-Pollution Hardening," in *11th Congress of the International for Heat Treatment and Surface Engineering*, Florence, Italy, 1998.

[10] V. Rudnev and G. E. Totten, Induction Heating and Heat Treating, Metals Park, OH: ASM International, 2014.

[11] V. S. Nemkov, "Computer Simulation of Induction Heating and Quenching Processes," in *2nd International Conference on Quenching and Control of Distortion, March 1999*, Prague, Czech Republic, 1999.

[12] N. Kobasko, Why Database for Cooling Capacity of Various Quenchants Should be Developed, vol. V, WSEAS, 2011, pp. 136-141.

[13] Z. Li, B. L. Ferguson and V. Nemkov, "Effect of Quenching Rate on Distortion and Residual Stresses During Induction Hardening of a Full-Float Truck Axle Shaft," *J. Mat. Eng. Perf.*, vol. 23, pp. 4170-4180, 2014.

[14] B. M. Whitley, J. G. Speer, R. L. Cryderman, K. O. Findley and D. K. Matlock, "Additions and Thermomechanical Processing on Austenite Grain Size Control in Induction-Hardenable Medium Carbon Steel Bar Rolling," *Mat. Sci. Forum,* vol. 879, no. International Conference on Processing & Manufacturing of Advanced Materials Processing, Fabrication, Properties & Applications (Thermec), pp. 2094-2099, 2016.

QDE 2025: Proceedings of the 3rd International Conference on
Quenching and Distortion Engineering
May 6–7, 2025, Vancouver, Canada
https://doi.org/10.31399/asm.cp.qde2025p0119

Understanding Principles of Quench Distortion of Carburizing Steel Parts using a Single Side Carburized Almen Strip and Computer Modeling

Zhichao (Charlie) Li, Lynn Ferguson, Stefan Habean, and Edward Lee
DANTE Solutions, Inc., Cleveland, Ohio, USA
Charlie.Li@Dante-Solutions.com

Abstract

Quench hardening of steel components is a complex problem because multiple factors are involved in the process, including the part geometry, quench uniformity and phase transformations, etc. The quench hardening of carburized steels is considered as a black box in general by the heat treatment industry mainly due to the complexities of microstructural changes, which are affected by the variation of chemical composition for a given steel grade, carbon profile, form and size of carbides, and quenching rate. To simplify the problem, a single side carburized Almen strip made of Pyrowear 675 is used to investigate the effect of phase transformations on distortion in this study. The Almen strip is low pressure carburized (LPC) on only the top surface by copper plating all the other surfaces. The conventional process for Pyrowear 675 is applied, including carburizing, double normalizing, quench hardening, deep freeze and tempering. The temperature gradient in the Almen strip during high pressure gas quench (HPGQ) is small due to its thin section thickness, and the thermal stress is negligible. The volume (strain) change caused by carburizing and phase transformation is considered as the main cause of bow distortions observed at different processing stages. Material models and material data required to describe the part responses during the heat treatment process steps are described. Computer modeling by DANTE is used to analyze the collected experimental data and demonstrate the principles of distortions and residual stress.

Introduction

Distortion caused by heat treatment is complicated and difficult to understand due to the combined effects of nonlinear material behavior, part geometry and complex heating and cooling conditions. The effects of part geometry and thermal boundary conditions on distortion are analyzed by general FEA using elastic-plastic material models, if there is no phase transformation involved during the process, such as for Aluminum alloy quench hardening processes [1]. For steel parts, carburizing and martensitic phase transformation are often the main mechanisms for hardening parts. The material volume changes during austenitizing and quenching have a significant effect on the distortion. With the computer modeling, the effects of different factors on distortion become more transparent [2, 3]. In this paper, the basic material models used in the commercial heat treatment simulation software DANTE (Distortion ANalysis for Thermal Engineering) are explained. The heat treatment process steps of an Almen strip sample made of Pyrowear 675 are modeled including: selective carburizing, double normalizing, quenching, deep freeze and double tempering. The modeling results are compared with the experiments, and the mechanisms of distortion are explained.

Material Models for Heat Treatment Process Modeling

Phase Transformation Models

Equation (1) describes the rate based austenitizing transformation model in DANTE. The transformation rate is a function of temperature, austenite fraction and initial phase fraction.

$$\frac{dA}{dt} = v_1 * \frac{1}{e^{(v2+v3*(ln\frac{(v4-T)}{v5}))}} * (A)^{v_6} * (1-A)^{v_7} * P \tag{1}$$

Where A is the volume fraction of austenite, t is time, v_1, v_2, v_3, v_4, v_5, v_6, and v_7 are constants, P is volume fraction of initial phase, and T is temperature.

A continuous heating transformation (CHT) diagram using equation (1) is plotted in Figure 1, with the variable values from v_1 to v_7 being *0.025, 9.0575, 15.965, 1025.0, 150.0, 0.15, 0.90*, respectively. At a slow heating rate, the transformation to austenite starts at about 750 °C and finishes at about 805 °C. At a high heating rate of 1000 °C/s, the transformation to austenite starts at about 850 °C, and finishes at about 1025 °C. The rate based austenitizing model can be used to describe the phase transformation for both slow heating (furnace heating) and fast heating (induction or laser heating) processes.

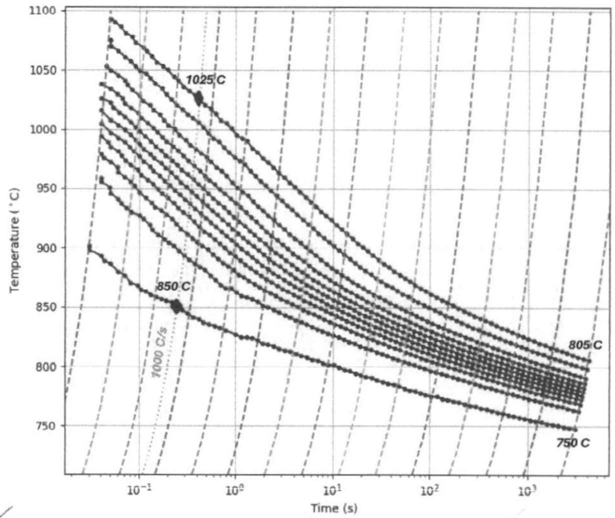

Figure 1: Example of CHT phase diagram generated using DANTE rate based austenitizing model

The material model for diffusive phase transformations during quenching is described in equation (2). The phase transformations include the austenite decomposing to ferrite, pearlite, and bainite.

$$\frac{dF}{dt} = v_1 * \frac{1}{e^{(v2+v3*(ln\frac{(v4-T)}{v5}))}} * (F)^{v_6} * (1-F)^{v_7} * A \qquad (2)$$

Where F is the volume fraction of diffusive phase formed from austenite, t is time, v_1, v_2, v_3, v_4, v_5, v_6, *and* v_7 are material constants, A is volume fraction of austenite phase, and T is temperature.

A time-temperature-transformation (TTT) diagram using equation (2) is plotted in Figure 2, with the variable values from v_1 to v_7 being: austenite to ferrite transformation: 0.10, 7.985, 0.690, 815.4, 153.5, 0.812, -0.759; austenite to pearlite transformation: 0.025, 10.50, 0.723, 793.1, 130.7, 0.796, -0.297, and austenite to bainite transformation: 0.067, 2.40, 2.05, 568.1, 118.5, 0.5562, -1.00.

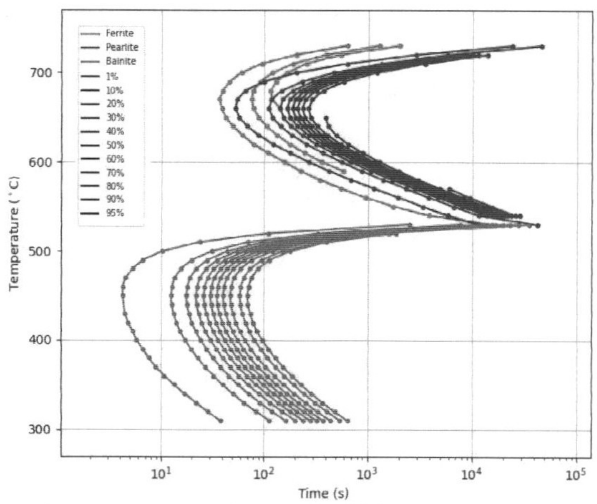

Figure 2: Example of TTT phase diagram generated using DANTE austenite decomposing model to diffusive phases

The austenite decomposing model to martensite is described in equation (3). Martensitic transformation is treated as diffusionless, and the transformation is temperature dependent instead of time dependent.

$$\frac{dM}{dT} = v_1 * (M + v_4 * D)^{v_2} * (1 - M)^{v_3} * A \qquad (3)$$

Where M is the volume fraction of martensite phase, v_1, v_2, v_3, and v_4 are material constants, A is volume fraction of austenite phase, D is volume fraction of diffusive phase, and T is temperature.

A calculated dilatometry curve for Pyrowear 675 from DANTE material database is shown in Figure 3, with the martensitic transformation model parameters v_1, v_2, v_3 and v_4 being 0.0285, 0.0, 0.658 and 0.20, respectively. The heating rate is 1.0 °C/s, and the cooling rate is 20 °C/s. The X-axis in Figure 3 is temperature, and the Y-axis is linear strain. With the starting phase being pearlite, the material volume increases during heating due to thermal expansion. When reaching the austenitizing temperature, material volume decreases with austenite transformation. During cooling, the material volume decreases due to thermal shrinkage, and the coefficient of thermal expansion (CTE) of austenite is higher than that of tempered martensite. When the temperature drops below the martensite start temperature (Ms), martensitic transformation starts with material volume expansion.

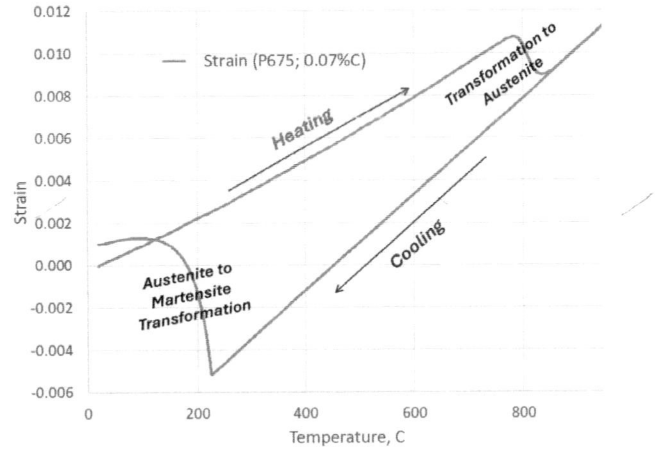

Figure 3: Calculated dilatometry curve for Pyrowear 675 from DANTE material database

The austenitizing transformation may lead to unexpected distortion, especially for parts with thin and thick sections if the heating rate is not well controlled. Figure 4 shows the simulated stress with concentration at the thin and thick section interaction region caused by nonuniform austenitizing transformation. At this snapshot time during heating, section A temperature is higher than that at section B. Section A has a volume shrinkage due to transformation to austenite, and a bending effect between section A and section B leads to concentrated stress at the connection region, which contributes to distortion.

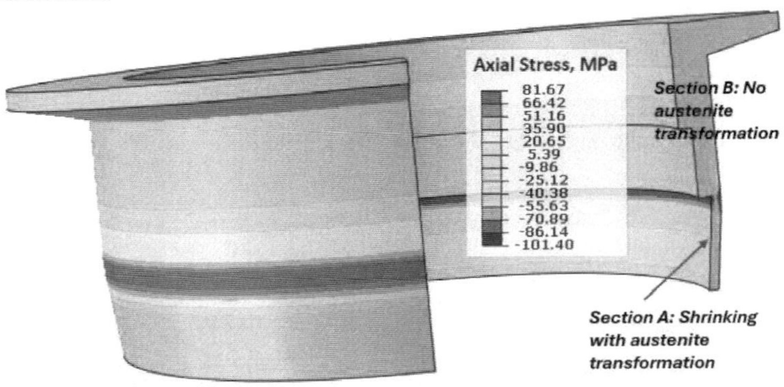

Figure 4: Example of distortion caused by concentrated stress during austenitizing process

Figure 5 shows two calculated dilatometry curves with two different initial phases: as-quenched martensite and fully tempered martensite. The green curve starts with as-quenched martensite. During heating, the martensite is tempered, and the iron carbide coarsens, which lead to a material volume shrinkage. Higher carbon content leads to a more significant volume shrinkage from the iron carbide coarsening. The brown curve starts with fully tempered martensite, so there is no noticeable volume shrinkage because the iron carbide is already coarsened. To more precisely control the distortion, the initial phase condition should be considered.

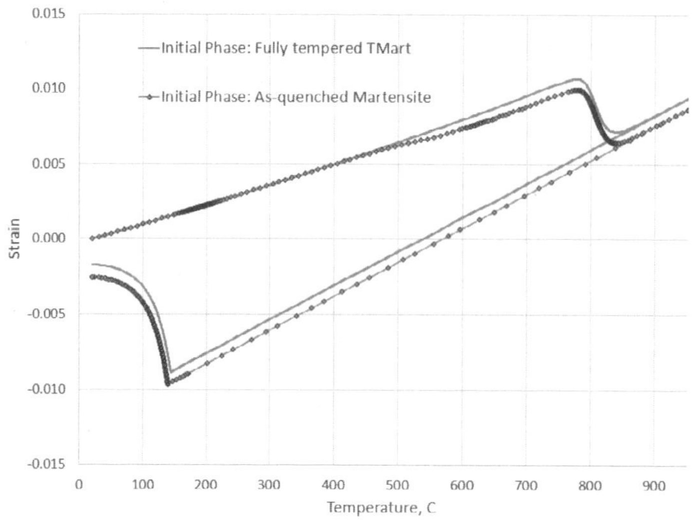

Figure 5: Calculated dilatometry strain curves for a higher carbon steel grade with fully tempered martensite and as-quenched martensite as the initial phases

ISV Plasticity Mechanical Model

Equation (4) is the stress constitutive formula in DANTE to describe the plasticity behavior of different phases. Instead of conventional effective strain, an internal state variable (ISV) model is used [4].

$$\sigma = \alpha + \kappa + Y + V * log(\frac{|\dot{\varepsilon}|+\sqrt{|\dot{\varepsilon}|^2+f^2}}{f}) \quad (4)$$

Where, Y contributes to yield, V and f describes the rate effect on yield, sigma (σ) is the stress tensor, with 6 components, alpha (α) is the directional ISV variable, with 6 components, kappa (κ) is the isotropic ISV variable, with 1 component, $|\dot{\varepsilon}|$ is strain rate magnitude.

The formula for Y, V and f are described in equations (5), (6) and (7)

$$Y = \frac{\{0.5*[1.0+tanh(C19*(C20-T)]\}*(C3+Carb*(C26-C27*T))}{C21+EXP(\frac{-C4}{T})} \quad (5)$$

$$V = C1 * EXP(\frac{-C2}{T}) \quad (6)$$

$$f = C5 * EXP(\frac{-C6}{T}) \quad (7)$$

Where Cx are plasticity model parameters, T is temperature, and $Carb$ is carbon.

The material yield criterion is defined in equation (8).

$$|\sigma - \alpha| > \left[\kappa + Y + V * log\left(\frac{|\dot{\varepsilon}|+\sqrt{|\dot{\varepsilon}|^2+f^2}}{f}\right)\right] \quad (8)$$

Using the plasticity model described above, the stress-strain curves of tempered martensite are calculated from DANTE material database for Pyrowear 675 at different temperatures, as shown in Figure 6. During a heat treatment process the internal stress may change from tension to compression due to the contributions from temperature gradient and phase transformations. To effectively describe the material conditions with stress reversals, the material model should be able to catch the Bauschinger effect schematically plotted in Figure 6(b).

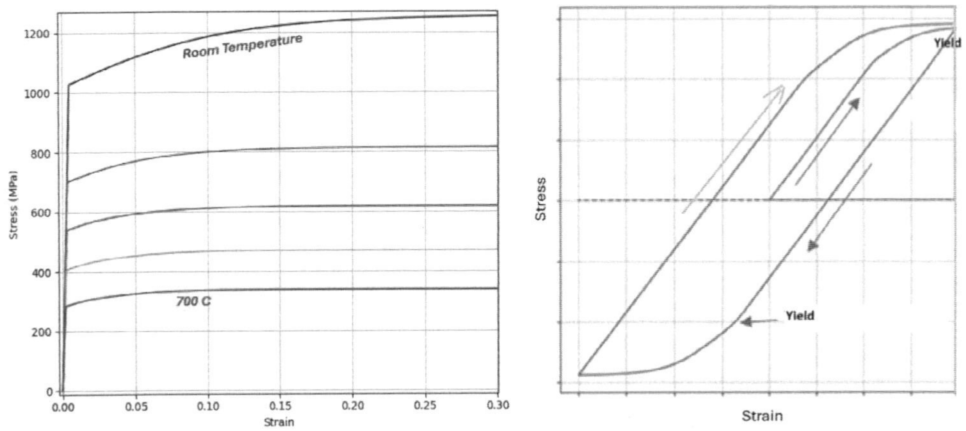

(a) (b)

Figure 6: (a) Pyrowear 675 tempered martensite stress-strain curves calculated from DANTE material database at different temperatures, and (b) Bauschinger effect with stress reversal plotted from the ISV model

During a low temperature tempering, carbon exits the steel matrix and forms iron carbide, so the as-quenched martensite transforms to tempered martensite. The hardness of tempered martensite is dependent on the conditions of iron carbide. The iron carbide coarsening behavior is modeled using equation (9).

$$\frac{dS}{dt} = c_1 + c_2 * e^{(T-c_4)/c_3} \quad (9)$$

Where S is the size of iron carbide or alloy precipitate, t is time, T is temperature, and c_1, c_2, c_3 and c_4 are material constants.

The same equation (9) is used to describe the alloy carbide coarsening during a high temperature tempering process. Figure 7(a) shows the Pyrowear 675 alloy carbide coarsening behavior during a high temperature tempering process calculated from DANTE material database using the models described above. The X-axis is tempering time, Y-axis is the alloy carbide volume fraction. Size Class-1 is the smallest size, and size class-11 is the largest size. For each size class, its volume fraction increases first when a smaller size class coarsens to this specific class. With further tempering, its volume fraction decreases with coarsening to the next size class. The in-situ alloy carbide size distribution condition is converted to the alloy factor, which is used in calculation of the hardness and mechanical properties of the tempered martensite. Figure 7(b) is an example of hardness evolution during austenitizing, quenching, deep freeze and tempering process for carburized Pyrowear 675. Early on during tempering, the hardness decreases due to the coarsening of iron carbide. The hardness increases with further tempering due to the alloy carbide precipitation and limited coarsening. When tempering time is too long, the hardness decreases due to more coarsening of alloy carbide.

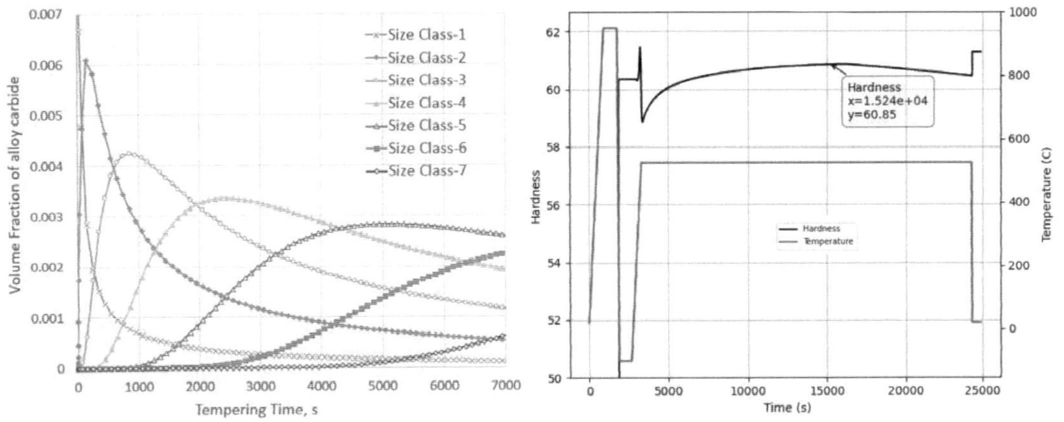

(a) (b)

Figure 7: (a) Pyrowear 675 alloy precipitate coarsening behavior during high temperature tempering, and (b) Simulated precipitation hardening effect during high temperature tempering of a carburized Pyrowear 675 steel

Experiment and Modeling Case Study

In this study, an Almen strip sample made of Pyrowear 675 is selectively carburized, double normalized, HPGQ quench hardened, deep freeze, and double tempered. The distortion after each process step is measured and compared with the modeling results. With the help of computer modeling, the mechanism of distortion is analyzed.

The Almen strip sample has dimensions: 100 mm length, 20 mm width, and 5 mm thickness. Before heat treatment, all surfaces are copper plated other than the top surface. With a plated copper layer, the carbon diffusion is blocked, so the copper plated surfaces are not carburized. Figure 8(a) shows the samples hung in the fixture for processing. Figure 8(b) shows the modeled carbon distribution profile, and the effective case depth (ECD) is about 1.75 mm using HRC 50 criterion.

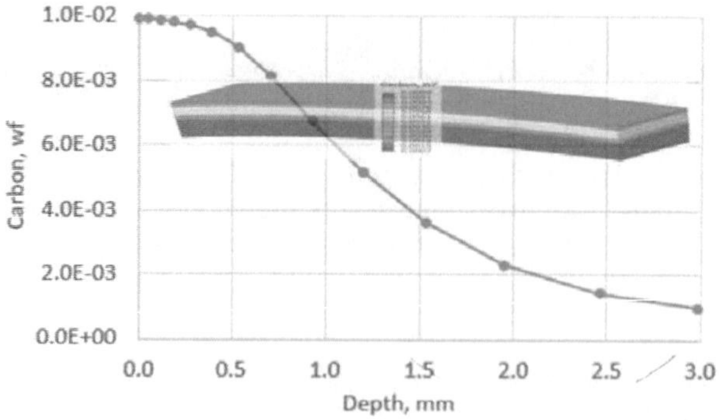

Figure 8: (a) Photo of samples in fixture process for process, and (b) simulated carbon distribution profile using LPC process for carburizing top surface selectively

After each processing step, two samples were pulled out to measure the distortion. All the processing steps were modeled to compare with the experimental results, as shown in Figure 9(a). For each process, two blue dots represent experimental measurements, and the brown dots represent modeling results. Figure 9(b) are photos of Almen strip samples after a specific process: machining (18), normalizing (02), LPC (04), 1st normalizing (06), 2nd normalizing (08), 10 bar HPGQ (10); 1st temper (12), deep freeze (14), and 2nd temper (16, 17). The two distortion measurements after each processing step are consistent, which is due to two main reasons: 1) samples were hung in the fixture, so the effect of the fixture on the sample cooling is minimized because there is no direct contact, and 2) very small thermal gradient in the thickness direction in the sample during quenching (<10 °C), so the distortion caused by the thermal stress is negligible.

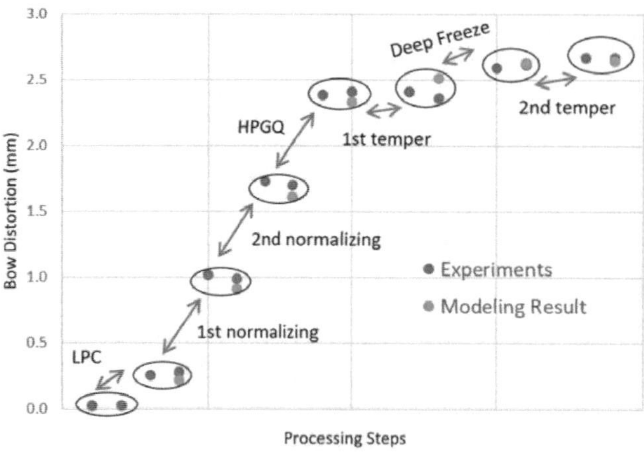

(a)

(b)

Figure 9: (a) Comparison of simulated and experimental bow distortion at the end of different processing steps, and (b) photo of Almen strip samples pulled out at the end of different process steps

For all the processing steps, it is expected the following factors contribute to distortion: 1) material volume expansion caused by carburizing, 2) stress relaxation during heating, carburizing and soaking at the austenitizing temperature, 3) gradient of martensitic transformation during cooling or quenching due to the carbon gradient, and 4) austenitizing difference in the thickness direction due to retained austenite in the carburized case.

The cooling rate in the carburizing step is the same as that of the double normalizing steps. However, the bow distortion caused by normalizing is much higher than that of the carburizing step, which means a large portion of the bow distortion is from the heating step. The distortion from each normalizing step is close to that of the 10bar quenching step, which means the quenching rate doesn't have a significant effect on distortion in this case.

Using the 10bar HPGQ hardening process step as an example, the mechanism of various factors contributing to distortions are analyzed by FEA modeling. Figure 10 shows the simulated temperature, retained austenite fraction, residual stress in the longitudinal direction, and bow distortion after carburizing and double normalizing process steps. The temperature is at room temperature of 20 °C. The retained austenite in the base carbon material is negligible, and the retained austenite in the carburized case is about 25%. The residual stress in the case (longitudinal direction) is about 250 MPa compression, the residual stress in the core is about 260 MPa in tension, and the residual stress in the opposite surface is about 360 MPa in compression. The compressive residual stress in the surface is caused by the delayed martensitic transformation. The tension in the core is to balance the surface compression in the case. The residual compression in the opposite surface is caused by the bow distortion. The bow distortion is about 2.3 mm before reheating.

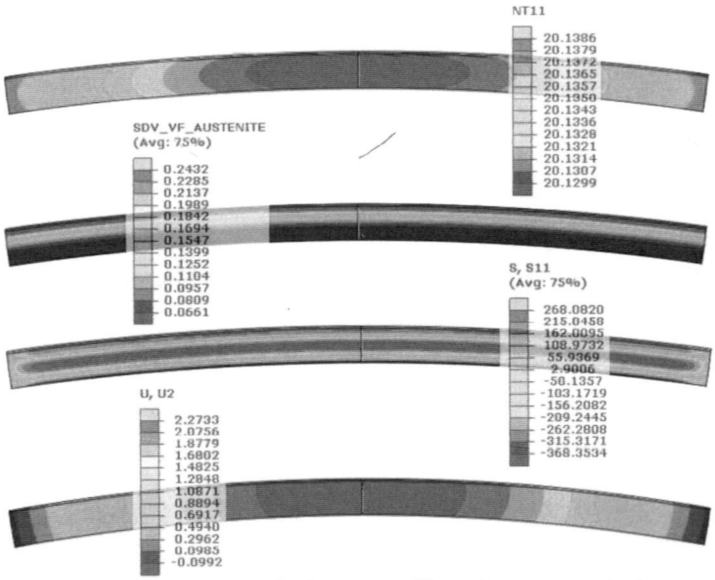

Figure 10: Temperature, retained austenite, residual stress and bow distortion at the beginning of reheating for 10 bar HPGQ hardening (after carburizing, double normalizing and cool down to room)

Figure 11 shows the temperature, austenite fraction, longitudinal stress and bow displacement at 185.5 seconds in the reheating step for quench hardening. The temperature in the mid-section is about 715 °C, with less than 1 °C variation, and the phase transformation to austenite has not started yet. The main contribution to the bow displacement prior to this processing time is the relaxation of the residual stress. The material strength decreases with the temperature increase, as shown in Figure 6(a), and the residual stress relaxes with plastic deformation and bow displacement. The thermal stress in the part has a negligible effect on the bow displacement.

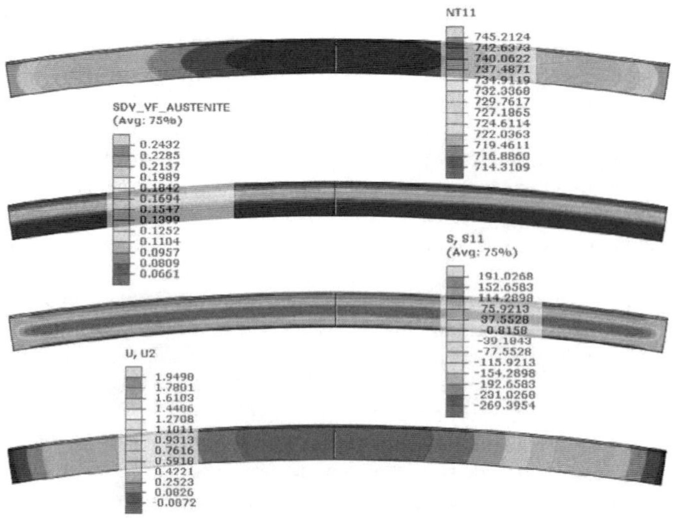

Figure 11: Temperature, austenite fraction, longitudinal stress, and bow displacement at 185.5 seconds in the reheating process step

With further heating, the austenitizing transformation occurs in the part, which has a significant effect on the bow displacement. The case has about 25% retained austenite prior to reheating, while the bottom surface without carburizing doesn't have retained austenite. Therefore, there is about 25% more transformation to austenite at the bottom surface, associated with more volume shrinkage at the bottom surface. The more volume shrinkage at the bottom leads to increased bow distortion at the end of the heating step, as shown in Figure 12.

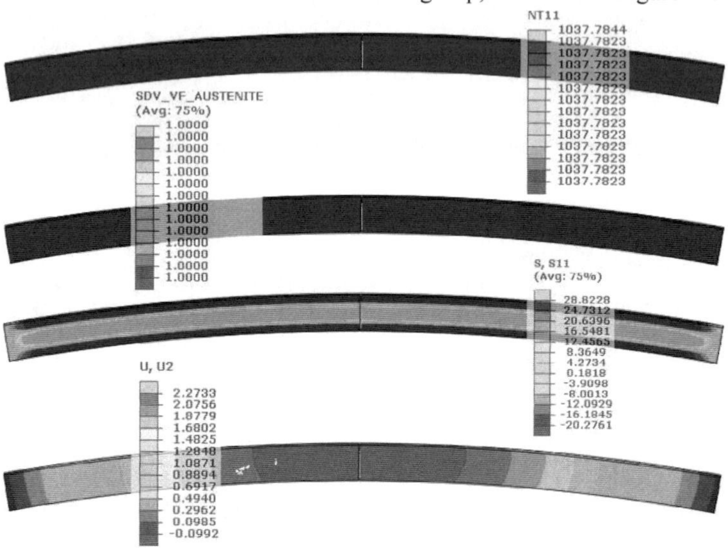

Figure 12: Temperature, austenite fraction, longitudinal stress, and bow displacement at the end of the reheating step

During quenching, the part cools with limited temperature difference < 10 °C in the thickness direction, so the distortion caused by the thermal stress is negligible. The simulation results show there is no noticeable bow displacement change before the 60 second in quench due to negligible thermal stress and no phase transformations. Martensitic phase transformation happened in the uncarburized region between 60s and 83s in the quenching step. The bow displacement changes significantly from ~2.2 mm to a negative value ~0.1 mm. The martensitic transformation and associated volume expansion leads to compressive in-process stress in the uncarburized region, and the bow displacement leads to tensile stress on the bottom surface with compressive stress in the core, as shown in Figure 13.

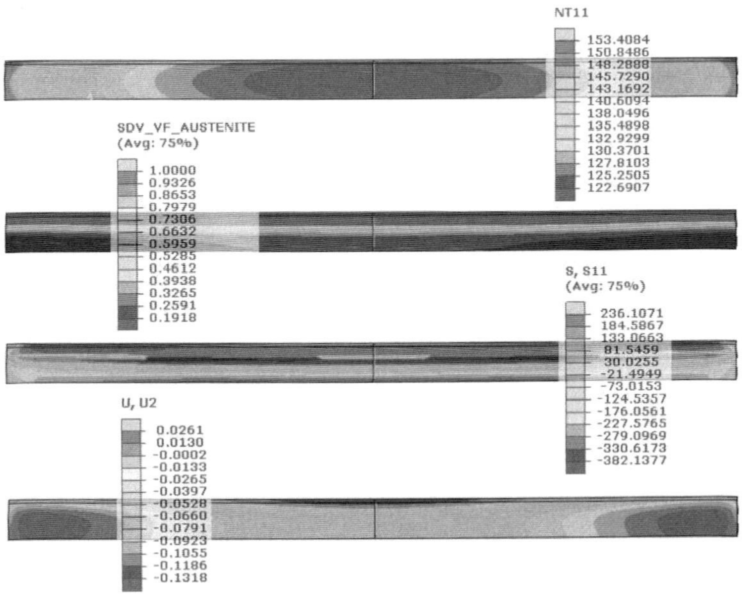

Figure 13: Temperature, austenite fraction, longitudinal stress, and bow displacement at 83 seconds in the quenching step

The martensite transformation in the carburized case is delayed because the martensite start temperature is decreased with high carbon. With further cooling, the martensite transformation in the case starts, which leads to material volume increase. A significant bow displacement is predicted due to this transformation, along with a shift of in-process stress from tension to compression in the carburized case. Figure 14 shows the simulated temperature, retained austenite, residual stress and bow distortion at the end of quenching process step.

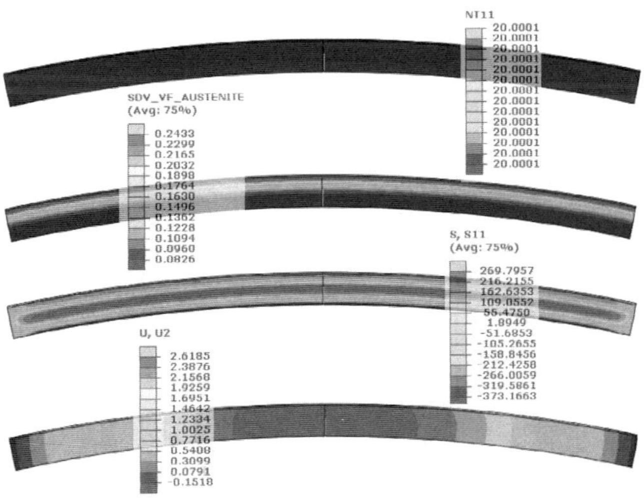

Figure 14: Temperature, austenite fraction, longitudinal stress, and bow displacement at the end of the quenching step

Conclusions

Using an Almen strip sample, the heat treatment process is modeled, and the simulated distortions after each processing step match well with the experimental measurements. With the help of computer modeling, the mechanisms or factors contributing to distortions are summarized: 1) residual stress relaxation during heating of the part, 2) nonuniform austenitizing transformation, 3) thermal stress, especially concentrated stress during heating and quenching, and 4) phase transformations during cooling or quenching. In this case study, the normalizing step causes a similar amount of distortion as the quench hardening step. For most heat treatment processes, each heating and cooling cycle will contribute to distortion or increase the distortion inconsistency. The common quench hardening process steps of aerospace parts include carburizing, cooling down to room temperature, then reheating and quench hardening. Reheating is used to refine the microstructure. However, from the distortion control aspect, the part should be quenched directly after carburizing.

References

[1] S. Masoudi, G. Amirian, E. Saeedi, and M. Ahmadi, "*The Effect of Quench-Induced Residual Stresses on the Distortion of Machined Thin-Walled Parts*", Journal of Materials Engineering and Performance (2015) 24: 3933–3941

[2] V.C. Prantil, M.L. Callabreski, J.F. Lathrop, G.S. Ramwaswamy, M.T. Lusk, "*Simulating Distortion and Residual Stresses in Carburized Thin Strips*," Journal of Engineering Materials and Technology, Vol. 125, April 2003, p 116-124.

[3] S.J. Lee, and Y.K. Lee, "*Finite element simulation of quench distortion in a low-alloy steel incorporating transformation kinetics*", Acta Materialia 56 (2008) 1482–1490

[4] D.J. Bammann, V.C. Prantil, et al., "*Development of a Carburizing and Quenching Simulation Tool: A Material Model for Carburizing Steels Undergoing Phase Transformations*", Proceeding of the Second Internal Conference on Quenching and the Control of Distortion, 4-7 November, 1996, Cleveland, Ohio, p 367-375

QDE 2025: Proceedings of the 3rd International Conference on
Quenching and Distortion Engineering
May 6–7, 2025, Vancouver, Canada
https://doi.org/10.31399/asm.cp.qde2025p0129

HEAT TREATMENT SIMULATION OF SAE 1060 STEEL USING POLYACRYLATE AQUEOUS SOLUTION AS QUENCHING MEDIA

Sergio Gallegos-Cantú, D. Scott MacKenzie
Quaker Houghton, Norristown, Pennsylvania, USA
sergio.gallegos@quakerhoughton.com; scott.mackenzie@quakerhoughton.com

Abstract

Simulations of heat treatment processes can help predict both mechanical and microstructural properties of metals. However, simulation results are dependent on process inputs such as, material, specimen shape, heating and cooling cycles and cooling curves characteristics, among others. Quenching fluids such as polymers are still under investigation and need to be characterized to improve the accuracy of the simulation results. This work aims to contribute to the optimization of the simulation process in the heat treatment industry. Cooling curves of a polyacrylate-based (ACR) polymer solution at a concentration of 9 and 12 %, using an axial flow rate of 1.30 L/min on an inmersion system and a fluid temperature of 45 °C were acquired and analyzed. Air quenching was also used to compare the polymer quenching conditions. The specimens made of SAE 1060 steel were designed for a laboratory-scale immersion quenching system; a 1 mm diameter thermocouple was introduced on the cross section to measure cooling curves at 2 mm from the bottom surface of the specimen. The results of the cooling curves were used as boundary conditions in the simulation. The commercial software JMatPro® (Version 15, Sente Software, Guildford, UK) was used to predict hardness and phases on the SAE 1060 steel. Finally, physical tests were performed on the specimens for the polymer and air quenching conditions to validate the results provided by the simulation.

Introduction

The quenching characteristics of polymer quenchants have been previously investigated using comparative techniques such as the determination of cooling curves by standard methods and observations of the effects of quenching. Polyacrylate-based (ACR) polymer has shown to be one of the slowest quenchants and therefore one of the candidates to replace mineral oil quenchants [1]. Additional, with the increased research, the creation of polymer quenchants has not only filled the gap between water and oil quenchants but also have achieved the slow quench rate required in many applications, for instance, high hardenability steel forgings, patenting high carbon wire, high carbon railroad, and steel castings [2].

ACR polymer has received particular attention to patenting process. Patenting depends on slow cooling rates to achieve a fine pearlite structure and therefore a ductile material. Such slow cooling rates were typically achieved in the extended film of vapor at high concentrations [1]. Additionally, ACR polymers have been used in accelerated cooling process such as modern railroad steel production; it enables phase transformation at lower temperature resulting in lower pearlite interlamellar spacing with low wear resistance [3].

One way to decrease cost and reduce time in product development can be to use simulation tools that can predict the final properties and shape of a component caused by the used manufacturing process. A decrease in cost and better knowledge of final properties already in product development can give the company a better market position and competitiveness [4].

In this project the hardness and phase transformation were predicted using JMatPro® and compared with physical data from test specimens made of SAE 1060 and quenched in air and ACR polymer solution at 9 and 12 %. The data obtained from the cooling curves was introduced to the software and compared with physical data.

Methodology

Quenching tests were conducted under laboratory-scale conditions using a test specimen made of SAE 1060 and having a chemical composition as follows: 0.58 %C, 0.23 %Si, 0.86 %Mn, 0.01 %P, 0.034 %S, 0.150 %Cr, 0.029 %Mo, 0.044 %V, 0.15 %Cu, 0.07 %Ni, and Fe as remainder. The test specimen was machined according to the sketch shown below on Fig.1. A thermocouple type K of 1 mm of diameter was introduced in the axial position at 2 mm from the bottom surface, the surface condition was 6 Ra and a total of 4 specimens were prepared for these runs.

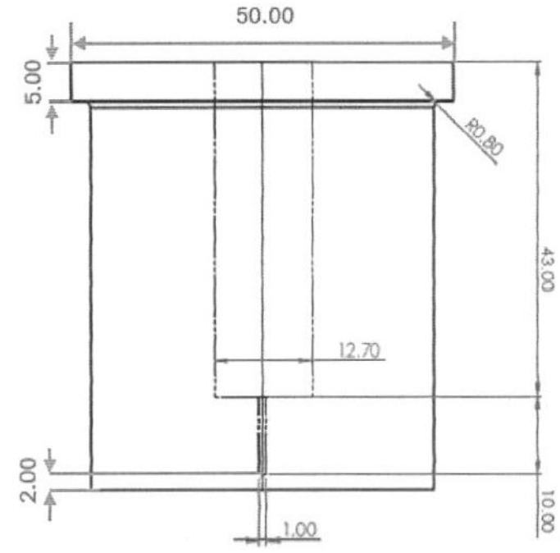

Figure 1: Test specimen design made of SAE 1060 steel.

The construction of the quenching system was carried out using different parts, a description of the components and test parameters are provided below. A 3D printed quench tank made

in ABS polymer was designed to provide an axial flow similar to the Jominy test, the test specimen was partially immersed on the geometric center of the quench tank to give a depth of ~ 50 mm; a picture of the setup is shown in Fig. 2. The temperature for the test specimen was 860 °C and it was heated in an electrical furnace, holding this temperature for 35 minutes, and the test specimen was quenched for 240 s. The quenchant used was an ACR polymer and two concentrations level were used on these experiments, the low level of concentration was 9 % and high level was 12 %, the flow rate was fixed at 1.30 L/min, a water flow meter was used for measuring the flow, however, the flow was determined using the time to fulfill the known volume of the 3D printed quench tank; another run was carried out using air quenching to compare against the ACR polymer solution. The fluid temperature was hold at 45 °C. The cooling power of the quenchant was characterized via cooling curve analysis using the data acquisition (called hand unit) from IVF SmartQuench equipment and its software to process the cooling curves data; the data record frequency of these experiments was 10 Hz. In Table 1 is shown the experimental testing.
Metallographic analysis and microhardness test were carried out on the cross section of the test specimens quenched at the experiment conditions. For metallography preparation, the specimens were cut, hot mount, grinded and polished, then Nital at 2 % was used as the etchant to revel the microstructure of the mounted samples; a digital microscope was used to obtain images at 500X at center position of the cross section at around 2 mm from the bottom surface of the specimen, near to the channel where the thermocouple was placed. Microhardness test was performed at the center of the cross section, at around 2 mm from the bottom surface.

Figure 2: Quenching system setup, a) quench tank, b) test specimen and c) recovery tank.

Table 1: Experimental testing.

Sample	Cooling curve analysis	Microhardness	Microstructure	Simulation
Air Quenching	✓	✓	✓	✓
ACR 9 %	✓	✓	✓	✓
ACR 12%	✓	✓	✓	✓

Simulation was performed using JMatPro®, the chemical composition used was the one described above. Cooling curves from the test specimens were introduced in the quench properties module from the software where hardness and phase transformation were predicted using a grain size ASTM 7 and 860 °C for the austenization temperature. Results were compared against physical data.

Results

The cooling curves were run in duplicate showing a consistency in the repeatability. One of the cooling curves was chosen to represent the test condition. The cooling curve for air quenching is shown in Fig. 3, where the adiabatic heat was observed at 672 °C; the maximum rate for this condition was found in 4.7 °C/s at 807 °C and the time to cool the specimen was found over 600 s (maximum recorded by IVF SmartQuench). Respect to the ACR polymer solutions, the results are shown in Fig. 4, where the 9 % of concentration had a maximum cooling rate of 12.3 °C/s at 808 °C and the vapor phase of polymer collapsed at a temperature of 337 °C. For the ACR polymer solution at 12 % of concentration, the maximum cooling rate was found at 10.44 °C/s and occurred at 808 °C; the vapor phase collapsed 334 °C; both concentrations gave a cooling time of 240 s to cooled down below 100 °C.

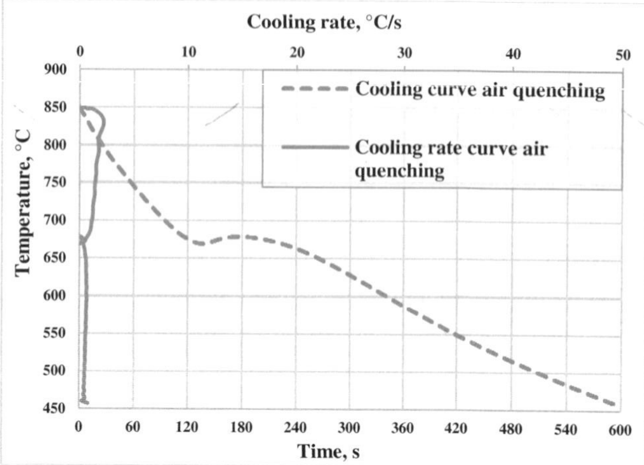

Figure 3: Cooling curve and cooling rate curve for air quenching condition.

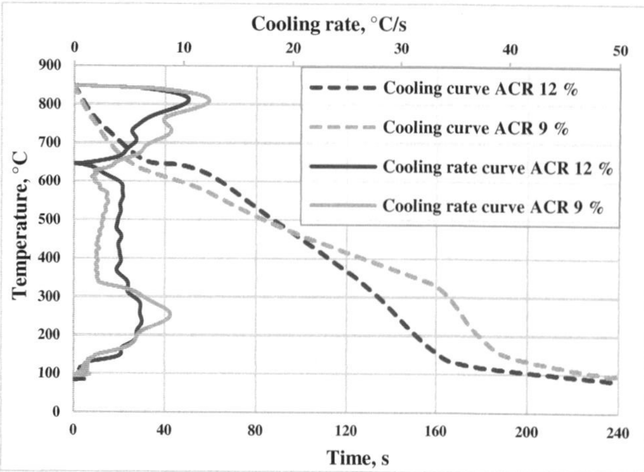

Figure 4: Cooling curves and cooling rate curves for ACR polymer solution; effect of concentration.

Hardness values from the test specimens were obtained using a microhardness equipment (Duramin-40 from Struers); the applied load was 300 gf for 20 s. Conversions to Rockwell C scale were also reported. In Table 2 are shown the results for hardness for all the conditions.

Table 2: Results of hardness for all the conditions.

Sample	Position	HV0.3	HRC (conversion)
Air Quenching	Center	262	24
ACR 9 %		296	29
ACR 12%		323	32

Micrographs are shown in Fig. 5 at 500X. Samples were grinded, polished and etched using Nital at 2 %, the images were taken using a Keyence digital microscope VHX-X1 series with a coaxial light. Image analysis also was performed to determine the amount of ferrite present on each condition.

In Table 3 are shown the results of the simulation provided by JMatPro®. The results were calculated based on the thermocouple position on the test specimens.

Table 3: Hardness simulation results and phases prediction.

Sample	Position	Microhardness, VPN	HRC	Phase prediction
Air Quenching	Center	278	25	Pearlite, 92.7 % Ferrite, 6.88 % Other, 0.42 %
ACR 9 %		325	32	Pearlite, 48.52 % Bainite, 45.84 % Ferrite, 5.28 % Other, 0.36 %
ACR 12%		319	32	Pearlite, 41.33 % Bainite, 52.80 % Ferrite, 5.50 % Other, 0.37 %

Continuous cooling transformation diagram also was calculated on this software (Fig. 6); cooling curves obtained from the test specimens for each condition were introduced in the diagram.

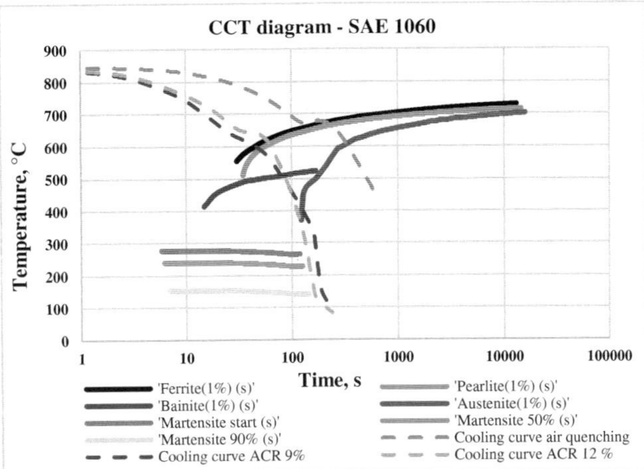

Figure 6: Continuous cooling transformation diagram for SAE 1060 steel with cooling curves from the test specimens.

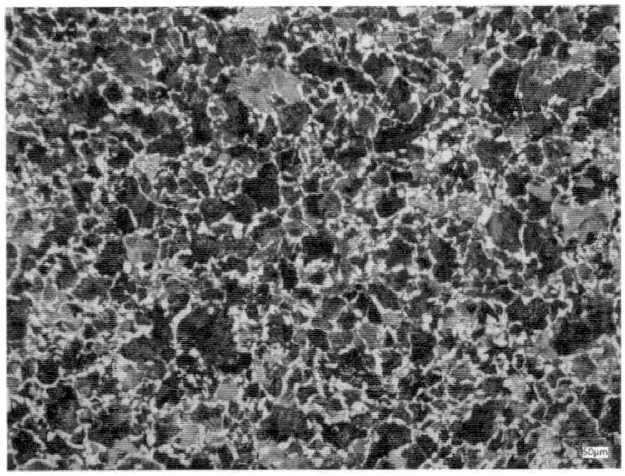

(a)

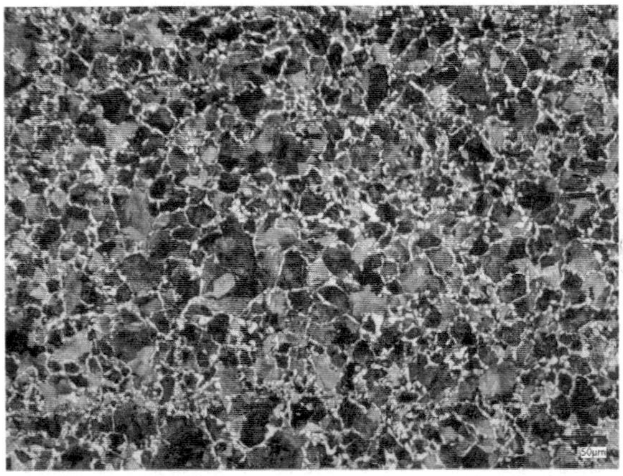

(b)

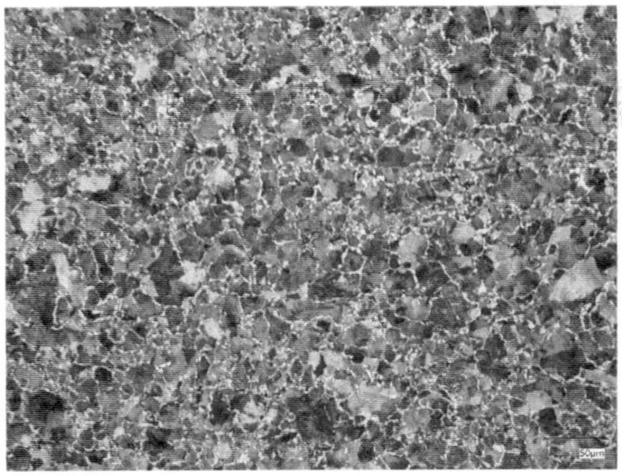

(c)

Figure 5: Micrographs of SAE 1060 steel at 500X (etchant Nital at 2 %): a) air quenching, b) ACR polymer solution at 9 % and c) ACR polymer solution at 12 %.

Discussion

In this paper, three quenching conditions were evaluated, air quenching, ACR polymer solution at 9 % and 12 %. From the cooling curves and cooling rate curves shown in Figures 3 and 4, differences between cooling profiles were observed: for air quenching cooling rates were slower than ARC polymer solution conditions. Adiabatic heat from the steel phase transformation was observed on the cooling curves in all the conditions, however it happened at different temperatures based on the quenching severity of the media. Air quenching condition also showed the longest quenching times. ACR polymer solution at 9 and 12 % showed a very long vapor phase, this vapor layer collapsed at 337 and 334 °C for 9 and 12 % respectively, this helped to promote a ferritic, pearlite and bainitic structures in the SAE 1060 steel. Based on the cooling curves for ACR polymer solution, the effect of the concentration did not show a significant difference, a similar cooling curve profile was found in both conditions; this effect will be discussed later.

Hardness test for air quenching condition showed the lowest value with a result of 262 HV0.3 (24 HRC), while the simulation showed a value of 278 VPN (25 HRC), having a slightly difference. The microstructure for air quenching condition was found in pearlite and ferrite mainly, only ferrite was quantified using the image analyzer on the digital microscope, giving a result of 19.62 %, while the prediction in the software was found in 6.88 %, considerably low for what was calculated by the software. ACR polymer solution condition at 9 % of concentration was found with a hardness value of 296 HV0.3 (29 HRC) and compared to the prediction was found slightly deviated with a value of 325 VPN (32 HRC); the microstructure was observed predominantly pearlite with ferrite. The amount of ferrite quantified was 17.37 % while the prediction by the software was 5.28 %. Based on the continuous cooling transformation diagram showed in the Fig. 6, there is a presence of bainite for the 9 % of concentration condition; on the Table 3 can be observed a 45.84 % of bainite on the prediction, however, on the micrograph was not completed clear the detection of bainite during the metallography evaluation. ACR polymer solution condition at 12 % of concentration was found with a hardness of 323 HV0.3 (32 HRC), the prediction for this condition was found in 319 VPN (32 HRC), having a very similar calculation; the microstructure was found with pearlite and ferrite. Similar case as the 9 % of concentration where the prediction of the phase's simulation gave a value of 52.80 % of bainite, while the presence of bainite was not cleared observed on the metallographies; ferrite was found with 15.49 % and the simulation gave a value of 5.50 %. The higher hardness value was found on the 12 % of concentration condition, it can be related to the smaller percentage of ferrite found in the microstructure and a smaller grain size in comparison with the other conditions.

Conclusions

In this paper, a quenching was performed on SAE 1060 steel using air and a polyacrylate-base (ACR) polymer solution. Additionally, hardness and metallography were evaluated finally, the simulation of this heat treatment process was compared against physical data. The conclusions of this work are:

- The vapor phase of the cooling curves with the ACR polymer solutions was stable up to a range of 334 to 337 °C. This was key to promote the presence of the ferritic, pearlitic and bainitic phases in the quenched specimens. The air quenched condition showed mainly pearlitic and ferritic microstructures, and this condition obtained a much longer cooling time compared to the cooling curves with the ACR polymer solutions.
- The ACR polymer solution at 12 % of concentration condition showed the highest hardness result, due to the lower presence of ferrite found in the micrograph compared to the other conditions. In addition, visually there is a finer grain size in the 12 % condition sample compared to a coarser grain size in the air quenched and 9 % ACR polymer solution specimen.
- The hardness results for the quenching simulation using the quench properties module on JMatPro® showed results very similar to the measured results. Regarding the prediction of the ferritic phase, deviated results were found in all the simulated conditions against the measured; this could have been due to the position of the metallographic evaluation; however, the hardness results were consistent and useful.

Acknowledgments

One of the authors (S. G.-C.) would like to thank Dr. Scott MacKenzie for his guidance, support and mentoring on this project, Conxita Suárez, Mireia Viure and Sebastian Kozdron for their contributions in this project, Joe Jankowski for all the work with the execution of the cooling curves, Dr. Michael Smith and the company Quaker Houghton for supporting me in publishing this article.

References

[1] W. D. Griffiths, *"The Quenching Characteristics of Sodium Polyacrylate Solutions", Sheffield Hallan University, 1989.*

[2] S. MacKenzie, *"Polymer Quenchants for Steel Heat Treatment"*, Gear Solutions, p 24-25, 2024.

[3] S. Sahay, G. Mohapatra and G. Totten, *"Overview of Pearlitic Rail Steel: Accelerated Cooling, Quenching, Microstructure, and Mechanical Properties"*, Journal of ASTM International, Vol. 6, No. 7, 2009.

[4] A. Henrik, *"Material Modelling for Simulation of Heat Treatment", Luleå: Luleå Tekniska Universitet, p 22, 2003.*

)E 2025: Proceedings of the 3rd International Conference on
enching and Distortion Engineering
ay 6–7, 2025, Vancouver, Canada
ps://doi.org/10.31399/asm.cp.qde2025p0133

Analysis of Geometry Orientation in the Calculation of Transient Heat Fluxes during Immersion Quenching of Hollow Cylinder

Gamaliel Salazar[1], Jean-Sebastien Lemyre-Baron[1], Mohammad Jahazi[1], Henri Champliaud[1], Antoine Tahan[1]
[1]Département de génie mécanique, École de technologie supérieure, Montréal, QC, Canada

Abstract

Quenching of large and complex geometries often involves surfaces with different orientation with regards to the quench bath. Previous work by the authors includes the characterization of industrial quenching facilities with specially design quench probes. The study herein explores the influence of quenching a hollow cylinder in a horizontal set up as it replicates a condition frequently faced in the industrial application of interest where large tubular geometries are quenched in a similar orientation. The study considers the required air transfer step before immersion that produces non-homogeneous condition of the heat-treated load even before its immersion. The work in this study includes the analysis of experimentally acquired thermal history at different angular and azimuthal locations of a hollow cylinder. Then, the inverse heat transfer problem is solved to compute the transient surface heat fluxes around the outer surface and bore surface of the tested geometry. Findings capture the evolving condition on both surfaces, including the gradient before immersion due to the air transfer as per process constraints and the thermal gradient once the body is immersed demonstrating the uneven cooling of the test specimen. Results in the form of transient heat fluxes are compared with previous findings from fully horizontal plates as well as the hollow cylinder condition quenched on a vertical orientation.

Introduction

Modeling of quenching processes requires the coupling of multiple physical fields: thermal, metallurgical, and mechanical. It is widely perceived that the thermal boundary condition, typically in the form of heat transfer coefficients (HTC), is one of the most influential parameters in the modelling of heat treatment and quenching processes [1] yet remains one of the biggest challenges. Accurate determination of the thermal evolution of the quenched body is crucial as the heat transfer triggers both the metallurgical and mechanical fields. For instance, the driving force for phase transformation is the transient temperature field. Similarly, thermal stresses due to uneven thermal shrinkage, even on transformation less cases, are caused by the temperature gradient across the thickness of the quenched body [2], [3].

Studies on the boiling conditions for horizontal tubes and surfaces can be traced back to the works of Zubber [4] and Bromley [5]. The former developed a correlation for boiling heat transfer, including a horizonal cylinder, and the latter focused on heat transfer during film boiling. Modern work on the determination and correlations for the critical heat flux (CHF) under pool boiling conditions have considered the influence of orientation [6], material and subcooling temperature [7], while a review of CHF predictive models has been reported by Liang et Mudawar [8]. Similarly, the pool film boiling regime has been studied for different geometries and fluids [9]. Notably, transient conditions, like the ones encountered during quenching by immersion under industrial environments, are barely addressed.

HTCs are often obtained for specific mediums and conditions [10] which might differ or misrepresent the industrial environment which limits the applicability of boiling curves obtained under steady-state conditions. Previous studies have explored the influence of decreasing the starting temperature of the load finding out its impact on the film boiling stage while shifting the critical heat flux towards lower temperatures [11], [12]. The influence of orientation during quenching has been reported to reduce the surface heat fluxes at the bottom of a horizontally quenched blade down to a range ≈15-25% when compared against the heat flux on the top surface [13]. For cylindrical geometries, the critical heat flux (CHF) presents a considerable variation around the surface when quenched horizontally under saturated and subcooled conditions [14], it has been proposed that the outer surface of a cylinder should consider at least three different regions due to the variable conditions [6]. Moreover, even vertically quenched standard quench probes might present non-uniform conditions in the axial and circumferential directions [15].

The authors of this study have identified the boundary condition for vertically quenched plates [16] and hollow cylinder [17] under an industrial environment. In both studies a thermal gradient was reported by the end of the air transfer step, which influences the remaining cooling once immersion occurs. Contrary to standard procedures, the immersion movement (vertical descend in this study) might take several seconds to complete due to the dimensions of the load. The interaction between tooling and quenched load, as well as the agitation system, influences the quenchant flow leading to different boundary conditions on the quenched geometry.

The study herein presents the determination of the boundary condition for a horizontally quenched hollow cylinder under an industrial environment. The experimental results of the quenching process are included first, followed by the methodology used to identify the surface boundary conditions. A comparison with vertically quenched geometries (plate and hollow cylinder) is included to highlight the variable conditions on the bore and outer surface of the quenched body.

Methodology

Quench Probe & Experimental Means

In this study a concentric hollow cylinder made of 304SS was quenched in a horizontal orientation so the quenchant flows perpendicular to the body. The quench probe has a length (L) of 304 mm, a thickness (δ) of 50.8 mm and an inner radius (r_{bore}) of 76.2 mm. The stations around the cylinder are labelled as A, C & D as shown in the schematic in Fig. 1. Station A is placed at the top of the cylinder, it is the first to exit the furnace but the last to enter the quenchant whereas Station D is the last to exit the furnace and first to enter the quenchant. These two stations are the main subject of interest in this study as they provide the most insightful results.

The instrumentation strategy considers K-type thermocouples inserted along the length of the body parallel to the main surfaces of the probe. Measurements are taken at mid-length of the quench probe. Each thermocouple is labelled according to their position: OutSurf & Bore are sub-surface thermocouples whose distance off the surface (γ) is 4.8 mm; the Mid thermocouple is placed at the mid-thickness of the body. A Phoenix TM PTM1220 data logger was used to record the thermal history during the test, it has an accuracy of 0.3°C & a resolution of 0.1°C. The test was carried out with a sampling rate of 5 Hz.

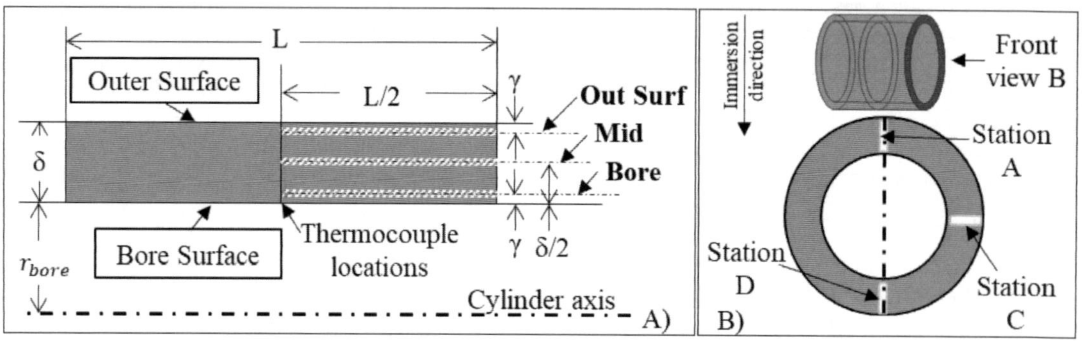

Figure 1: A) Schematic of typical instrumentation per station & B) stations labelling

The testing was carried out at the industrial partner site, thus accounting for the heterogeneities of the process as well as the transferring through air form the furnace to the quench bath. The schematic in Fig. 2 depicts the typical movements of such transfer: extraction of the system from the furnace (1), lateral transition until the system is placed on top of the quench bath (2-4), descend step as immersion (5) occurs, the load remains in the bath to complete its cooling (6). The probe was mounted on an industrial heat treatment rig as shown in Fig. 2 while placed around the mid-height of the assembly. Both ends of the quench probe are open, thus the quenchant can flow in and out of the bore freely. Due to the orientation of the body, the outer surface of the hollow cylinder is subject to crossflow during immersion and the remaining cooling time in the bath. The quenchant flow in the bath is assumed to move upwards (contrary to the immersion direction) due to the agitation system. The comparison cases included in the results section were tested under similar bath conditions and parameters such as air transfer step, bath temperature and agitation.

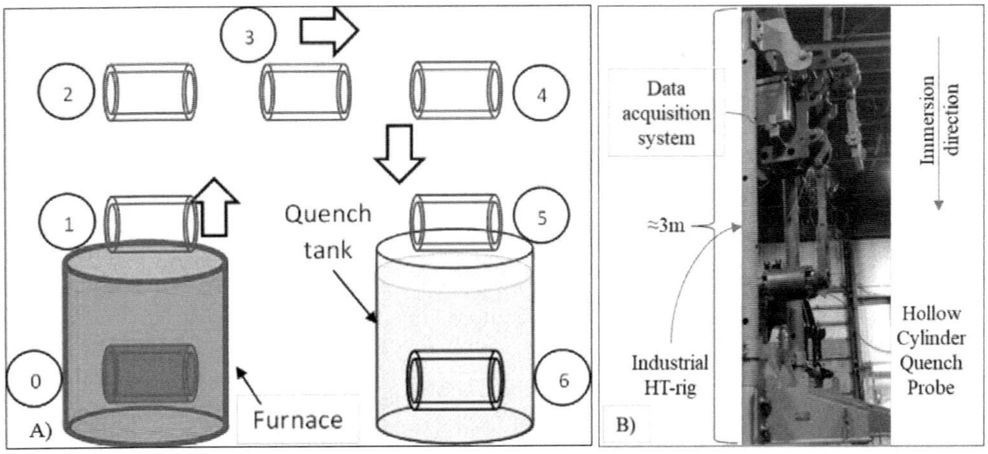

Figure 2: A) Schematic of the quenching process at the industrial partner site and B) test setup on industrial HT-rig with the quench probe in a horizontal setup.

Surface Boundary Condition Identification

The identification of the boundary condition requires solving the inverse heat transfer problem. Methods and strategies to solve the inverse problem are well documented in the literature [18]-[20]. Recent works combine commercially available software with optimization algorithms to identify the surface boundary conditions in the form of transient heat fluxes and HTCs varying in time and space [21], [22]. In this study, the boundary conditions are identified on a station-by-station basis, thus solving a 1D axisymmetric case. The solution is verified running the direct problem for a 2D case (half cylinder) using the identified BCs as input. Overall, the identification strategy follows the methodology applied on a plate probe [16] and a vertically quenched hollow cylinder [17].

The main steps of the methodology are shown in the schematic in Fig. 3. Firstly, the surface temperature profile is reconstructed using a damped Newton-Raphson algorithm [23]. The sum of square residuals between simulation and test data is optimized following a linear piecewise temperature evolution for sequential time intervals. Numerical differentiation based on sensitivity simulations of the thermal field using the FE method are used to build up the Hessian matrix and gradient vector. Temperature dependent thermophysical properties of the material are considered in the FE simulations using the commercially available FE package Abaqus. An axisymmetric FE model was employed with element type DCAX4, and element size of ≈ 1 mm.

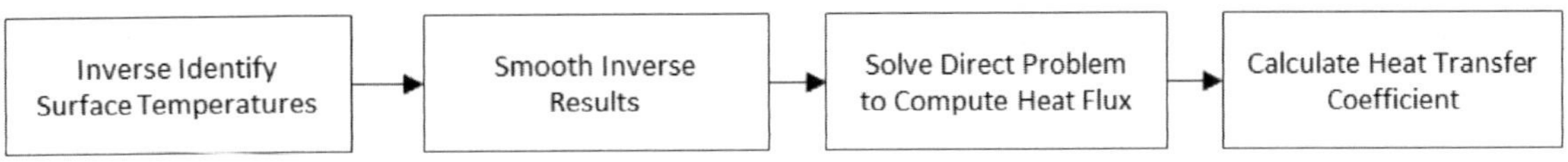

Figure 3: Main steps of inverse identification strategy [16]

The surface temperature profiles are smoothed out using a Least Squares B-Spline, inspired from a point distance minimization algorithm, to diminish the impact of the characteristic «saw teeth» [24]. The smoothed thermal profiles are used as boundary condition in a FE simulation from which the reactive fluxes are extracted. This step uses a refined axisymmetric FE model with a refined size ≈ 0.1 mm and same element type (DCAX4). The results of this simulation can be used to compute the heat transfer coefficient following the well-known Newton's Law of cooling.

A verification simulation using the identified HTC as boundary condition was run for a 2D geometry (half cylinder, element type DC2D4). The bore and outer surfaces were split into sections to assign a local HTC per station and compared against test readings. The relative difference (%) between simulated temperatures and the experimental values is computed as:

$$T_{RelDiff} = \frac{(T_{i_{sim}} - T_{i_{test}})}{T_{i_{test}}} \tag{1}$$

The relative difference always remained below ±5% throughout the entire dataset, including air transfer and immersion quenching, thus confirming that the identified boundary conditions adequately match the experimental readings. Note that the results section is centered on the temperature evolution and heat flux profiles per station though, the former due to the experimental nature of the values, and the latter due to the physical significance in terms of cooling regimes.

Results & Discussion

This section presents the experimental data followed by the identified surface boundary condition. Most of the discussion is centered on Stations A & D as they present the most insightful results. Comparison with previous studies considering a plate and vertically quenched hollow cylinder [16], [17]. The results from the comparison cases were obtained under similar conditions, thus comparison can be done directly.

Experimental Data

The complete thermal history of the different sensors is plotted in Fig. 4. The labelling of each sensor follows the convention outlined in Fig. 1. The temperature evolution during the air transfer is also included in Fig. 4 (right plot). The experimental readings indicate that the bore surface enters the oil at ≈863°C, barely decreasing from the starting temperature and even slightly higher than the readings at the MID position (854°C - 857°C). The high temperature in the bore is mainly attributed to the radiation exchange between the faces of the hollow cylinder. The outer surfaces enter the oil at temperatures in the range of 822°C to 830°C, with Station A being the coolest and Station D the hottest. These readings clearly demonstrate that there is a thermal gradient across the cross section of the body by the end of the air transfer.

The temperature evolution and cooling rates after immersion are shown in Fig. 5 for Stations A & D. The influence of the body orientation can be clearly appreciated in both plots. Station D is immersed first as it is placed in the lower side of the horizontally quenched body. The data shows that Station D outer surface cools down faster than its bore, yet considerably more similar than the readings for Station A which is placed on the upper side of the body. The outer surface at Station A cools down abruptly, even faster than Station D outer surface, down to the ≈400°C. This is attributed to the interaction between the fluid and body that enhances the cooling during the boiling regimes in the top side of the body (A – at top), when compared with the lower side (D – at bottom). However, when the influence of the convective stage becomes more evident (below 400°C), then the lower side (D-at bottom) has a better heat exchange than the top side as it is more susceptible to the oil flow coming from the agitation underneath. This is also influenced by the conditions in the bore surface at the top (D – bore), which has very low cooling rates during most of the process. During the boiling stages, this can be attributed to a combination of quenchant bubbles moving upwards (vertical) inside the bore, which functions as a prolongated vapor film in the upper station A, thus reducing the heat extraction in the bore surface in that station, and the flow stagnation that may occur due to the concave geometry itself in the top half of the body.

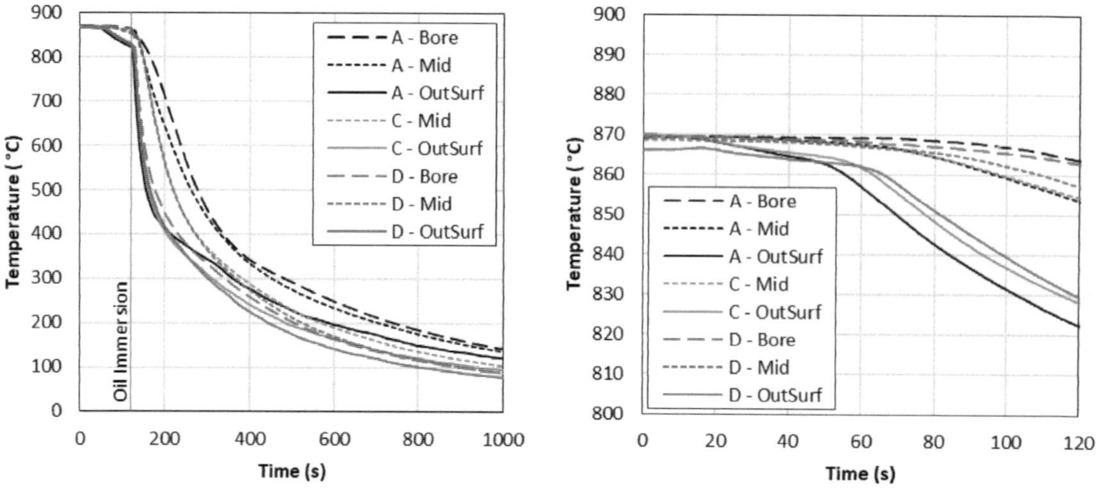

Figure 4: Experimental data: complete thermal history (left) and air transfer only (right)

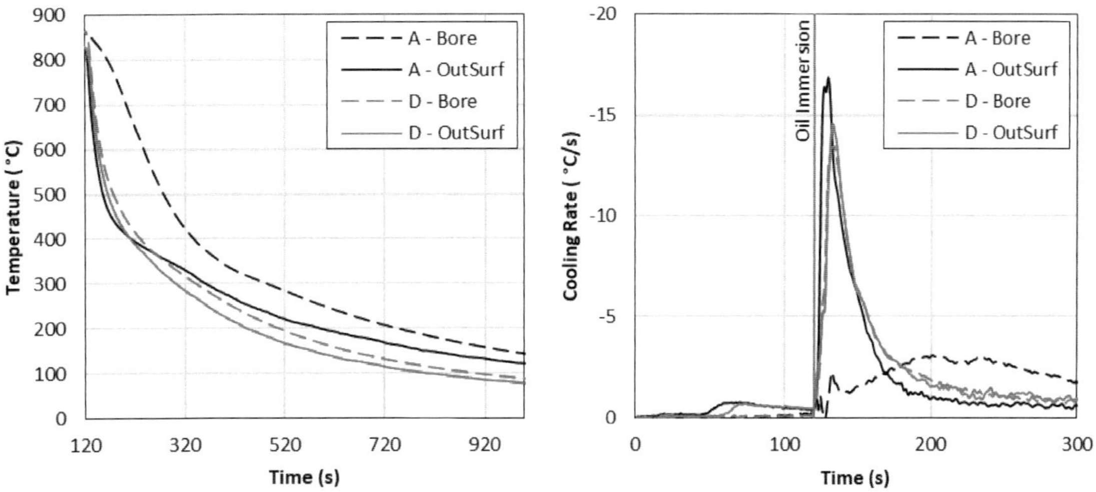

Figure 5: Experimental data for Stations A & D: temperature evolution at sub-surface locations for immersion only (left) and cooling rates (right).

Surface Identification Results

The resulting surface temperature profiles for the entire cooling are shown in Fig. 6 for both stations A & D. Clearly, both stations present a thermal gradient but entirely different shape. At Station D, the temperature gradient, although more pronounced towards the outer surface, appears towards both surfaces as they cool down faster than the mid location. However, the temperature gradient in Station A has a clear shift towards the outer surface as the bore surface remains at high temperatures. In fact, the highest temperature gradient in Station A can be computed at a location ≈12 mm off the bore surface, not the mid thickness, as most of the heat is extracted through the outer surface thus the mid location cools faster than the reference location. The max temperature difference at Station A is computed at ≈445°C around the 38 s mark after immersion (t=158s). Conversely, in station D the max temperature difference is ≈25% less in magnitude (≈350°C) while both surfaces present a large gradient with regards to the mid body temperature.

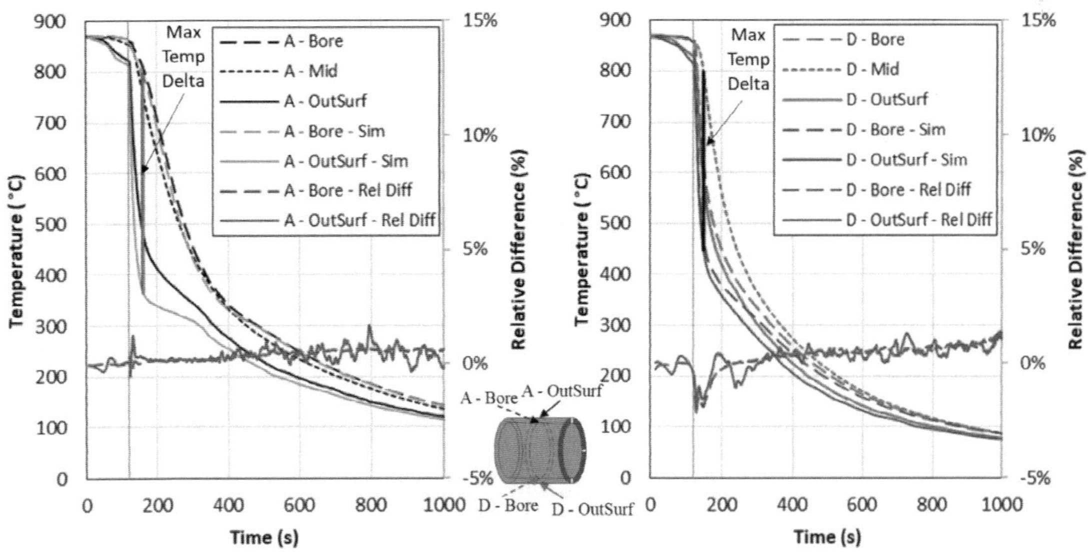

Figure 6: Surface temperatures for Station A and Station D - surface temperature (main Y-axis) and relative error (secondary Y-axis). Evolution of thermal difference at max delta.

Figure 6 includes the estimated relative difference (secondary axis) that compares the simulation results at the sensors location. Every case remains within the ±5% relative difference during both the identification stage and the verification step, thus confirming the accuracy of the proposed methodology.

Surface Heat Flux

The temperature dependent heat flux profiles are plotted in Fig. 7, bore (7A) and outer (7B) surfaces are presented separately. Figure 7 also contains the surface fluxes for a vertically quenched hollow cylinder [17], and the flux profile for the lower surface of horizontally quenched plate (7A), based on the thermal profile previously reported [16]. The following lines address the current test setup (horizontal) first, followed by a comparison against the vertically quenched hollow cylinder and the horizontal plate during the different cooling stages.

- Hollow cylinder - horizontal only

The resulting flux profiles at the bore (Fig. 7A, dashed lines) clearly depict the radically different conditions between D-Bore, which has higher flux rates, when compared with A-Bore, whose surface flux magnitude is only ≈18% of the fluxes at D-Bore. Once the convection stage is reached, the fluxes in the bore surface remain lower in magnitude than those in the outer surface (Fig. 7A vs Fig. 7B), which can be attributed to the poor flow circulation inside the bore.

The heat flux profiles along the outer surfaces (Fig. 7B, continuous lines) are more similar than those in the bore surface, yet they present clear differences in the temperature range between 800°C to 300°C, during the boiling stages. The external surface at the top (A-OutSurf) has a higher extraction rate than the lower side D-OutSurf until the convective stage appears below 300°C. The CHF varies around 10% between both locations, although the difference in magnitude is more evident in the full nucleate boiling towards the convective stage. Such condition is attributed to the interaction between liquid and vapor boundary layers flowing upwards next to the quenched body [14].

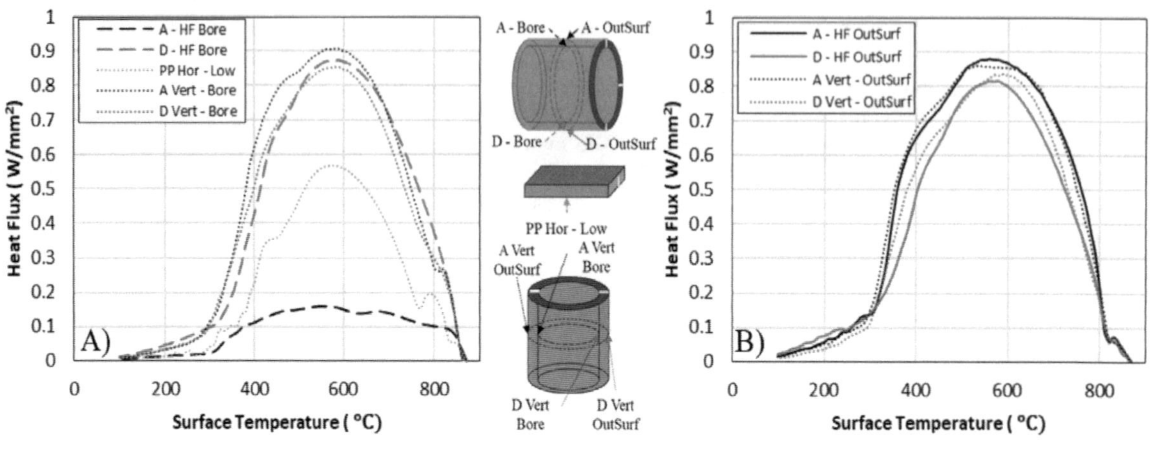

Figure 7: Comparison of surface heat flux: A) Bore surface (horizontal & vertical) vs horizontal plate, B) Outer surface (horizontal & vertical). geometry schematics for the three geometries shown between plots.

- Influence of orientation: horizontal vs vertical

Data corresponding to the vertically quenched body is shown as dotted lines in Fig. 7. Clearly, the body orientation has a strong effect on the location A-Bore (black dashed line, Fig. 7A). The profile shape and magnitude are entirely different from those for the vertical setup. Conversely, the D-Bore flux profile exhibits similar shape like the vertically quenched cases, its CHF is almost the same in magnitude (0.8-0.9 W/mm^2) and appears at similar temperatures (580-590°C). Although the heat extraction during the nucleate boiling at lower temperatures after the CHF is more effective for the vertical configuration (dotted lines, Fig. 7A). Similarly, the transition to convection is more drastic for the vertical setup and begins at slightly lower temperatures than the horizontal setup.

The outer surfaces have similar flux profiles for both orientations (Fig. 7B). Peak heat fluxes (CHF) appear at slightly higher temperatures for the vertical setup although the magnitude remains almost the same. Once convection appears, the horizontal setup has higher fluxes (Fig. 7B) than the vertical configuration. This could be attributed to the interaction of the body with the oil flow as the horizontal setup is more sensitive to the oil flow.

- Hollow cylinder horizontal vs horizontal plate

The lower surface heat flux profile of a horizontally quenched plate (orange dotted line, Fig. 7A) is around 60-65% lesser in magnitude than the D-Bore flux profile during transition and full nucleate boiling. Notably, the stagnant flow under the plate's surface, although reducing the heat extraction, is less insulating as previously reported for plate-like horizontal surfaces [13]. Notably, the horizontally quenched plate has a higher flux magnitude than the A–Bore, which is located at the top of the hollow cylinder. Such difference is attributed to the concave geometry that restrains any displacement of any bubbling coming from the bore surfaces thus decreasing the heat extraction more drastically. Conversely, once the convective stage begins, both plate and A-Bore have practically the same heat flux profile, which is still lower than the other profiles for the vertical case and D-Bore.

Conclusions

A hollow cylinder quenched horizontally under the industrial environment presents different boundary conditions for the bore and outer surface. Notably, the condition in the hollow cylinder bore varies drastically, the predicted boundary condition at Station A Bore is only ≈18% of the flux profiles for the other locations. The variable conditions caused by the quenched orientation create a non-symmetrical thermal gradient inside the thickness of the body. Such gradient is around 25% higher at the top station than the lower station. In fact, at the top station most of the heat is extracted through the outer surface, which causes the max thermal gradient at a different location other than the mid thickness.

The outer surface of the hollow cylinder has higher heat extraction rates at the top of the body compared with the lower side. This phenomenon is attributed to the interaction between flow and body that enhances the heat exchange at such location. The boiling regime is more effective when quenched vertically as the heat flux profiles are higher in magnitude than those for the horizontal configuration, yet differences remain within a range of 10-15%. Conversely, once convection starts, the horizontal setup is more efficient than the vertical setup along the outer surface. Findings confirm the influence of orientation and geometry that should be considered to adequately model a quenching process.

Acknowledgments

This study is financially supported by the Natural Sciences and Engineering Research Council of Canada (NSERC), the Metal Transformation Research and Innovation Consortium of Quebec (CRITM) and Safran Landing Systems. Immense gratitude to the board of academy and industrial advisors for their guidance and advice.

References

[1] C. Simsir, "Modeling and Simulation of Steel Heat Treatment - Prediction of Microstructure, Distortion, Residual Stresses, and Cracking," in Steel Heat Treating Technologies, vol. 4B, J. Dosset and G. E. Totten, Eds. ASM International, 2014, pp. 409–466.

[2] A. Samuel and K. N. Prabhu, "Residual Stress and Distortion during Quench Hardening of Steels: A Review," J. Mater. Eng. Perform., vol. 31, no. 7, pp. 5161–5188, 2022, doi: 10.1007/s11665-022-06667-x.

[3] R. A. Wallis, "Modeling of Quenching, Residual-Stress Formation, and Quench Cracking," in Metals Process Simulation, .

[4] N. Zuber and M. Tribus, "Further Remarks on the Stability of Boiling Heat Transfer," Oak Ridge, TN, 1958.

[5] L. A. Bromley, "Heat Transfer in Stable Film Boiling," 1949.

[6] A. H. Howard and I. Mudawar, "Orientation Effects on Pool Boiling Critical Heat Fux (CHF) and Modeling of CHF for Near-Vertical Surfaces," Int. J. Heat Mass Transf., vol. 42, 1999.

[7] B. Pattanayak and H. Kothadia, "Experimental Study of Critical Heat Flux During Pool Boiling on Mini Tubes: Effect of Subcooling, Orientation, and Dimensions," Heat Transf. Eng., vol. 43, no. 11, pp. 896–921, 2022, doi: 10.1080/01457632.2021.1919971.

[8] G. Liang and I. Mudawar, "Pool Boiling Critical Heat Flux (CHF) – Part 1: Review of Mechanisms, Models, and Correlations," Int. J. Heat Mass Transf., vol. 117, pp. 1352–1367, 2018, doi: 10.1016/j.ijheatmasstransfer.2017.09.134.

[9] F. Ahmad, M. Meyer, J. Hartwig, I. Mudawar, K. Ave, and W. Lafayette, "Saturated Pool Film Boiling of Cryogenic Fluids: Review of Databases, Assessment of Existing Models and Correlations, and Development of New Universal Correlation," Int. J. Heat Mass Transf., vol. 235, p. 126190, 2024, doi: 10.1016/j.ijheatmasstransfer.2024.126190.

[10] R. A. Wallis, "Modeling of Quenching , Residual-Stress Formation , and Quench Cracking," in Metals Process Simulation, vol. 22B, D. U. Furrer and S. L. Semiatin, Eds. ASM International, 2010, pp. 547–585.
[11] B. Hernandez-Morales, F. Lopez-Sosa, and L. Cabrera-Herrera, "A New Methodology for Estimating Heat Transfer Boundary Conditions during Quenching of Steel Probes," in Proceedings from the 6th International Quenching and Control of Distortion Conference, 2012, pp. 81–92.
[12] M. Maniruzzaman, X. Dai, and R. Sisson, "Effect of Quench Start Temperature on Surface Heat Transfer Coefficients," in Proceedings from the 6th International Quenching and Control of Distortion Conference, 2012, pp. 57–68.
[13] D. S. MacKenzie and A. L. Banka, "Determination of Heat Transfer Coefficients for Thermal Modeling," in ASM Handbook, vol. 22A, ASM International, 2009, pp. 106–126.
[14] J. Thibault, "Boiling Heat Transfer Around a Horizontal Cylinder and in Tube Bundles," McMaster University, 1978.
[15] G. Ramesh and K. N. Prabhu, "Assessment of Axial and Radial Heat Transfer during Immersion Quenching of Inconel 600 Probe," Exp. Therm. Fluid Sci., vol. 54, pp. 158–170, 2014, doi: 10.1016/j.expthermflusci.2014.01.016.
[16] G. Salazar, J.-S. Lemyre-Baron, M. Jahazi, H. Champliaud, and A. Tahan, "Analysis of Industrial Quenching (Air Transfer + Oil Immersion) and the Cooling Regimes after Immersion," in Proceedings of the 32nd ASM Heat Treating Society Conference, 2023, pp. 106–113, doi: https://doi.org/10.31399/asm.cp.ht2023p0106 Copyright.
[17] G. Salazar, J.-S. Lemyre-Baron, M. Jahazi, H. Champliaud, and A. Tahan, "Identification of Thermal Boundary Conditions during Industrial Quenching (Air Transfer + Mineralized Oil Immersion) of a Hollow Cylinder," in Proceedings of the 29th International Federation for Heat Treatment and Surface Engineering World Congress, 2024, pp. 152–159.
[18] J. V. Beck and K. A. Woodbury, "Inverse Heat Conduction Problem: Sensitivity Coefficient Insights, Filter Coefficients, and Intrinsic Verification," Int. J. Heat Mass Transf., vol. 97, pp. 578–588, 2016, doi: 10.1016/j.ijheatmasstransfer.2016.02.034.
[19] K. A. Woodbury, H. Najafi, F. de Monte, and J. V Beck, Inverse Heat Conduction: Ill-Posed Problems, 2nd Editio. John Wiley & Sons Inc, 2023.
[20] F. Samadi, K. Woodbury, and F. Kowsary, "Optimal Combinations of Tikhonov Regularization Orders for IHCPs," Int. J. Therm. Sci., vol. 161, 2021, doi: 10.1016/j.ijthermalsci.2020.106697.
[21] P. Duda and M. Konieczny, "An Iterative Algorithm for the Estimation of Thermal Boundary Conditions Varying in Both Time and Space," Energies, vol. 15, 2022.
[22] S. Behrem and B. Hrnjica, "Estimation of Heat Transfer Coefficient during Quenching in Water," Trans. FAMENA XLII, vol. 1, no. 1, pp. 61–73, 2018.
[23] S. Chapra and R. Canale, Numerical Methods for Engineers, 6th ed. McGraw-Hill Higher Education, 2010.
[24] J. V Beck, "Non-linear Estimation Applied to the Non-linear Inverse Heat Conduction Problem," Int. J. Heat Mass Transf., vol. 13, pp. 703–716, 1970.

QDE 2025: Proceedings of the 3rd International Conference on
Quenching and Distortion Engineering
May 6–7, 2025, Vancouver, Canada
https://doi.org/10.31399/asm.cp.qde2025p0141

Residual stress and precipitation behavior during heat treatment of FGH96 alloy

Dingge Fan[1], Xiongxi Zhao[1], Jian Zhang[1], Jiantang Jiang[*1,2]
[1]School of Materials Science and Engineering, Harbin Institute of Technology, Harbin, 150001, China
[2]National Key Laboratory of Precision Hot Processing of Metals, Harbin Institute of Technology, Harbin, 150001, China
Email of communicating author: DinggeFan1@163.com

Abstract

The impact of heat treatment on residual stress and precipitation behavior in FGH96 alloy was investigated. Results obtained from the contour method for evaluating residual stress in FGH96 alloy demonstrate that aging conducted at 760 °C enhances mechanical properties and reduces residual stress levels. The size of secondary γ' precipitates gradually increases with prolonged aging duration. However, the increase in hardness of FGH96 alloy after a 1-hour aging period is found to be insignificantly significant, indicating a gradual decrease in the contribution of precipitate strengthening as the size of secondary γ' precipitates increases.

Introduction

Due to their high strength and excellent damage tolerance, nickel-base superalloys have been extensively employed in turbine discs, which are critical safety components in modern aero engines [1,2]. In general, heat treatment is necessarily important for excavating the potential mechanical properties of nickel-base superalloys [3,4]. Particles of the secondary/tertiary γ' phase precipitate in nickel-base superalloys during the quenching and the subsequent aging process [5] and contribute to the strengthening of this series of alloys [6]. The quenching, on the other hand, induces violent stress and eventually brings about significant residual stress, which may lead to cracking in quenching or distortion in the following processes in the components that are fabricated [7,8]. Therefore, the quenching-induced residual stress (Q-RS) should be controlled or reduced in the subsequent aging process. Since aging time is the dominating factor that affects the reduction of residual stress and the properties during quenching, it is necessary to study the heat treatment method and precipitation behavior of nickel-based superalloys to reduce the residual stress efficiently and to improve the mechanical properties of nickel-based superalloys.

The precipitation strengthening related to γ' precipitation together with the reduction of Q-RS remains unclear in nickel-based superalloys [9,10]. Zhang et al. [11] examined the microstructure and the RS in FGH96 alloy turbine disks of different conditions using the neutron diffraction method and found that the reduction of matrix supersaturation contributed to increased compressive radial stress in the rim after the disc was aged. D.M. Collins et al. [12] designed a heat treatment process to achieve optimized microstructure using the modeling software PrecipiCalc. The optimized thermal cycle for achieving high mechanical properties was determined through the simulation of an extensive matrix of heat treatments: a solution heat treatment at 1105 °C for 4 hours, followed by controlled cooling at a rate of 40 °C/s, and finally aging for 16 hours at 798 °C. However, due to the effect of composition on the precipitation behavior of γ', the precipitation strengthening of other nickel-based superalloys remains unclear [13,14].

The FGH96 alloys need to be quenched in oil and then aged properly to achieve high strength for being applied in turbine discs. The current study then focuses on the oil quenching and the subsequent aging treatment on the FGH96 alloy aiming to investigate the reduction of Q-RS and the accompanying precipitation behavior and to explore the potential for maximizing the strength of FGH96 alloy while minimizing quenching residual stress.

Experiments Methods

The test material utilized is FGH96, a powder metallurgy superalloy developed in China. Its chemical composition is presented in Table 1.

Table 1: Chemical composition of FGH96 alloy (mass fraction, %).

Element	Mass fraction
Ni	Bal.
Al	1.9
Ti	3.7
Cr	16.3
Co	10.9
Nb	0.7
Mo	4
W	4

The strengthening of FGH96 alloy is achieved by coherent γ'-Ni3(Al, Ti) precipitates in the face-centered cubic (fcc) matrix [15]. Based on the average size, morphology, and number density, the γ' precipitates can be divided into two classes: the secondary γ' precipitates with relatively large size, irregular shape, and low quantity density, and the tertiary γ' precipitates with small size, spherical shape, and high quantity density [16].

The theoretical phase equilibrium of FGH96 alloy in the range of 600-1400°C was predicted by Thermo-Calc. The mass fraction of γ, γ', and the liquid phase in the FGH96 alloy over this temperature range are calculated and the results are illustrated in Fig. 1. It can be seen that for FGH96 alloy the predicted γ' solvus temperature is around 1100 °C. The super solvus heat treatments were usually conducted at temperatures ranging from 20 to 50 °C above the γ' solvus temperature of the respective alloy [17]. The solid solution process of FGH96 alloy is conducted at a temperature of 1150°C for a duration of 2 h, which was then followed by oil quenching at 160°C. The cooling rate during oil quenching is around 338 °C/ min. The aging treatment process of FGH96 alloy involves air cooling after being held at 760°C for 16 hours.

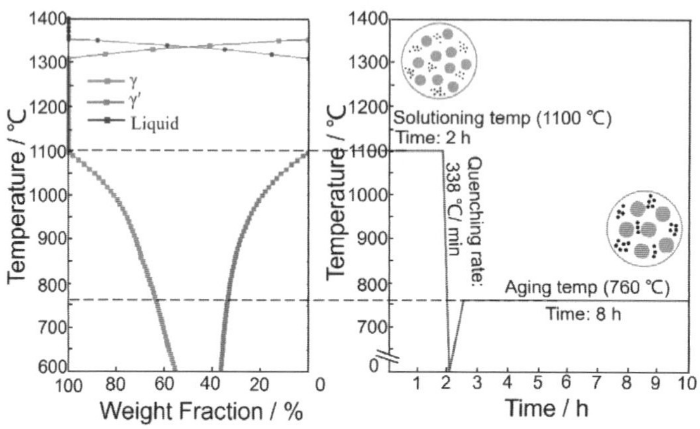

Figure 1: Predicted equilibrium phase composition of FGH96 alloy with weight fraction of phase present as a function of temperature.

The residual stress of FGH96 alloy specimens of various conditions was measured using the contour method. Initially, the sample was cut using the Swiss CUT200P slow-walking wire-cutting machine. The cutting process involved utilizing a 0.15 mm copper wire at a wire speed of 0.5 mm/min while ensuring symmetrical clamping on both sides of the cutting surface. Subsequently, the Leitz PMM-C 12107 coordinate measuring instrument was employed to measure the deformation profile of the FGH96 alloy sample with a measurement point spacing of 1 mm × 1 mm. Finally, by employing the deformation contour data as displacement boundary conditions through finite element software ABAQUS, we calculated the residual stress after quenching and aging of FGH96 alloy.

Results and Discussion

The residual stress in the bulk sample (47×50×75 mm3) of FGH96 alloy, resulting from the oil-quenching or from aging for 1 h, was quantified using the contour method and the results are illustrated in Fig. 2. The profiles of the quenched and aged samples, obtained through measurements conducted with a coordinate measuring instrument, are depicted in Figs. 2 (a) and (b) respectively. The surface profile of the sample exhibits a concave inner and convex outer morphology, with a vertical distance between the highest and lowest points measuring 0.1mm. The deformation contour data were inverted and incorporated into the sample model as displacement boundary conditions, devoid of any initial stress. Consequently, the stress result graphs presented in Figs. 2 (c) and (d) were obtained. The central region of the sample experiences tensile stress, while the surface undergoes compressive stress. A comparison was made between the residual stress values of quenched and aged specimens along the central path in the Z direction, as depicted in Fig. 3. The maximum tensile residual stress observed in the quenched sample is 1087 MPa. Generally, aging treatment can effectively reduce the quenched residual stress in bulk samples.

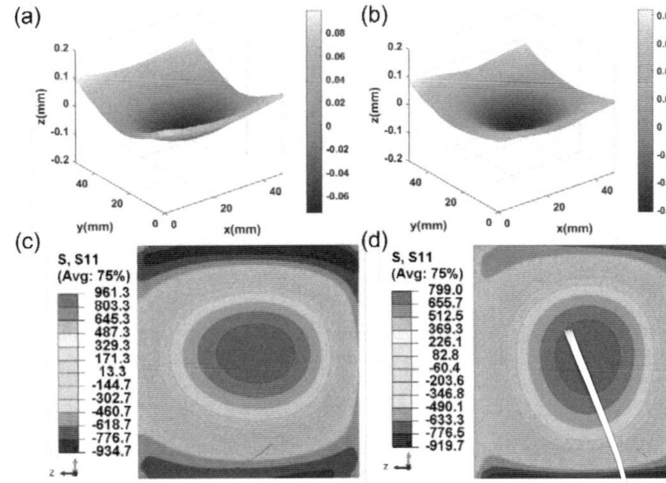

Figure 2: (a)The contour map of oil-quenched FGH96 alloy. (b) The contour map of aging FGH96 alloy. (c) Test results of quenching residual stress of FGH96 alloy. (d) Test results of aging residual stress of FGH96 alloy.

During the initial stage of aging, the Q-RS remains confined within the sample. Subsequently, in the heating stage, there is a more rapid increase in surface temperature compared to its center, resulting in thermal expansion on the surface. However, this expansion is constrained by tensile deformation at the center, leading to tensile stress at the center and compressive stress on the surface. The original compressive stress induced by quenching is further enhanced during the aging process, resulting in amplified compressive stress and accelerated formation of compressive deformation. As the heating duration prolongs, the temperature at the center rises, leading to constrained thermal expansion due to the presence of compressive deformation on the outer surface. The sample center undergoes further elastomeric expansion during the subsequent aging insulation process, surpassing the limitations imposed by the outer layer. Ultimately, an equilibrium state is achieved where the surface layer compresses the sample core inward while the core expands outward due to tensile stress. In conclusion, novel stresses are generated throughout the aging process, effectively counteracting the initial quenching stress and resulting in a reduced residual stress level. Specifically, the core stress decreases from 961 MPa to 799 MPa.

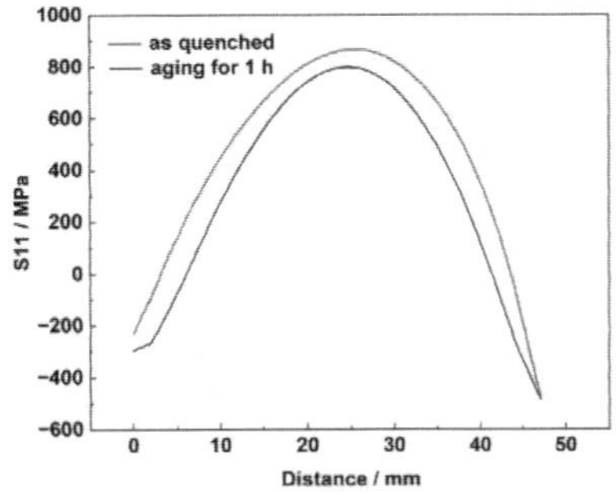

Figure 3: The results of measured residual stress values along the Z direction.

Fig. 4 (a)-(c) and (d)-(f) depict the microstructures of the as-quenched and aged FGH96 alloy, respectively. The size of secondary γ' in the quenched sample is approximately 34 nm. During aging at 760℃, growth of the secondary γ' is observed concomitant with precipitation of tertiary γ'. The uniformly dispersed secondary γ' precipitates within the γ matrix can be seen in Fig. 4 (e), while the tertiary γ' segregates around them. Specifically, the size of the secondary γ' is approximately 100 nm, whereas that of the tertiary γ' measures around 19 nm.

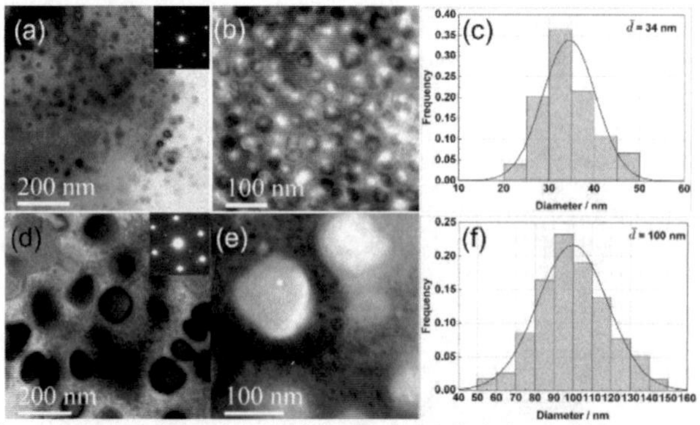

Figure 4: Microstructure of quenched and aged FGH96 alloy: (a) Bright-field image and the corresponding SAED pattern for quenched sample; (b) HAADF-STEM image for the quenched sample; (c) Size statistics of secondary γ' precipitates for the quenched sample; (d) Bright-field image and the corresponding SAED pattern for the sample aged for 16 h; (e) HAADF-STEM image for the sample aged for 16 h; (f) Size statistics of secondary γ' precipitates for the sample aged for 16 h.

The surface and line sweep maps of the secondary and tertiary γ' precipitates in the FGH96 sample aged for 16 hours are presented in Figs. 5 (a1) - (h1). The line sweep maps reveal that the secondary γ' is enriched with Ti, Al, Ni, and Nb elements while the matrix γ exhibits enrichment in Co, Cr, and Mo elements. In Fig. 5 (b1) and (d1), it can be observed that the contents of Ti and Al drop sharply around the γ'/γ phase interface. The composition of Cr and Co elements near the γ'/γ interface is significantly higher compared to that in the γ matrix away from the interface (Fig. 5 (g1)). This disparity can be attributed to the presence of a secondary γ' phase at elevated temperatures, which exhibits enhanced diffusion kinetics. Consequently, this facilitates rapid elemental redistribution between the secondary γ' phase and its surrounding γ matrix. Specifically, the elements Al and Ti are attributed to the ordered γ' phase, while Cr and Co are ascribed to the γ matrix. These secondary γ' phases exhibit rapid growth due to swift element allocation. Once the anisotropic elastic strain energy surpasses the isotropic interface energy, a transition in the morphology of the precipitated phase occurs from spherical to cubed or butterfly-shaped.

Generally, during the cooling process of the FGH96 alloy from the super solution line temperature, there is a continuous increase in the chemical volume free energy (∇Gv) as the temperature decreases below the solution line temperature (1120 °C). This increase occurs with an increasing degree of supercooling until it reaches a critical activation energy barrier (∇Gv) for nucleation that can be overcome, leading to the initiation of nucleation and subsequent formation of secondary γ'. The second nucleation explosion takes place when Al and Ti, which are elements forming γ', reach saturation within the γ matrix. This is accompanied by a sufficiently low degree of supercooling and a critical energy barrier (∇Gv) that is also sufficiently low. Consequently, tertiary γ' precipitates are formed within the γ matrix, positioned between existing secondary γ' precipitates. The tertiary γ' precipitates exhibit a high number density and small size due to limited atomic diffusivity at high supercooling and significant thermodynamic driving force for nucleation. Figs. 5 (a2) - (h2) illustrates the line sweep and surface sweep of tertiary γ' precipitates. It is evident from the figure that the enrichment of Ni, Al, and Ti elements in tertiary γ' is not as pronounced as that in secondary γ'. The reason for this can be attributed to the fact that secondary γ' contains fewer elements compared to the newly nucleated tertiary γ'. Additionally, Singh et al. [18] investigated the nucleation and growth of γ' in Rene 88DT alloy, which exhibits a remarkable similarity to FGH96. They discovered that at a cooling rate of 24°C/min, tertiary γ' forms within a relatively narrow temperature range between 700 °C and 800 °C.

The FGH96 alloy, employed as the turbine disc in the critical high-temperature component of an aero engine, necessitates exceptional mechanical properties at elevated temperatures. In general, the microstructure of nickel-based superalloys, including the size distribution, morphology, and volume fraction of γ', can significantly influence their mechanical properties. The mechanical properties of FGH96 alloy were assessed through high-temperature tensile tests and room-temperature hardness tests in various heat treatment states. Fig. 6 illustrates the high-temperature tensile curves of FGH96 alloy under different heat treatment conditions. Additionally, Table 2 presents the yield strength, tensile strength, and fracture elongation values for FGH96 alloy in different heat treatment states. Following aging treatment, there is an increase in the strength while a decrease in elongation is observed.

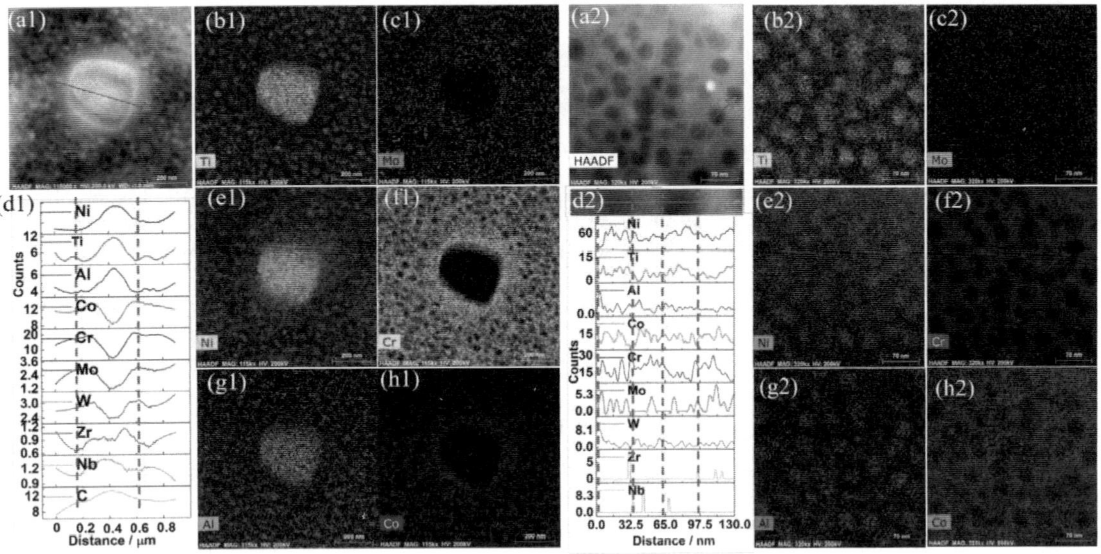

Figure 5: The surface and line sweep maps of (a1-h1) for the secondary γ' precipitates and (a2-h2) for the tertiary γ' precipitates for the aged sample.

The aging hardness curve (Fig. 6 (b)) exhibits a diminished hardness for quenched samples, followed by a subsequent increase after 1 hour of aging. Subsequently, the hardness value remains relatively stable after a 2-hour aging period. The precipitation strengthening mechanism plays a pivotal role in enhancing the strength of FGH96 alloy during its aging process, with the intensity being determined by both the size and volume fraction of γ' particles. Therefore, achieving optimal mechanical properties relies on attaining optimum precipitation sizes for γ'. Furthermore, an increase in the size of secondary γ' particles leads to a gradual reduction in their contribution to precipitate strengthening.

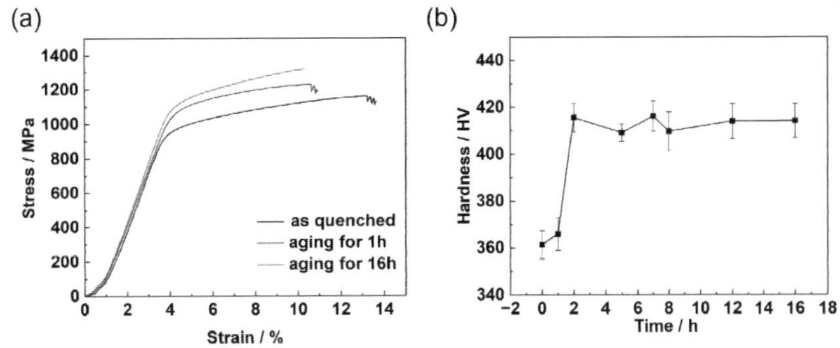

Figure 6: (a)Engineering stress-strain curves of the quenched and aged samples at 650 ℃. (b) The variation in hardness of FGH96 alloy with aging time.

In conclusion, the aging treatment has been demonstrated to effectively reduce quenching residual stress and enhance the strength of FGH96 alloy. The quenching residual stress in FGH96 alloy is primarily influenced by its thermal conductivity, thermal expansion coefficient, and constitutive behavior. The determination of these factors depends on the composition and microstructure of the alloy, thus establishing a correlation between quenching residual stress and the alloy's composition and microstructure.

Table 2. Mechanical properties of FGH96 alloy under different heat treatment conditions.

Sample condition	Yield strength/MPa	Tensile strength/MPa	Elongation/%
As quenched	942	1161	8.8
Aging for 1 h	1054	1228	5.8

Aging for 16 h	1102	1322	5.5

Conclusions

1. The tensile strength of FGH96 alloy increases while the elongation decreases during the aging process. Tensile tests conducted at 650°C demonstrate that FGH96 alloy exhibits excellent mechanical ductility after being aged for 16 hours. Furthermore, the hardness of the FGH96 alloy increases with prolonged aging time, reaching a plateau after 2 hours.
2. The size of the secondary γ' precipitates in the quenched FGH96 alloy is small, approximately 34 nm. During the aging process, these secondary γ' precipitates grow to around 100nm and are accompanied by the precipitation of tertiary γ'. This phenomenon significantly enhances the mechanical properties of the FGH96 alloy.
3. The quenched sample of FGH96 alloy exhibits tensile stress at its center and compressive stress on its surface in the Q-RS profile. The aging process significantly reduces the magnitude of Q-RS. The maximum tensile Q-RS measures 961 MPa, which decreases to 799 MPa after a 1-hour aging treatment.

Acknowledgments

This work was performed under the auspices of the National Key Research and Development Program of China (Grant No. 2023YFB3711900，2022YFB3400245_01) and National Key Research and Development Program of China (Grant No. 2023YFB3711900，2022YFB3404700).

References

[1] H. Ge, W. Xing, B. Chen, X. Hu, S. Zheng, K. Liu, X. Ma, Acta Mater., 214 (2021) 116985.
[2] E.I. Galindo-Nava, L.D. Connor, C.M.F. Rae, Acta Mater., 98 (2015) 377–390.
[3] Y. Mishima, S. Ochiai, T. Suzuki, Acta Metall., 33 (1985) 1161–1169.
[4] U. Bru¨ckner, A. Epishin, T. Link, Acta Mater., 45 (1997) 5223–5231.
[5] H. Song, H. Gao, Q. Wu, Y. Zhang, J. Alloys Compd., 886 (2021) 161269.
[6] M. Li, J. Coakley, D. Isheim, G. Tian, B. Shollock, J. Alloys Compd.,732 (2018) 765–776.
[7] R.W. Kozar, A. Suzuki, W.W. Milligan, J.J. Schirra, M.F. Savage, T.M. Pollock, Metall. Mater. Trans. A, 40 (2009) 1588–1603.
[8] S.S. Babu, M.K. Miller, J.M. Vitek, S.A. David, Acta Mater., 49 (2001) 4149–4160.
[9] R. Wang, H. Jiang, W. Shao, S. Yang, X. Sun, J. Cao, L. Zhen, J. Mater. Res. Technol., 24 (2023) 5298–5308.
[10] M. Strantza, B. Vrancken, M.B. Prime, C.E. Truman, M. Rombouts, D.W. Brown, P. Guillaume, D. Van Hemelrijck, Acta Mater., 168 (2019) 299–308.
[11] Z. Zhang, Y. Feng, Q. Tan, J. Zou, J. Li, X. Zhou, G. Sun, Y. Wang, Mater. Des., 166 (2019) 107603.
[12] D.M. Collins, H.J. Stone, Int. J. Plast., 54 (2014) 96–112.
[13] Y.K. Xiao, Z.Y. Bian, Y. Wu, G. Ji, Q. Lian, H.Z. Wang, Z. Chen, H.W. Wang, Mater. Charact., 178 (2021) 111242.
[14] S. Goel, M. Neikter, J. Capek, E. Polatidis, M.H. Colliander, S. Joshi, R. Pederson, Mater. Des., 195 (2020) 109045.
[15] M. Hou, W. Mou, G. Yan, G. Song, Y. Wu, W. Ji, Z. Jiang, W. Wang, C. Qian, Z. Cai, Surf. Coat. Technol., 397 (2020) 125972.
[16] P.E. Aba-Perea, T. Pirling, M. Preuss, Mater. Des., 110 (2016) 925–931.
[17] R.J. Williams, F. Vecchiato, J. Kelleher, M.R. Wenman, P.A. Hooper, C.M. Davies, J. Manuf. Process., 57 (2020) 641–653.
[18] A.R.P. Singh, S. Nag, S. Chattopadhyay, Y. Ren, J. Tiley, G.B. Viswanathan, H.L. Fraser, R. Banerjee. Acta Mater., 2013, 61(1): 280-293.

QDE 2025: Proceedings of the 3rd International Conference on
Quenching and Distortion Engineering
May 6–7, 2025, Vancouver, Canada
https://doi.org/10.31399/asm.cp.qde2025p0147

Effect of fluid-probe interaction on heat extraction in laboratory-scale forced convective quench tests

B. Hernández-Morales, R. Cruces-Reséndez
Facultad de Química, Depto. de Ingeniería Metalúrgica, Universidad Nacional Autónoma de México, Ciudad de México, México
bernie@unam.mx

E.I. Durán-García
Fairbanks Morse, Totoltepec, Estado de México, México

Abstract

Quenching is conducted with agitation, which produces a non-uniform velocity field around the components. To assess the effect of quenchant velocity on cooling curves in laboratory-scale tests, special devices have been developed to produce one-dimensional fluid flow parallel to the probe. However, the international standards for these tests recommend probes that are flat-end cylinders which promote non-uniform flow around them.

In this work, we report the evolution of heat extraction from probes of three different geometries: flat-end, conical-end and hemispherical-end cylinders, characterized through video-recordings and cooling curves. The experimental set up assures one-dimensional fully developed flow, parallel to the probe. The probes were machined from AISI 304 stainless steel and instrumented with four thermocouples (one at its geometric center and the others near the surface). The initial test temperature was 850 °C; water was flowing at 0.2 and 0.6 m/s, at 60 °C.

From the video recordings, the flat-end probe favors a distorted vapor film near the probe end which was barely seen in the hemispherical-end probe and absent in the conical-end probe. This observation could be explained through the results of a previous paper in which an isothermal CFD model was developed to compute the velocity and dynamic pressure fields around the probes at room temperature. The model showed a region of low pressure near the base of the flat-end probe which is consistent in shape and position with the vapor film observed at the beginning of the quench of this probe. Because of the distorted vapor film, the cooling curves showed more scatter from test to test when the flat-end probe was used. The wetting front velocity was constant for each one of the experimental conditions except for the flat-end probe cooled with water at 0.6 m/s in which case a secondary wetting front was formed.

Introduction

Cooling curve analysis is the preferred methodology to assess the cooling power of new and used quenchants [1]. It is based on recording a cooling curve at a position within a probe or actual part and calculating the cooling rate curve to obtain parameters such as the maximum cooling rate, the cooling rate al 300 °C and the time to reach 600, 400 and 200 °C [2]. The probes are usually cylinders that are quenched in an apparatus that simulates quenchant agitation in a quench tank and are machined from a material that does not transform upon cooling [3, 4].

An important piece of information derived from cooling curve analysis is the characterization of the heat transfer modes [1]. When a probe is quenched from the austenitizing temperature in a vaporizable liquid, three distinct stages can be distinguished along the cooling curve. First, a vapor film is formed around the probe which results in a nearly linear segment in the cooling curve; the slope of the line is small because the vapor film insulates the surface from the quenchant. Once the surface temperature has dropped to a level that cannot sustain the vapor film, the liquid starts to boil, generating bubbles; this second stage is known as the nucleate boiling stage and, because of the phase change in the liquid side results in a very efficient heat extraction and the cooling curve decreases sharply. Finally, boiling ceases altogether and the only heat transfer mechanism available is pure convection which reduces the efficiency of the cooling, resulting in a soft descent of the cooling curve. The vapor film does not disappear at once; instead, a wetting front (the locus of the interface between the vapor film and the nucleate boiling region) is formed and advances.

Although a flat-end cylinder is recommended in the international standards for cooling curve analysis [1], it has been shown [5] that a conical-end cylinder probe produces a more symmetrical and repeatable wetting front which would result in more reproducible cooling curves. Also, having the thermocouple tip located at the geometric center of the probe – as indicated in the standards - produces cooling curves that are not representative of the events occurring at the probe surface because of the thermal resistance – by conduction - offered by the probe material. Thus, in this work we compare cooling curves, cooling rate curves and wetting front velocities obtained with three different geometries: flat-end, conical-end and hemispherical-end cylindrical probes, quenched in water flowing parallel to the probe axis at two free-stream water velocities: 0.2 and 0.6 m/s. The probes are instrumented with thermocouples near the probe surface, at three heights along the probe length, and a fourth thermocouple located at the geometric center of the probe.

Methodology

Experimental

Austenitic stainless steel (AISI 304) cylindrical probes with three different end geometries were quenched by submerging them in water flowing parallel to the axis probe. The main feature of the experimental set-up is that fully developed flow is ensured at the position within the flowing fluid where the probe is immersed. Details of the experimental set-up are given in [6]. The probes had three different geometries: flat-end, conical-end and hemispherical-end, and were instrumented with four 1/16-in type K thermocouples as shown in Figs. 1, 2 and 3, respectively; the subsurface thermocouples are labelled as TC1, TC2 and TC3, starting from the uppermost position while TC4 is located at the geometric center. The thermocouples were connected to a data acquisition system (Data Translation, model DT9828) controlled by the software QuickDAQ; the data acquisition frequency was 10 Hz. Videos were recorded using a reflex camera (Samsung, model NX30, with an 18–55 mm lens) in full 1080 × 1920 HD resolution with a frequency of 60 fps.

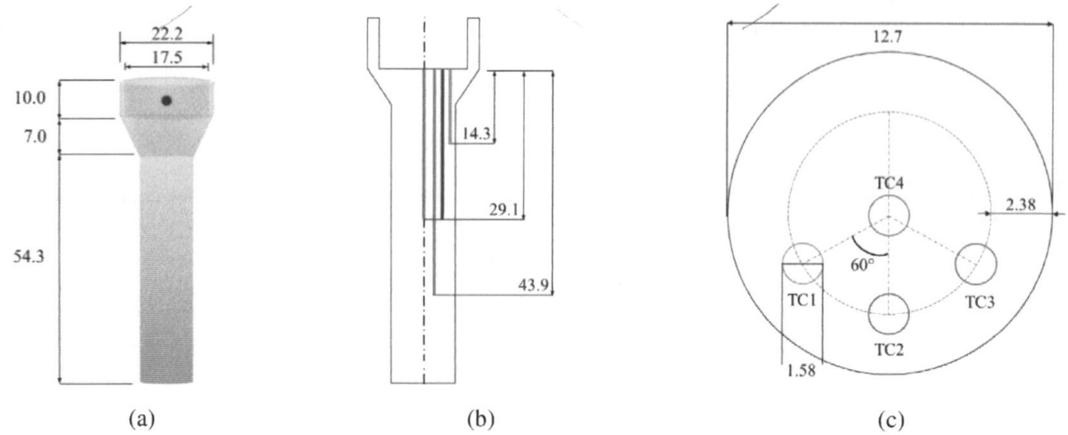

Figure 1: Schematic representation of the flat-end cylindrical probe. Distances are in mm.

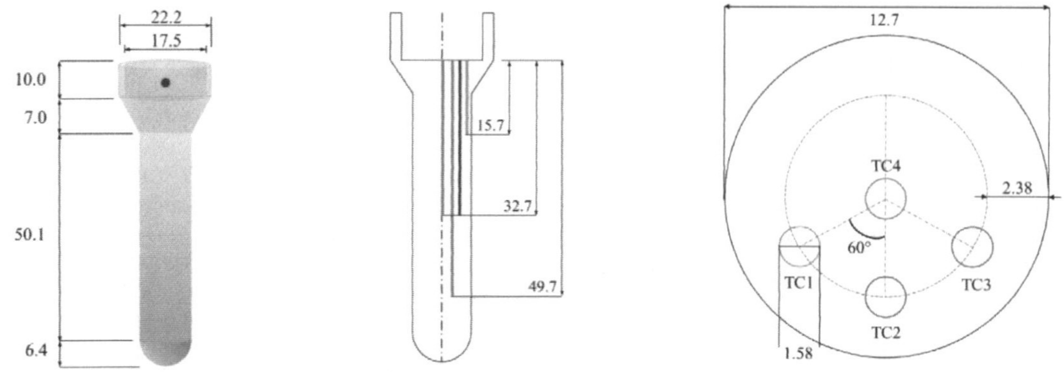

Figure 2: Schematic representation of the conical-end cylindrical probe. Distances are in mm.

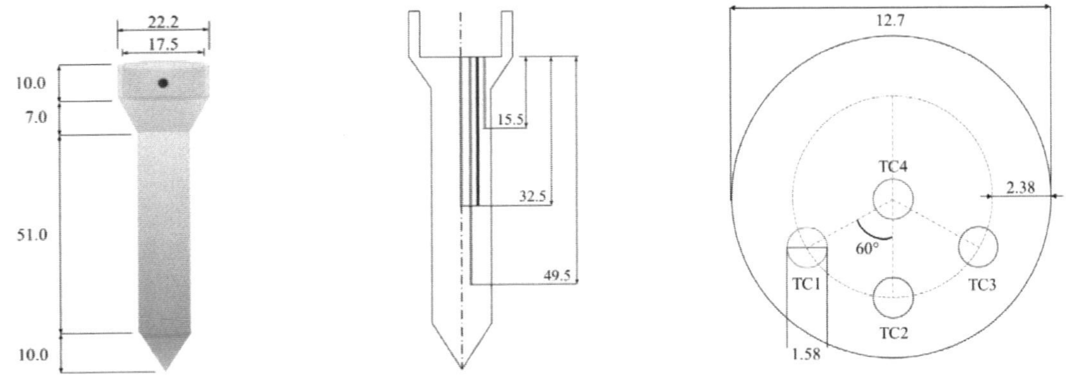

(a) (b) (c)

Figure 3: Schematic representation of the hemispherical-end cylindrical probe. Distances are in mm.

To extend the duration of the vapor film (thus allowing us to observe in detail the different phenomena taking place at the surface) water at 60 °C was used; two free-stream velocities were studied: 0.2 and 0.6 m/s (equivalent to Reynolds numbers of 20678 and 62035, respectively. The initial probe temperature was 850 °C.

Computational

The local cooling rate was calculated form the respective cooling curve using a forward finite difference. On the other hand, the videos were analyzed to track the wetting front using the 1-D Autotracker module included in Tracker v. 5.0 video analysis software [7] as detailed in [6]. The result is a plot of the wetting front position along the probe length as a function of time; the wetting front velocity was estimated as the slope of a linear regression of the data, which is valid if the coefficient of determination, R^2, is close to 1.

Results and discussion

Cooling curves

Measured cooling curves – at the near surface thermocouple positions - for each one of the three geometries studied are shown in Figs. 4, 5 and 6, for a free-stream water velocity of 0.2 m/s. Images at selected times during the quench are also shown in each figure.

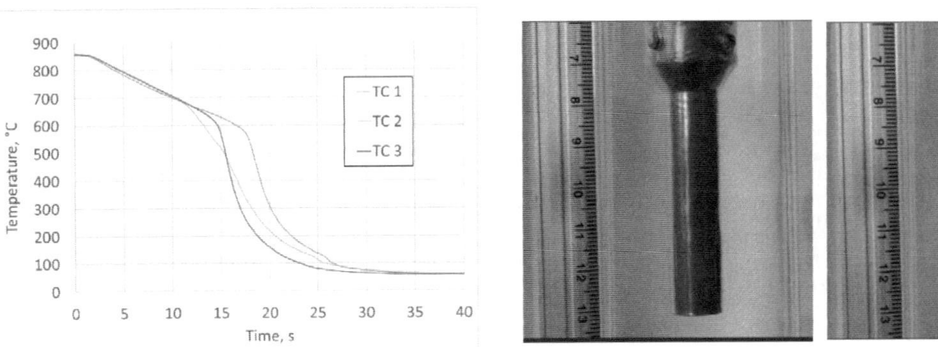

(a) (b) (c)

Figure 4: a) Measured cooling curves at the near-surface thermocouple positions during quenching of a flat-end cylindrical probe in water flowing at 0.2 m/s; b) image at the time at which the probe reaches the test position; and c) image 3 s after the probe reaches the test position. The rule shown has its scale in cm.

The probe geometry results in markedly different heat extraction which is reflected in the cooling curves: the flat-end cylindrical probe (Fig. 4) favors a very steady vapor film as reflected by the duration of the corresponding times at thermocouples TC3 and TC 2. While the cooling curves at those positions indicate the same sequence of boiling stages, the curve measured by TC 1 shows a very distinct behavior: after following the same pattern as TC2 and TC3 for the first 12 s, it suddenly decreases at a significantly higher rate. This was caused by a second wetting front being formed at the top of the probe which quickly displaces downward, thus causing the change observed at TC 1. There is still vapor film surrounding the entire probe 3 s after it reached the test position.

In contrast, the three cooling curves measured near the surface in the conical-end probe (see Fig. 5) all follow the pattern expected based on their position along the length of the probe. Also, the duration of the vapor film stage at the position of TC3 and TC2 is markedly shorter than those observed with the flat-end probe. Three seconds after the probe has reached the test position the wetting front has already formed and is moving upward, nearly reaching the base of the inverted cone; this is a result of the tip being an efficient heat extractor.

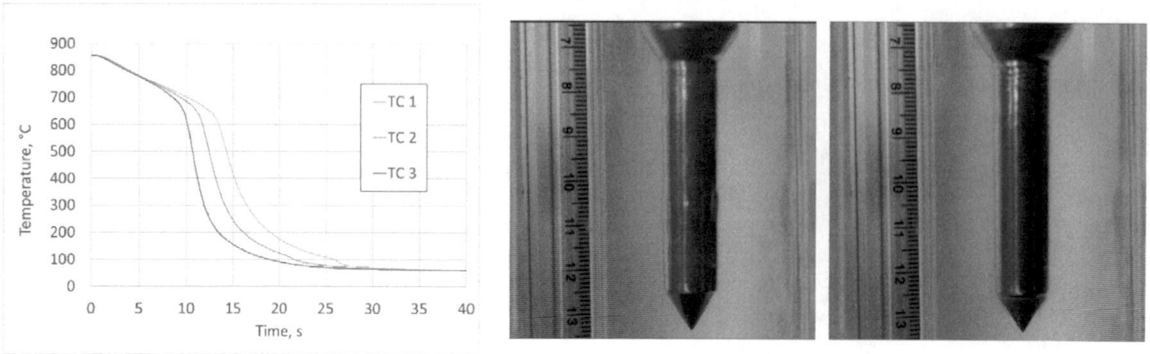

(a) (b) (c)

Figure 5: a) Measured cooling curves at the near-surface thermocouple positions during quenching of a conical-end cylindrical probe in water flowing at 0.2 m/s; b) image at the time at which the probe reaches the test position; and c) image 3 s after the probe reaches the test position. The rule shown has its scale in cm.

The behavior of the hemispherical-end probe (refer to Fig. 6) is somewhat intermediate between the other two. The duration of the vapor film stage as registered at the positions of TC3 and TC2 is long but shorter than those for the flat-end probe. The cooling curve measured at TC1 shows, as was the case with the flat-end probe, a faster rate of change after some time but it was not as marked as with the flat-end probe; it was also caused by a second wetting front formed at the top of the probe which descends quickly. The probe is surrounded by a vapor film even 3 s after the probe has reached the test position.

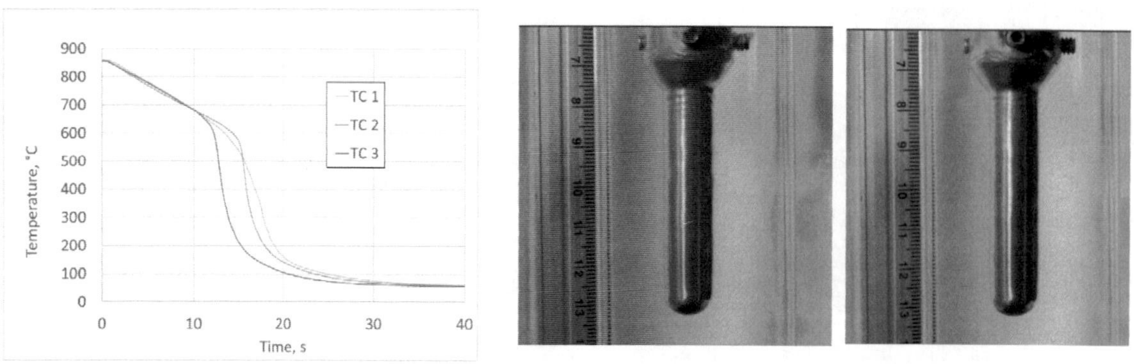

(a) (b) (c)

Figure 6: a) Measured cooling curves at the near-surface thermocouple positions during quenching of a hemispherical-end cylindrical probe in water flowing at 0.2 m/s; b) image at the time at which the probe reaches the test position; and c) image 3 s after the probe reaches the test position. The rule shown has its scale in cm.

Cooling curves registered during the experiments conducted with a water free-stream velocity of 0.6 m/s are shown in Figs. 7, 8 and 9, for the flat-end, conical-end and hemispherical-end probes, respectively. In all cases, the duration of the film boiling stage and the total cooling time decrease with respect to their values for water flowing at 0.2 m/s. Also, the three cooling curves for each probe are now ordered as expected.

It is noteworthy that a distorted vapor film was observed for the flat-end probe (Fig. 7 (a)) when it reached the test position and was still present 3 s later (Fig.7 (b); the reason for this will be discussed later on in the paper.

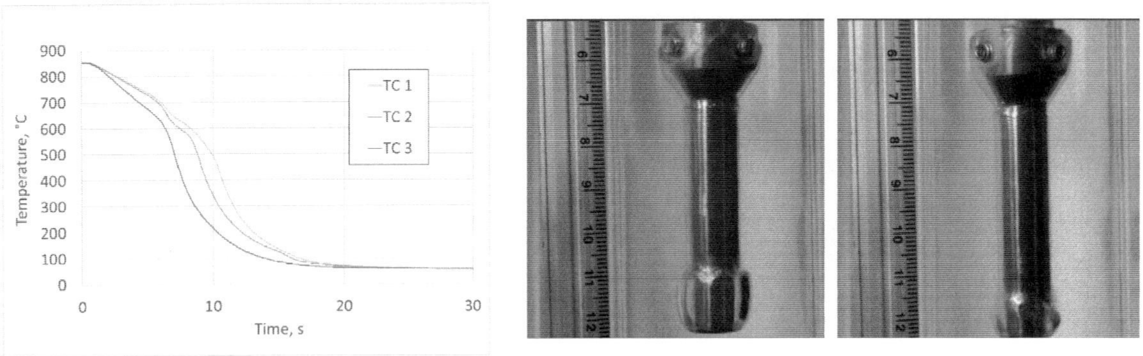

(a) (b) (c)

Figure 7: a) Measured cooling curves at the near-surface thermocouple positions during quenching of a flat-end cylindrical probe in water flowing at 0.6 m/s; b) image at the time at which the probe reaches the test position; and c) image 3 s after the probe reaches the test position.

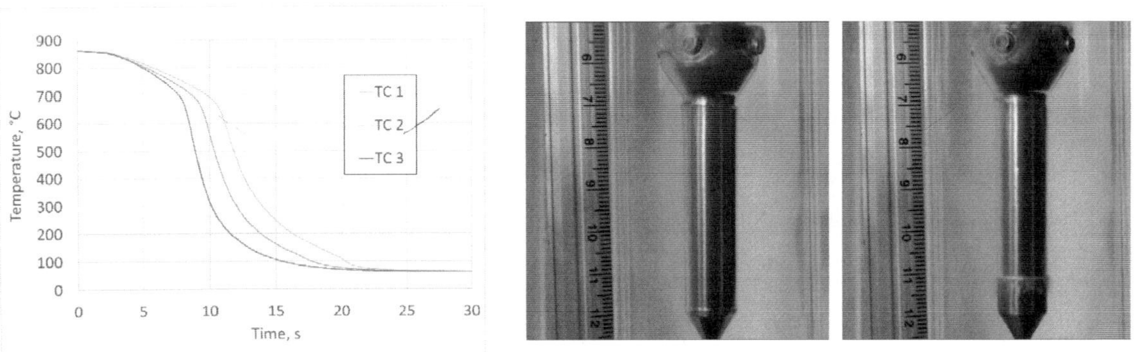

(a) (b) (c)

Figure 8: a) Measured cooling curves at the near-surface thermocouple positions during quenching of a conical-end cylindrical probe in water flowing at 0.6 m/s; b) image at the time at which the probe reaches the test position; and c) image 3 s after the probe reaches the test position.

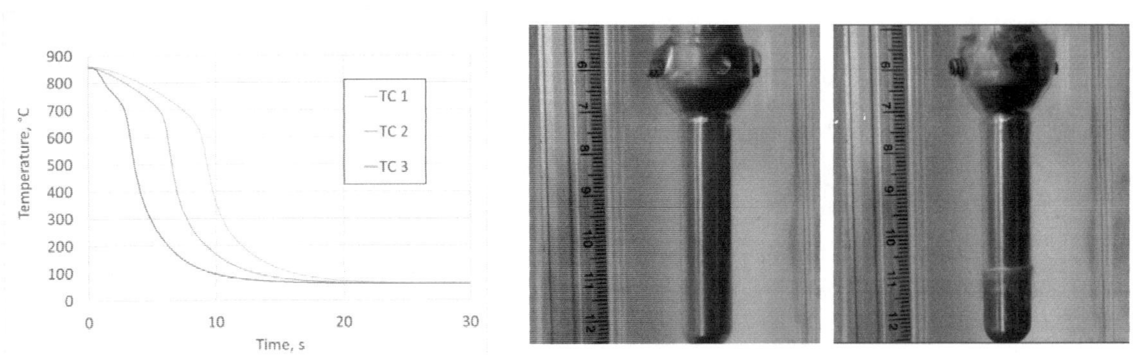

(a) (b) (c)

Figure 9: a) Measured cooling curves at the near-surface thermocouple positions during quenching of a hemispherical-end cylindrical probe in water flowing at 0.6 m/s; b) image at the time at which the probe reaches the test position; and c) image 3 s after the probe reaches the test position.

The wetting front at the surface of the conical-end probe has advanced more after 3 s of cooling with respect to the experiments with water flowing at 0.2 m/s; in the case of the hemispherical-end probe, the wetting front has now formed and advanced while when the water flowing at 0.2 m/s there was no evidence of the wetting front. These latter results are in line with the fact that an increase in the water free-stream velocity results in a larger convective component of the heat extraction [8] which causes the wetting front to form earlier and advance at a higher velocity.

Cooling rate curves

Cooling rate curves were computed from the cooling curves measured at the thermocouple positions. They are plotted, as a function of time, in Figs. 10 and 11, for a water free-stream velocity of 0.2 and 0.6 m/s, respectively. In all cases, the presence of a vapor film at the surface is evidenced by a horizontal line that is apparent almost from the beginning of the test; the fact that it is not observable at the start of the test is because of the thermal resistance by conduction offered by the probe material which causes a lag in the response even though the thermocouples TC1, TC2 and TC3 are located near the surface. The highest cooling rates are observed for the hemispherical-end probe, followed by the conical end probe and the flat-end probe.

In Fig. 10, a double peak is evident for the flat-end and the hemispherical-end probes. In the former case, the maximum value at TC 2 (near the surface, at mid-height) occurs at a time between those observed for TC1 and TC3 and is a consequence of a second wetting front that is formed at the top of the probe and descends quickly.

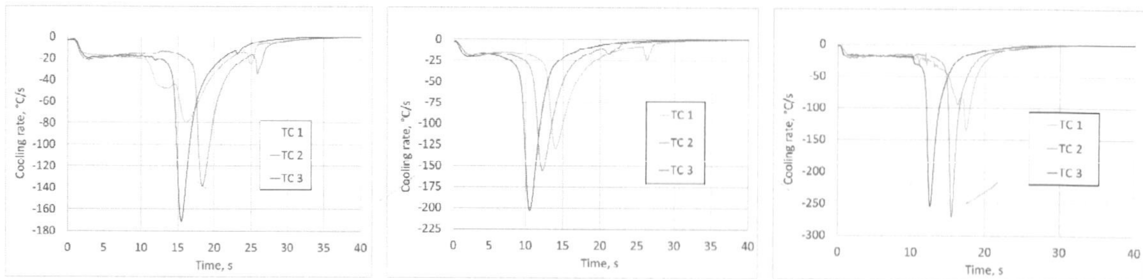

(a) (b) (c)

Figure 10: Calculated cooling rate curves for probes rapidly cooled in water flowing at 0.2 m/s: a) flat-end probe, b) conical-end probe and c) hemispherical-end probe.

An increase in water free-stream velocity to 0.6 m/s produces higher cooling rates associated with a more efficient heat extraction (see Fig. 11). The curves are now ordered as expected, *i.e.*, the curve at TC3 shows the end of the vapor film stage first, followed by the curve at TC2 and then that one at TC1. However, the hemispherical-end probe produced the largest maximum cooling rate at TC2. Only the curves at TC2 and TC1 for the flat-end probe show a double peak; in the videos, a secondary wetting front was observed near TC2 while the primary one was ascending. Also, note that the flat-end probe was the only one that produced a horizontal line associated with the boiling stage; this implies that there is more insulation between the surface probe and the quenchant with this type of probe.

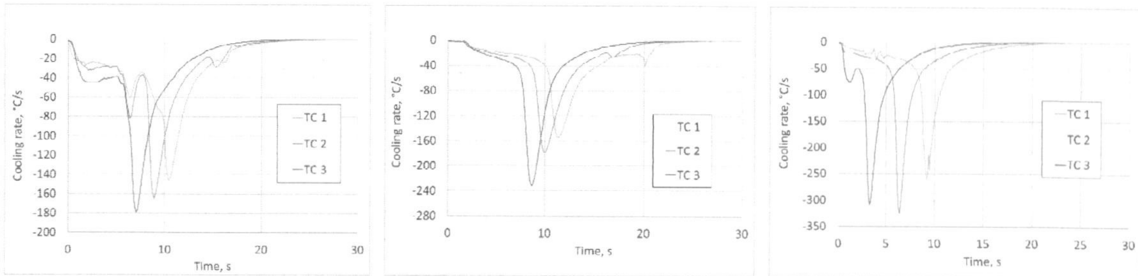

(a) (b) (c)

Figure 11: Calculated cooling rate curves for probes rapidly cooled in water flowing at 0.6 m/s: a) flat-end probe, b) conical-end probe and c) hemispherical-end probe.

One of the parameters usually reported in cooling curve analysis is the maximum cooling rate; according to the standards it must be measured at the probe geometric center but that value may be different to the one at or near the surface. In Fig. 12, the cooling rate curves calculated for TC2 and TC4 are plotted, as a function of time, for the three probe geometries quenched with water flowing at 0.2 m/s. In all cases, the cooling rate curve at the center (TC4) lags the one calculated at the subsurface (TC2) and shows smaller values while mirroring its shape.

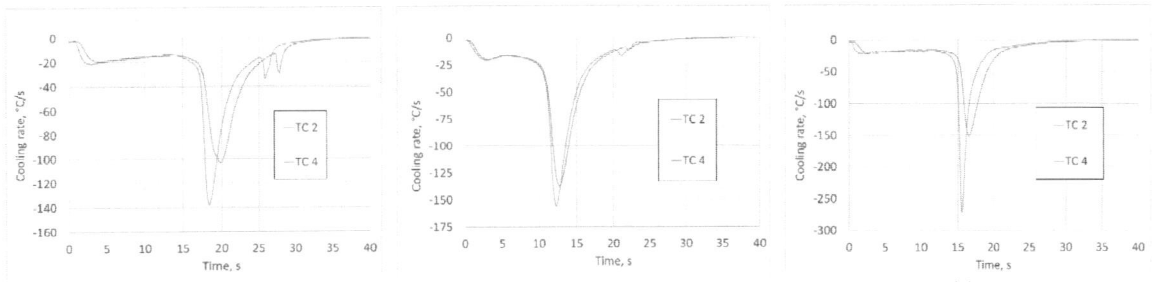

(a) (b) (c)

Figure 12: Calculated cooling rate curves, at TC2 and TC4, for probes rapidly cooled in water flowing at 0.2 m/s: a) flat-end probe, b) conical-end probe and c) hemispherical-end probe.

The corresponding curves for water flowing at 0.6 m/s are shown in Fig.13; a similar behavior is observed but with higher values of the cooling rates, as expected. Also, a second peak is shown in the curves for the flat-end probe.

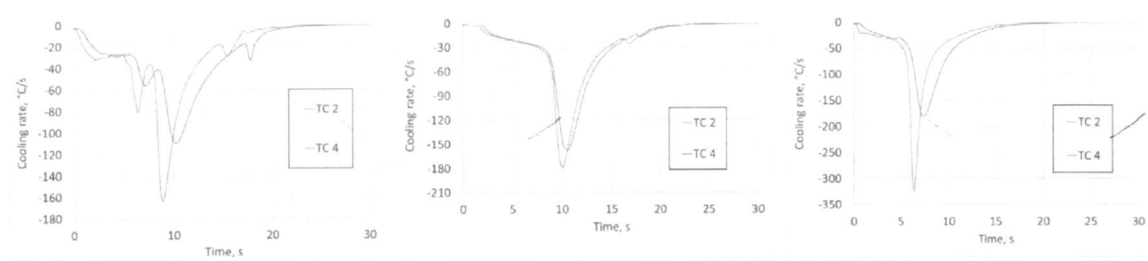

(a) (b) (c)

Figure 13: Calculated cooling rate curves, at TC2 and TC4, for probes rapidly cooled in water flowing at 0.6 m/s: a) flat-end probe, b) conical-end probe and c) hemispherical-end probe.

The maximum cooling rate at the positions of TC2 and TC4 as well as the difference between them (in value and as percentage with respect to the value at TC4) are shown in Table 1 for the three probe geometries and the two free-stream water velocities. The difference is the smallest for the conical-end probe followed by the flat-end probe and is the largest for the hemispherical-end probe. In all cases, the difference increases as the water free-stream velocity increases which is consistent with the fact that and increase in water velocity promotes a more efficient heat extraction which is not immediately felt at the center due to the conductive thermal resistance within the probe. It is interesting to note that the difference in maximum cooling rate between the positions of TC 2 and TC4 expressed as percentage (last column in Table 1) barely changes for the conical-end and the hemispherical-end probes while there is a significant difference for the flat-end probe when the free-stream velocity changes.

Table 1: Maximum cooling rate (CR) at the positions of TC2 and TC4 and the difference between them (in value and as percentage with respect to the value at TC4) for all experimental conditions.

Water free-stream velocity, m/s	Probe geometry	Max. CR at TC 2 °C/s	Max. CR at TC 4 °C/s	Difference °C/s	Difference % (w.r.t. TC4)
0.2	Flat-end	-137.96	-103.01	-34.95	33.9
	Conical-end	-155.81	-137.39	-18.42	13.4
	Hemispherical-end	-271.31	-151.08	-120.23	79.6
0.6	Flat-end	-164.46	-109.94	-54.52	49.6
	Conical-end	-178.47	-158.07	-20.41	12.9
	Hemispherical-end	-325.2	-180.4	-144.74	80.2

From the video recordings, the wetting front velocity for all the experimental conditions was estimated. The results, including the value of the coefficient of determination, R^2, are given in Table 2. The fact that the coefficient of determination is close to 1 indicates that the behavior of the data can be explained with a linear regression [9] and, therefore, the wetting front velocity in each case is valid and constant. For water flowing at 0.2 m/s, the wetting front velocity was largest for the hemispherical-end probe, followed by the flat-end probe and the conical-end probe. An increase in the water free-stream velocity to 0.6 m/s increased the wetting front velocity while reversing the order between the hemispherical-end and the conical-end probe; a value for the flat-end probe could not be determined because a secondary wetting front was formed near the position of TC2 which interfered with the primary wetting front.

Table 2: Wetting front velocity and correlation coefficient for all experimental conditions.

Water free-stream velocity, m/s	Probe geometry	Wetting front velocity	R^2
0.2	Flat-end	5.39	0.9981
	Conical-end	4.17	0.9985
	Hemispherical-end	6.03	0.9985
0.6	Flat-end	--	--
	Conical-end	6.21	0.998
	Hemispherical-end	5.81	0.998

At the beginning of the experiments with the flat-end conical probe quenched in water flowing at 0.6 m/s, a distorted vapor film was observed (refer to Figure 7(b) and (c)); later on, it gave way to a "normal" vapor film. This behavior is the responsible for the duration of the vapor film and the formation of a second wetting front near the probe base, and can be explained based on the results of a CFD model that was developed in [5] to compute the velocity and dynamic pressure fields in the neighborhood of each one of the probes of this study. It must be pointed out that the model did not include boiling and the system (the probe plus the fluid) was assumed isothermal, at ambient temperature; nonetheless, the results shed light on the effects of the probe-fluid interaction. Because the flat-end probe generates a region of stagnant fluid at its base when the probe interacts with water flowing at 0.6 m/s, the computed dynamic pressure field (see Fig. 14) clearly showed a region of very low pressure near the base of the flat-end probe. This allows the expansion of the vapor film during the current set of experiments and helps explaining the observed geometry of the vapor film at the beginning of the quench when water was flowing at 0.6 m/s. Given that the expanded vapor film may distort its shape as it diminishes its size (refer to Fig. 7(c)), the wetting front may not be symmetrical, causing a low repeatability in the measured cooling curves, as was reported in [5]. In contrast, even for this water velocity, the conical-end probe does not produce a region of low pressure and its tip acts as an efficient heat extractor, which explains the cooling curves, cooling rate curves and wetting front behavior observed when using this probe. The computed dynamic pressure field for the hemispherical-end probe is similar to the one for the conical-end probe but its blunt shape does not promote the rapid formation of the wetting front and, therefore, the vapor film lasts a long time.

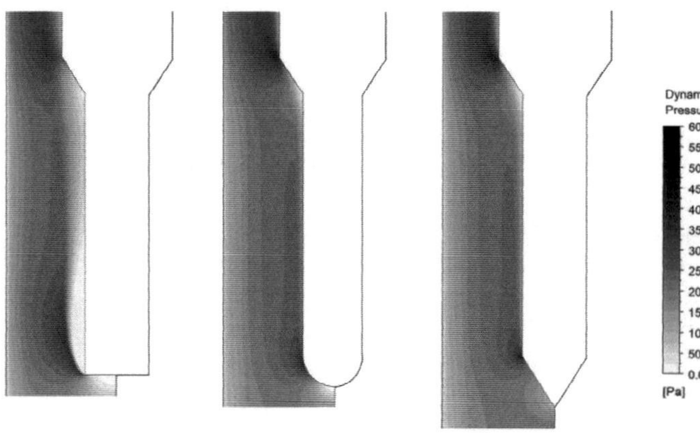

Figure 14: Computed dynamic pressure field (Pa) around: flat-end probe (left column), hemispherical-end probe (middle column) and conical-end probe (right column). Water at 60°C, flowing at 0.6 m/s, and probe at room temperature [5].

Conclusions

The effect of probe geometry on cooling curves, cooling rate curves and wetting front velocity was studied for two free-stream water velocities: 0.2 and 0.6 m/s; in all cases, the three boiling stages could be identified. However, each one of the three probe geometries produced significantly different cooling curves. The conical-end probe generates cooling curves measured near the surface which are ordered (in the sense that the start of the nucleate boiling stage was observed first in the cooling curve measured at the position of TC3, followed by TC 2 and TC1) for both water velocities while the cooling curves measured at the upper-most subsurface thermocouple (TC1) with the flat-end and the hemispherical-end probes quenched in water flowing at 0.2 m/s suddenly showed a higher rate of change than at the other two thermocouple positions. This was due to the formation of a second wetting front at the top of the probe which descended rapidly, thus increasing heat extraction at the position of TC1. Increasing the water velocity to 0.6 m/s reduced the duration of the vapor film stage and the total duration of the test and resulted in all the cooling curves now ordered following the subsurface thermocouple positions along the probe length.

The maximum cooling rate at the geometric center of the probe (the position where a thermocouple is located in the standard probes) increases in the following order: flat-end, conical-end and hemispherical end probe, while the difference in maximum cooling rate between the two thermocouples located at mid-height is largest for the hemispherical-end probe, followed by the flat-end probe and the conical-end probe. This implies that very different values of cooling rate at the geometric center would be reported, depending on the probe used.

When the higher water velocity was used in the experiments, the flat-end probe generated an expanded vapor film near the probe end which increases the duration of the vapor film. The computed dynamic pressure field in the neighborhood of the probe end - reported in a previous paper – explains this behavior. Because of the distorted vapor film, the cooling curves showed more scatter from test to test when the flat-end probe was used. The short duration of the vapor film when conical-end probes were used results from both a more uniform dynamic pressure field and the probe tip being an efficient heat extractor; although the hemispherical-end probe produces a similar dynamic pressure film, its blunt shape does not promote the rapid formation of the wetting front.

The wetting front was rather symmetrical around the probe for the hemispherical-end and the conical-end probe, which promotes repetitive cooling curves while scatter values are obtained when the flat-end probe is used. The wetting front velocity was constant for each one of the experimental conditions except for the flat-end probe cooled with water at 0.6 m/s in which case a secondary wetting front was formed which impeded the estimation of the wetting front velocity.

References

[1] B. Hernández-Morales, *Cooling Curve Analysis*, in *Encyclopedia of Iron, Steel and Their Alloys*, R. Colás and G.E. Totten, Eds., CRC Press/Taylor & Francis Group, (Boca Ratón, Florida, 2016), p 854–874.

[2] D. S. MacKenzie, "Understanding the Cooling Curve Test", *Thermal Processing*, 2017, January/February, p 28-32.

[3] International Organization for Standardization. (1995). *Industrial quenching oils—Determination of cooling characteristics—Nickel alloy probe test method (ISO Standard No. 9950:1995).* https://www.iso.org/standard/17848.html, accessed December 2024.

[4] ASTM International. (2021). *ASTM D6482-21, Standard Test Method for Determination of Cooling Characteristics of Aqueous Polymer Quenchants by Cooling Curve Analysis with Agitation (Tensi Method).* https://www.astm.org/d6482-21.html, accessed December 2024.

[5] R. Cruces-Reséndez and B. Hernández-Morales, "Effect of probe geometry on fluid flow near laboratory-scale quench probes", *Can. Metall. Q.*, 2020, 59(1), p 85–100. https://doi.org/10.1080/00084433.2019.1696605

[6] R. Cruces-Reséndez, B. Hernández-Morales and J.E.V. Guzmán. "A Detailed Study of Boiling Phenomena in Forced Convective Quenching Experiments". *J. of Materi Eng and Perform*, 2024, 33, p 4210–4225. https://doi.org/10.1007/s11665-023-09095-7.

[7] ComPADRE Digitital Library. *Tracker - Video Analysis and Modeling Tool*, https://physlets.org/tracker/. Accessed October 2024

[8] M.A. González-Melo, O.A. Rodríguez-Rodríguez, B. Hernández-Morales and F.A. Acosta-González. "Numerical Model of Simultaneous Multi-Regime Boiling Quenching of Metals". *J. Manuf. Mater. Process.*, 2024, 8, p 31. https://doi.org/10.3390/jmmp8010031

[9] Frost, J., *How To Interpret R-squared in Regression Analysis*, Mar 2018, https://statisticsbyjim.com/regression/interpret-r-squared-regression/, accessed December 2024.

QDE 2025: Proceedings of the 3rd International Conference on
Quenching and Distortion Engineering
May 6–7, 2025, Vancouver, Canada
https://doi.org/10.31399/asm.cp.qde2025p0157

Residual Stress and Microstructure Evolution in TC4 Titanium Alloy During Heat Treatment: A Combined Simulation and Experimental Approach

*Xiongxi Zhao[1], Dingge Fan[1], Jian Zhang[1], Jiantang Jiang[*1,2]*
[1]School of Materials Science and Engineering, Harbin Institute of Technology, Harbin, 150001, China
[2]National Key Laboratory of Precision Hot Processing of Metals, Harbin Institute of Technology, Harbin, 150001, China
Email of communicating author: 871456624@qq.com

Abstract

This study investigates the stress evolution and microstructural changes in TC4 titanium alloy through isothermal gradient solution quenching (900~1000°C solution treatment, water quenching at room temperature and 60°C), followed by annealing at 700~800°C and aging at 500~550°C. A series of hot compression and creep tests were conducted to establish the elastic-plastic and creep constitutive equations for TC4 titanium alloy in different treatment states. Finite element analysis (FEA) was employed to study stress evolution and residual stress distribution. An ABAQUS subroutine was developed to account for the $\alpha \leftrightarrow \beta$ phase transformation during heat treatment. The influence of this transformation on residual stress distribution was determined through comparative analysis. Residual stresses in the sample sections were experimentally validated using the contour method, showing strong agreement with the simulation results. By optimizing the heat treatment process, residual stresses were significantly reduced without compromising the alloy's mechanical properties, demonstrating the effectiveness of the proposed approach.

Introduction

Titanium alloy structural components in aerospace applications are often thin-walled with complex, variable-thickness geometries. Among these, TC4 titanium alloy parts, designed for direct assembly, undergo multiple processing steps including heat treatment, machining, and welding. Due to TC4 titanium alloy's poor thermal conductivity compared to aluminum and steel, and its phase transformation behavior during heat treatment, residual stresses accumulate, leading to potential deformation and cracking during manufacturing. Therefore, it is necessary to study the stress evolution behavior and the microstructure evolution of TC4 titanium alloy in the key process. [1,2,3]

Machining deformation caused by residual stress is one of the main challenges in the production of thin-walled monolithic structural parts. One of the reasons for the distortion is the release and relaxation of the initial residual stress inside the original part due to the removal of materials during the machining process, and the initial residual stress mainly depends on the manufacturing process of the part and the subsequent heat treatment process. M. Landwehr et al. [4] studied the effect of heat treatment on residual stress-related machining deformation of Ti-6Al-4V titanium alloy. And in this paper, the effects of heat treatment on the residual stress-related processing deformation of Ti-6Al-4V alloy were summarized through experimental and computational results.

To sum up, it is not only of great academic significance but also of practical engineering significance to carry out research on the corresponding scientific problems of residual stress and subsequent failure in TC4 titanium alloy and reveal the corresponding evolutionary law and mechanism. This paper takes the residual stress in TC4 titanium alloy as the main research object, pays attention to the evolution process and law of the residual stress in the process of heat treatment and machining, closely combines the heat treatment, machining and other manufacturing processes to carry out the research on the stress/shape coupling evolution law and mechanism, and fully considers and coupling the phase transition process characteristics and microstructure characteristics of TC4 titanium alloy. In order to reveal the structure/stress cross-scale coupling evolution, shape response law.

Experiments Methods

The selected material in this paper is TC4 titanium alloy, and the material state is rolling state, which is the initial state for subsequent basic parameter testing and microstructure observation. For TC4 titanium alloy ($\alpha + \beta$ type dual phase titanium alloy), the strengthening heat treatment is generally adopted, that is, solid solution quenching & aging treatment, and annealing treatment. In addition, the phase transition temperature T_β of TC4 titanium alloy is also an important factor in formulating its heat treatment scheme. The heat treatment scheme of TC4 titanium alloy in this paper is shown in the Table 1.

Table 1: The heat treatment scheme of TC4 titanium alloy in this paper.

Solid solution quenching	**Annealing treatment**	**Aging treatment**
950°C×1h+WQ(20°C)	750°C×1h+AC	/
1000°C×1h+WQ(20°C)	750°C×1h+AC	/
950°C×1h+WQ(20°C)	/	500°C×2h/8h+AC
1000°C×1h+WQ(20°C)	/	500°C×2h/8h+AC

The small cylindrical specimens of Φ8 mm × 12 mm or Φ6 mm × 9 mm were cut by wire cutting. All the small cylindrical specimens were treated by solution quenching. The hot compression test was carried out on a Gleeble-3500D dynamic thermodynamic simulation testing machine. The small cylindrical sample needs to use 320 # and 2000 # water grinding sandpaper to smooth the wire cutting marks on the surface and end face of the small sample in turn, and ensure that the size of the sample is basically unchanged. In the process of thermal compression test, in order to reduce the friction between the end face and the indenter, graphite sheets need to be placed on both ends. And the test method of TC4 titanium alloy basic thermal physical property parameters is shown in the Table 2.

Table 2: The test method of TC4 titanium alloy basic thermal physical property parameters

Thermal parameter	**Test equipment**
Thermal conductivity λ and specific heat capacity c	Laser Flash Analyzer-457
Thermal expansion coefficient α	Thermal Dilatometer
Elastic modulus E and Poisson 's ratio γ	RFDA-HTVP1750C

As for the residual stress of TC4 titanium alloy specimens of various conditions, it was measured using the contour method. Initially, the sample was cut using the Swiss CUT200P slow-walking wire-cutting machine. The cutting process involved utilizing a 0.15 mm copper wire at a wire speed of 0.5 mm/min while ensuring symmetrical clamping on both sides of the cutting surface.

And for the contour method, the Leitz PMM-C 12107 coordinate measuring instrument was employed to measure the deformation profile of the TC4 titanium alloy sample with a measurement point spacing of 1 mm x 1 mm. Finally, by employing the deformation contour data as displacement boundary conditions through finite element software ABAQUS , we can calculate the residual stress after quenching and annealing & aging of TC4 titanium alloy.

Results and Discussion

The thermal parameters required for the finite element simulation of residual stress include but are not limited to : density ρ, thermal expansion coefficient α, thermal conductivity λ and specific heat capacity c, elastic modulus E and Poisson 's ratio γ. The density ρ of TC4 titanium alloy is 4430 kg/m^3. Therefore, through a series of experimental tests, the basic thermal physical parameters of TC4 titanium alloy are obtained as shown in Table 3.

Table 3: The basic thermal physical parameters of TC4 titanium alloy.

Temperature (°C)	200	400	600	800	1000
λ(W/(m·K))	9.59	12.93	17.04	20.56	24.53
c(J/(g·K))	0.578	0.646	0.740	0.785	0.791
$\alpha\times10^{-6}$(1/K)	10.60	10.90	12.10	12.45	7.31
E(GPa)	103.1	93.22	82.85	74.59	72.07
γ	0.310	0.320	0.328	0.338	0.348

The constitutive equation refers to the equation describing the properties and response characteristics of a specific material or material. It is usually an equation describing the functional relationship between stress and strain rate or between stress tensor and strain tensor. In the process of heat treatment, the strong temperature change inside the material caused by the heat exchange between the material and the outside world has a great influence on the evolution of temperature and stress distribution inside the material. TC4 titanium alloy, as mentioned above, its thermal conductivity is relatively poor, so in the process of heat treatment, the stress evolution caused by temperature change is more significant. The residual stress of the material is the final residual stress with the evolution of the stress and strain inside the material during the development of this thermal-mechanical coupling factor. Therefore, it is necessary to know the law of stress-strain evolution for the simulation analysis of residual stress evolution of a specific material. For TC4 titanium alloy, the construction of its constitutive equation is the basis for analyzing the stress-strain law, and it is also the basis for the simulation of residual stress. The hot compression test of TC4 titanium alloy was carried out. Through the data processing and fitting of the hot compression test, the hot deformation constitutive equation of TC4 titanium alloy material can be constructed. The true stress-strain curves of TC4 titanium alloy at different deformation temperatures and different deformation rates are shown in the Figure 1.

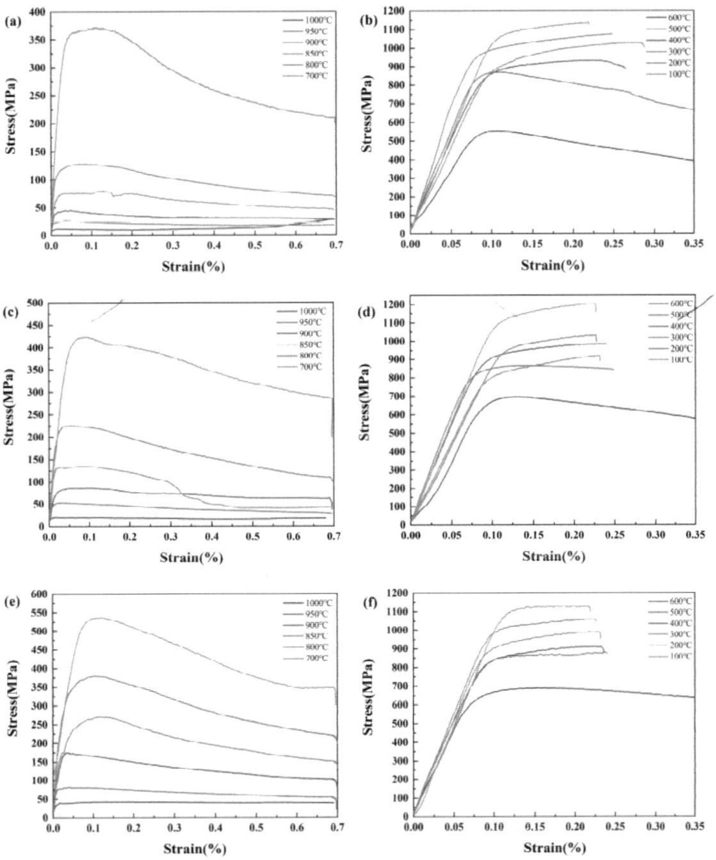

Figure 1: The true stress-strain curves of TC4 titanium alloy (solid solution quenched) at the same strain rate (different temperatures).
(a) (b) $0.001s^{-1}$; (c) (d) $0.01s^{-1}$; (e) (f) $0.1s^{-1}$.

According to the curves and related data obtained from the hot compression test, the hot deformation constitutive equation of TC4 titanium alloy can be constructed. The functional relationship between the deformation rate, temperature and flow stress of TC4 titanium alloy can be described and characterized by Arrhenius relationship. Therefore, the hot deformation constitutive equation of TC4 titanium alloy at high temperature and medium and low temperature can be obtained :

(1) The constitutive equation of TC4 titanium alloy at high temperature (600°C~1000°C) :

$$\dot{\varepsilon}=2.9939\times10^{18}[sinh(0.0077972\sigma)]^{3.6}\,exp\left(-\frac{440738}{RT}\right)$$

(2) The constitutive equation of TC4 titanium alloy at low temperature (100°C~500°C) :

$$\dot{\varepsilon}=432483.736[sinh(0.0010345\sigma)]^{31.64}\,exp\left(-\frac{75982}{RT}\right)$$

After obtaining the basic thermophysical parameters and material constitutive equations of TC4 titanium alloy, the stress evolution of TC4 titanium alloy during heat treatment can be simulated and analyzed. The following assumptions are made for the material before the finite element simulation of residual stress,:

(1) Assuming that the sample is isotropic ;

(2) Assume that the material is continuous and uniform ;

(3) It is assumed that in the process of heat treatment, the sample is small deformation, that is, the deformation can be ignored; and there is no initial stress inside the material.

The size of the simulation calculation and verification sample was 20×30×50 mm square sample, which met the ratio of 1:1.5:2. The temperature field simulation analysis of TC4 titanium alloy solution quenching (1000°C) process was carried out by using the finite element software ABAQUS, and the temperature field evolution cloud map was shown in Figure 2. The analysis shows that when the TC4 titanium alloy sample is subjected to water quenching (WQ), the temperature drop gradually develops from the outer layer of the sample to the inside; the temperature of the four corners and the outer layer of the sample decreased rapidly, and the temperature of the outer layer of the sample decreased to about 150°C in only 8s. At this time, the center temperature of the sample was still as high as 600°C, and the temperature gradient between the inner and outer layers was very large. When the time reaches 230s, the whole sample has been reduced to 20°C, which is consistent with the temperature of the quenching medium. In addition, through the cloud image results, the temperature drop gradually develops from the four corners of the sample to the outer layer of the sample and then to the center of the sample. Therefore, the cooling rate of different parts of the sample is different. The cooling rate from fast to slow can be summarized as: four corners-edge-outer layer-center. The solution quenching (950°C) process is similar.[5~8]

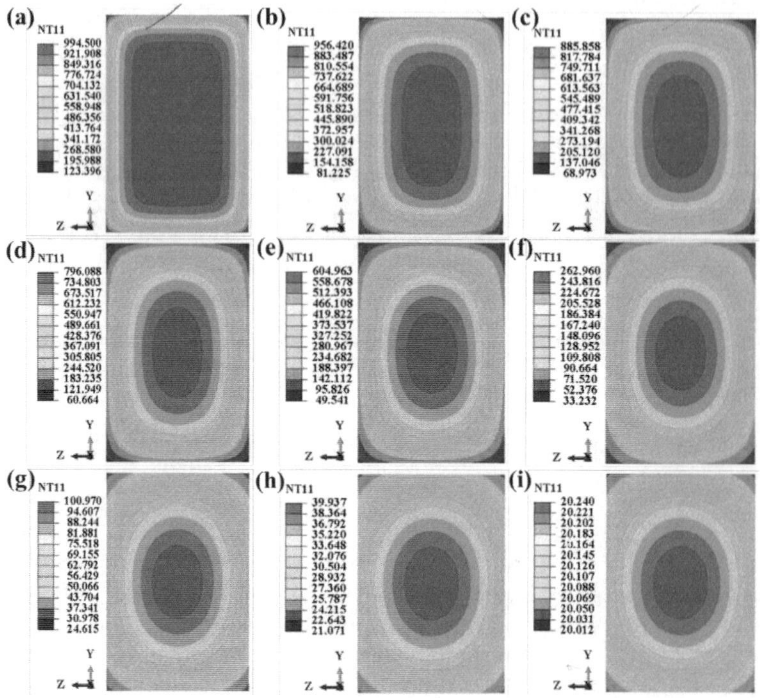

Figure 2: Temperature field evolution of TC4 titanium alloy during solution quenching (1000°C). (a) 0.70s; (b) 1.96s; (c) 3.04s; (d) 4.31s; (e) 7.36s; (f) 17.39s; (g) 32.44s; (h) 54.92s; (i) 126.90s.

The stress field evolution cloud diagram of TC4 titanium alloy during solution quenching (1000°C) is shown in Figure 3. The selected stress tracking direction is S11 direction-X axis direction; according to the residual stress evolution cloud map in the S11 direction, the analysis shows that the tensile stress evolution develops from the four corners of the sample to the edges, surface and center. Combined with the evolution results of the 1000°C temperature field, the temperature gradient evolution trend in the temperature field is basically coincident with the tensile stress evolution trend. The four corners of the sample and the surface layer are first exposed to the quenching medium, resulting in the first drop in temperature and the phenomenon of thermal expansion and cold shrinkage. That is, the volume shrinkage during cooling produces tensile stress at the surface layer of the sample. In order to achieve the force balance of the whole sample, the compressive stress state is generated in most areas of the sample, as shown in Figure 3(a)~(c), and the tensile stress value of the surface layer and the compressive stress value of the center increase with time. As the temperature of the sample core decreases, the volume shrinkage phenomenon gradually occurs in the core area of the sample, and the tensile stress phenomenon gradually shifts to the core area of the sample, so that in the subsequent cooling process, the tensile stress area develops from the four corners near the core area of the sample to the core area, as shown in Figure 3(d)~(f). With the passage of time, the tensile stress area is transferred to the center of the sample, and the tensile stress value is gradually increasing. A compressive stress area is formed in the surface area of the sample, and the value of the compressive stress area also increases, as shown in Figure 3(g)~(i). At the end of the stress evolution, the residual stress distribution characteristics of the surface compressive stress and the core tensile stress are formed. For the solution quenching process of 1000°C×1h+WQ(20°C), the maximum tensile stress in the core of the sample is 265 MPa, and the maximum compressive stress in the surface layer is 421 MPa, which transits from the maximum tensile stress area in the core to the maximum compressive stress area in the surface layer.

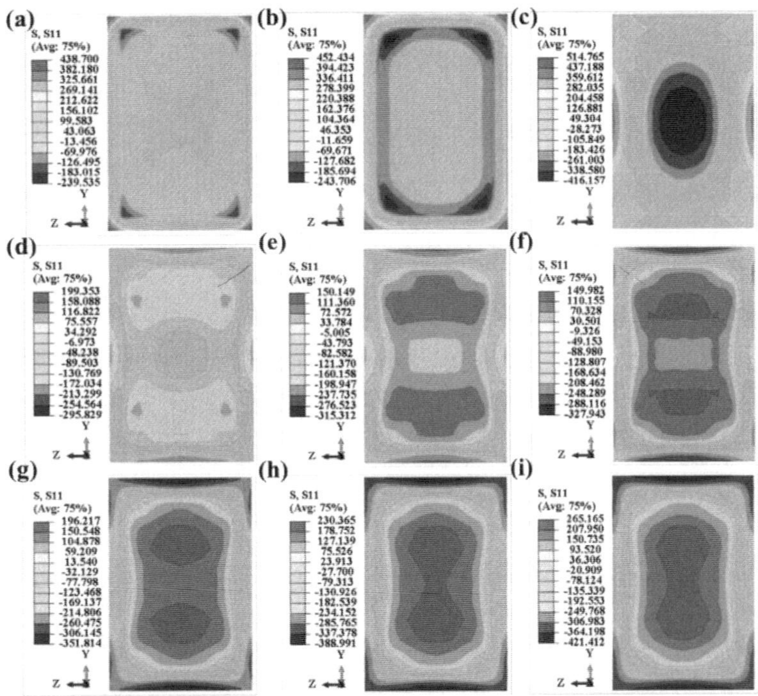

Figure 3: Stress field evolution of TC4 titanium alloy during solution quenching (1000°C).
(a) 0.96s; (b) 1.92s; (c) 9.97s; (d) 24.97s; (e) 30.97s; (f) 36.97s; (g) 51.97s; (h) 91.97s; (i) 375s.

Similarly, the stress evolution of TC4 titanium alloy during solution quenching (950°C) is similar. The selected stress direction is also S11 direction-X axis direction. According to the residual stress evolution cloud map in the S11 direction, the analysis shows that the residual stress evolution process at 950°C is similar to that at 1000°C. The evolution of tensile stress is developed from the four corners of the sample to the edge, surface and center of the sample, which is highly coincident with the evolution trend of temperature field. When the tensile stress is generated at the surface of the sample, most of the internal area of the sample is in a compressive stress state, and the tensile stress value of the surface layer and the compressive stress value of the core increase with time. As the temperature of the core area of the sample decreases, the tensile stress area gradually shifts to the core of the sample, so that in the subsequent cooling process, the tensile stress area develops from the four corners near the core area of the sample to the core area. As time goes on, the tensile stress area shifts to the center of the sample, and the tensile stress value gradually increases. The surface layer of the sample is a compressive stress area, and the compressive stress value also increases. The residual stress distribution characteristics of surface compressive stress and core tensile stress are formed in the final sample. The maximum tensile stress in the center of the sample is 256 MPa, and the maximum compressive stress in the surface layer is 355 MPa. The absolute value of stress is lower than that at 1000°C.
Compared with the residual stress cloud diagram of the contour method and the simulation results after solution quenching at 1000°C, it can be seen that as shown in Figure 4, the measured residual stress cloud distribution of the

contour method is basically consistent with the simulation results, but there is a slight difference. It is speculated that the possible reason is that the smaller volume of the sample will reduce the accuracy of the test, and the slow wire cutting has an effect on the stress distribution. In the simulation results, the maximum tensile stress in the center is 265 MPa, and the maximum compressive stress in the surface layer is 421 MPa; in the measured results of the contour method, the maximum tensile stress in the center is 227 MPa, and the maximum compressive stress in the surface layer (excluding the error area of the inlet and outlet) is about 400 MPa. The maximum tensile stress difference between the two is only 38 MPa, and the maximum compressive stress difference is about 20 MPa. It can be seen that the simulation results are not much different from the measured results. Considering that the Y-axis is the long side direction, which is consistent with the inlet and outlet of the slow wire cutting, the Y-axis path is selected to obtain data and draw to compare the evolution trend of residual stress on the path, as shown in Figure 5. The evolution law of residual stress on the Y-axis path, the simulation results and the measured results of the contour method are basically the same in the residual stress value and evolution trend.

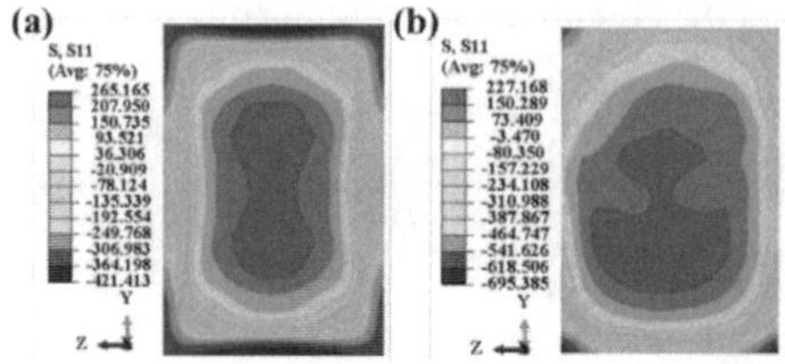

Figure 4: Comparison of simulation of solution quenching stress field of TC4 titanium alloy with contour method (1000°C).

(a) Simulation results ; (b) The measured results of contour method.

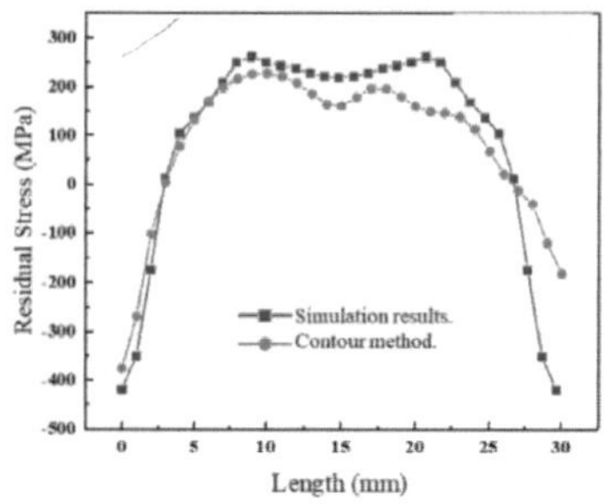

Figure 5: The curve of residual stress in Y direction of TC4 titanium alloy solution quenching stress field with length.

In summary, the material constitutive equations of TC4 titanium alloy at high temperature and medium-low temperature were constructed by using the thermal compression test data. At the same time, the relevant thermal parameters and yield strength obtained from the test are used as the basic parameters of TC4 titanium alloy simulation. The temperature field and residual stress field evolution of TC4 titanium alloy during solution quenching process are simulated by using ABAQUS finite element software and sequential coupling thermal stress analysis method. Limited by the length of this paper, the detailed information about the microstructure evolution of TC4 titanium alloy is not listed, and the stress evolution simulation analysis of coupled phase transformation is not listed. [9]

Conclusions

1. In the true stress-strain curve of TC4 titanium alloy, at the same temperature and different strain rates, the flow stress value and its peak value in the high temperature section increase significantly with the increase of strain rate, while the flow stress value in the middle and low temperature section tends to be relatively stable. At the same strain rate and different temperatures, as the test temperature decreases, the flow stress value increases significantly, and as the strain rate increases, the peak flow stress also increases.
2. In the simulation results of residual stress field, the basic law of residual stress field evolution at 1000°C and 950°C is almost the same, and finally the overall stress distribution of surface compressive stress and core tensile stress is formed. The difference in solution temperature leads to the difference in the absolute value of the maximum tensile stress in the final core and the maximum compressive stress in the surface layer. At 1000°C, the maximum tensile stress in the core is 265 MPa, and the maximum compressive stress in the surface layer is 421 MPa. At 950°C, the maximum tensile stress in the core is 256 MPa, and the maximum compressive stress in the surface layer is 355 MPa.

The contour method is used to verify the simulation results, which can basically verify the accuracy and credibility of the simulation results.

3. The microstructure of TC4 titanium alloy after solution quenching, aging and annealing heat treatment was studied. It was found that when the solution temperature (1000°C) was in the β single phase region, there would be obvious martensite relief after quenching. In the subsequent aging and annealing heat treatment, many intermediate phase transformations will be involved. When the solution temperature (950°C) is in the α+β two-phase region, because the temperature is close to the phase transition temperature of TC4 titanium alloy, a small amount of martensite relief phenomenon also appears in the microstructure.

Acknowledgments

This work was performed under the auspices of the National Key Research and Development Program of China (Grant No. 2023YFB3711900).

References

[1] Tabatabaeian A, Ghasemi A R, Shokrieh M M, et al. Residual Stress in Engineering Materials: A Review [J]. Advanced Engineering Materials, 2022, 24(3): 2100786.

[2] Li Y, Gan W, Zhou W, et al. Review on residual stress and its effects on manufacturing of aluminium alloy structural panels with typical multi-processes[J]. Chinese Journal of Aeronautics, 2023, 36(5): 96-124.

[3] Withers P J. Residual Stress Part 1-Measurement Techniques[J]. Materials Science and Technology, 2001, 17(4): 354-365.

[4] Landwehr M, Oehler F, Behnken H, et al. Influence of heat treatment on the residual stress-related machining distortion of Ti-6Al-4V alloy monolithic parts[J]. Procedia CIRP, 2021, 104: 1328-1333.

[5] Cerutti X, Mocellin K. Influence of the machining sequence on the residual stress redistribution and machining quality: analysis and improvement using numerical simulations. Int J Adv Manuf Technol 2016;83(1):489–503. 6.

[6] Bi Zhong Nan, Qin Hai Long, Liu Pei, et al. Research progress on quantitative characterization and control technology of residual stress in superalloy forgings [J]. Journal of Metals, 2023, 59 (09): 1144-1158.

[7] Guo J, Fu H, Pan B, et al. Recent progress of residual stress measurement methods: A review [J]. Chinese Journal of Aeronautics, 2021, 34(2): 54-78.

[8] Zhao Q, Sun Q, Xin S, et al. High-strength titanium alloys for aerospace engineering applications: A review on melting-forging process [J]. Materials Science and Engineering: A, 2022, 845: 143260.

[9] Liu C Chang, Li Y Huan Zi, Gu J, et al. Phase transformation in titanium alloys: A review[J]. Transactions of Nonferrous Metals Society of China, 2024, 34(10): 3093-3117.

QDE 2025: Proceedings of the 3rd International Conference on
Quenching and Distortion Engineering
May 6–7, 2025, Vancouver, Canada
https://doi.org/10.31399/asm.cp.qde2025p0164

Distortion of gears due to induction hardening with the gap-to-gap technique

Prof. Rainer Fechte-Heinen, Dr. Holger Surm, Grace Babb, Dr. Matthias Steinbacher
Leibniz Institute for Materials Engineering - IWT, Bremen, Germany

Abstract

The main goal of the European Commission Research project "WINDUCTION" is the design of an eco-friendly and low-energy consuming production route for the planet gears of the wind turbine gearboxes, by the replacement of classical case hardening process with induction hardening. Quantifying the distortions that occur during heat treatment is an important criterion when evaluating process chains. The purpose of this paper is to contribute more information about the type and severity of gear distortions that occur on each tooth during gap-to-gap hardening. The distortion of gears from different manufacturing processes after induction hardening and conventional case hardening are compared. From the distortion analysis, a difference between the induction hardened and reference steel was noticeable for profile slope deviation, adjacent pitch deviation, and runout error, which can primarily be attributed to the procedure during the induction hardening process.

Introduction

As wind turbines are both large and remote, potential repairs are very costly and should be minimized as much as possible. This can be achieved by ensuring a high quality of components susceptible to wear, as the quality of the gears directly impacts the longevity of the wind turbine. The purpose of the WINDUCTION project is to investigate the use of induction heating for surface hardening the planet gears, which are currently being fabricated with case hardening.

Induction heating offers many advantages over the conventional carburizing, such as the reduction in both energy consumption and manufacturing time (and therefore cost), as well as the elimination of pollutant production. While smaller gears can be more evenly heated as they are spun around inside an inductor coil, larger gears, such as the planet gears used in this project, are induction hardened using a gap-to-gap technique (also known as the gap-by-gap or tooth-by-tooth technique) where the inductor coil fits into the gap between two gear teeth and heats the flanks of each gap individually. This method allows the induction coil to better fit the shape of the gear and eliminates the need to spin such a large mass, which would require a large amount of energy and be costly. Some disadvantages to this asynchronous heating method are the potential back tempering as heat is conducted through the tooth to previously hardened areas, as well as the size and shape distortions that result from nonuniform volumetric changes. These effects are particularly noticeable at the last heating position [1], as the first tooth experiences back tempering on both flanks [2] and has the largest amount of time between heating on each flank.

Gears with a module of 8 mm and a tip diameter of 220 mm were used in this experiment. The gears to be induction hardened were manufactured from the materials 50CrMo4, 56NiCrMoV7, and 56NiCrMoVBi7. The influence of the initial condition was adjusted by selecting different tempering conditions when tempering the gears. This results in the variants LH (low hardness) and HH (high hardness). The gap-to-gap induction hardening technique was applied, in which each gear was iteratively rotated and aligned to allow the inductor coil to vertically scan each gap. The design of the coil and quenching unit went through multiple iterations before settling on a design that ensured a sufficient surface hardened depth of 2.4mm at the flank and 1.8mm at the root, and a minimum surface hardness level of 700HV. An Independent Frequency and Power (IFP) machine with an applied coil frequency of 10kHz and applied maximum power of 50kW was used.

In addition, gears with identical geometry made of the case hardening steel 18CrNiMo7-6 were used as a reference, and their hardening process consisted of controlled gas carburizing in a box furnace at 930°C, direct quenching in a mild quenching oil and subsequent tempering at 170°C. All variants investigated are listed in Table 1.

Table 1: List of the variants investigated

Item	Material	Acronym	Heat treatment	Hardness level	Core Hardness in HV1
1	18CrNiMo7-6	18ref	Case hardening	-	442
2	50CrMo4	50LH	Surface hardening through induction	LH	287
3	50CrMo4	50HH		HH	325
4	56NiCrMoV7	56LH		LH	289
5	56NiCrMoV7	56HH		HH	329
6	56NiCrMoVBi7	56Bi		HH	366

Results

Metallographic investigations

Figure 1 shows the case hardened reference (left) labeled with the positions at which the microstructure was evaluated by light optical metallography (LOM) (shown in Figure 2 and Figure 3), as well as an induction hardened variant (right) labeled with the positions where the hardness profiles were measured (shown in Figure 4, Figure 5, and Figure 6). A classic gas carburizing with direct quenching in mild quenching oil was used as a reference process. In the overview image of Figure 1 (left), the case hardened layer follows the contour of the tooth flank and root as a continuous dark seam. Figure 1 (right) shows that, in accordance with the procedure during induction hardening (gap-to-gap induction heating), there is an area at the tooth tip that has not been induction hardened. In the induction hardened variant, the back tempering effect can also be observed at the tip of tooth flank A. In each variant, the hardening layer is continuously hardened along the tooth flank and the tooth root. The hardening layer varies in thickness along the hardening contour. In addition, the etching behavior is different: In the area of the tooth root, the hardness layer appears lighter.

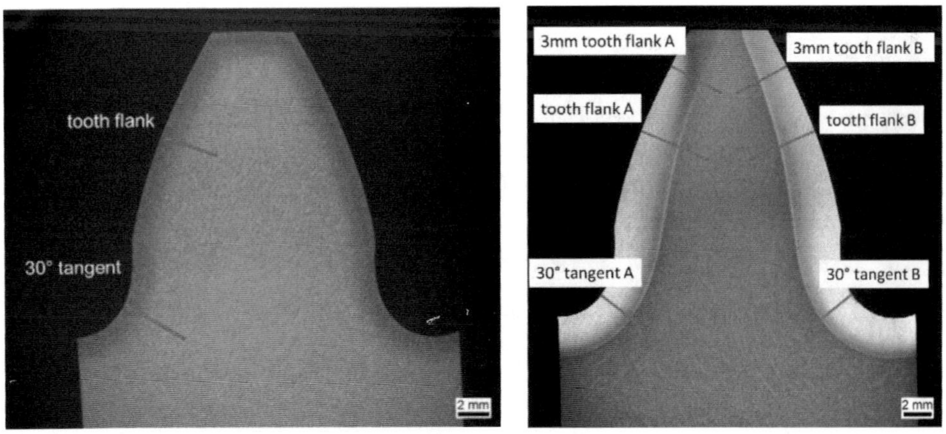

Figure 1: Positions where the microstructure was evaluated and where the hardness profiles were measured on the flank and 30° tangent of the gear tooth, as shown on the reference (left) and one induction hardened variant (right)

Figure 2 shows the microstructure of the reference (18CrNiMo7-6, case hardened) in more detail. Some differences between the tooth flank and the 30° tangent can be observed, which can be primarily explained by the geometric boundary conditions during carburizing. The microstructure consists of martensite and retained austenite. A slightly higher proportion of retained austenite can be expected at the tooth flank in comparison to the 30° tangent, without exceeding 25 % by mass.

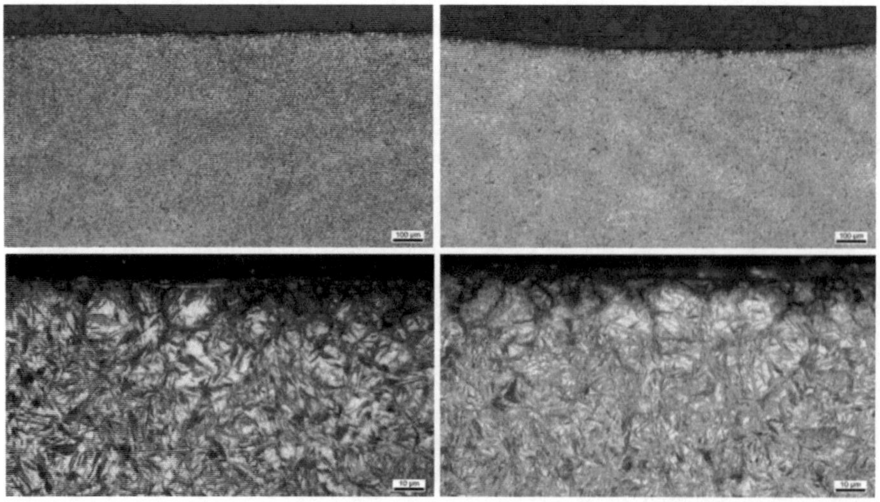

Figure 2: Microstructure at different positions after carburizing the 18ref: tooth flank (left) and 30° tangent (right). Sampling: cross section of tooth, half of the tooth width. Etching: 45 s, 3 % alc. HNO3.

An example of the microstructure after induction hardening is shown in Figure 3, with the condition at the tooth flank on the left-hand side and the microstructure at the 30° tangent on the right-hand side. In general, a martensitic microstructure is present in the hardened surface layer. The local structure of the microstructure is influenced by segregation. Despite these effects, the hardened surface layer can be defined as uniform in general. Depending on the material, only slight differences in the microstructure can be identified.

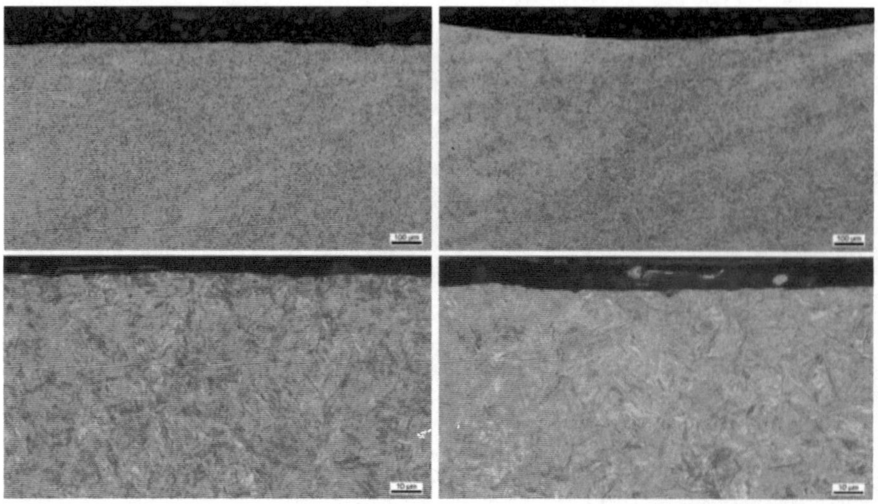

Figure 3: Microstructure at different positions after induction hardening 50LH: tooth flank (left) and 30° tangent (right). Sampling: cross section of tooth, half of the tooth width. Etching: 35 s, 3% alc. HNO3

The further characterization of the surface layer after induction hardening is based on hardness measurements. A total of six hardness profiles per tooth were measured at the following positions on both flanks (as shown in Figure 1, right): on the tooth flank approximately 3 mm from the tooth tip (Figure 4), on the tooth flank in the area of the pitch circle (Figure 5), and on the 30° tangent (Figure 6). For the evaluation of the hardness profiles on the tooth flank near the tooth tip, it is also important to note that the flank on which the back tempering effect occurs is always referred to as flank A.

At flank A (Figure 4, left), the hardness at the surface scatters between 620 HV1 and 690 HV1. The deviations appear to be purely process-related, as there is no apparent correlation with the material or the initial condition (HH, LH).

There is a continuous decrease in hardness up to the end of the surface hardened layer. In the transition zone, there is the typical sharp drop in hardness towards the untreated region. There are clear differences in the SHD (Surface Hardened Depth) between the variants (1.45 mm to 2.44 mm). The hardness in the untreated area corresponds to the hardness after tempering. No minimum in the hardness profile can be identified in the heat-affected zone. The hardness profile on flank B (Figure 4, right) is in principle comparable to flank A. However, there are small differences: The hardness on the surface of flank B is higher. It reaches values of around 715 HV1 for the variants with the HH treatment, while the LH variants are below 700 HV1. The drop in hardness towards the transition zone is not as pronounced. The SHD values vary between 1.99 mm and 2.59 mm. In addition to the reduced scattering width, the SHD range is significantly lower for flank B than for flank A. It is also noticeable that the order of the variants has changed in relation to the SHD. The hardness values in the untreated area are in line with expectations.

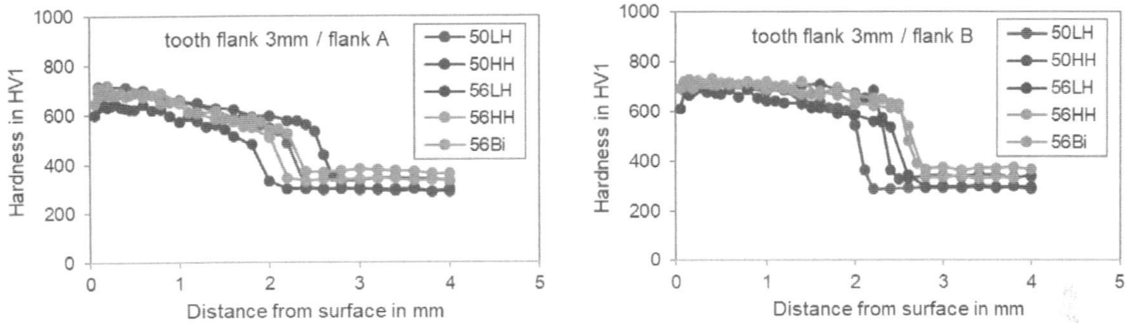

Figure 4: Hardness profile after induction hardening; position: tooth flank 3 mm; left diagram: flank A (with back tempering effect), right diagram: flank B (without back tempering effect)

There are also differences between the hardness profiles of flank A and B in the second position (tooth flank, pitch circle) in Figure 5. In particular, for the two variants made of the material 56NiCrMoV7 (56LH and 56HH), the hardness at the surface differs or is below 700 HV1. An early decrease in hardness towards the transition zone can also be seen for these two variants. In addition, the SHD of the two variants is significantly lower than the other three variants. The differences are particularly clear at flank A (Figure 5, left).

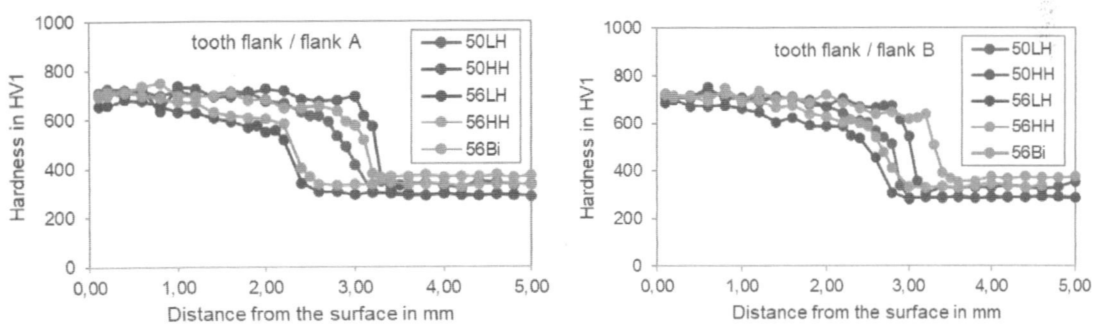

Figure 5: Hardness profile after induction hardening; position: tooth flank; left diagram: flank A, right diagram: flank B

The smallest differences between the hardness profiles of the five variants can be found at the 30° tangent (Figure 6). The hardness at the surface is between 300 HV1 and 720 HV1. A hardness plateau exists at the surface hardness level for almost all variants. Only for variant 56HH does the hardness decrease slightly from a depth of approximately 1.5 mm. The differences between the two flanks A and B in terms of SHD are significantly smaller. The exception in this case is variant 56Bi with a difference of 0.3 mm.

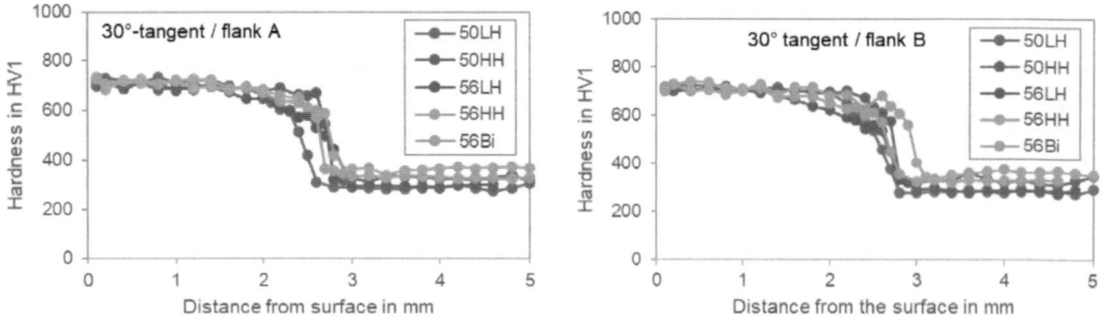

Figure 6: Hardness profile after induction hardening at the 30°-tangent; left diagram: flank A, right diagram: flank B

Distortion Behavior

Measurements were carried out to investigate the distortion behavior of the gears. These measurements were conducted on a coordinate measuring machine (CMM), type PMM 654 (Hexagon Metrology GmbH), using a probe with a diameter of 5 mm. For each measurement, a gear was fixed to the CMM using a chuck. The coordinate system was aligned via the bore and a face of the gear.

The following parameters were measured to evaluate distortion: profile slope deviation ($f_{H\alpha}$), helix slope deviation ($f_{H\beta}$), total cumulative pitch deviation (F_P), adjacent pitch deviation (f_u), and runout error (F_r). Figure 7 shows these parameters in more detail. As part of the measurement, three flank lines and three profile lines were measured on both flanks of each tooth. The measurement points were recorded when measuring the flank and profile lines in scan mode. For pitch and runout measurement, single point measurement was used.

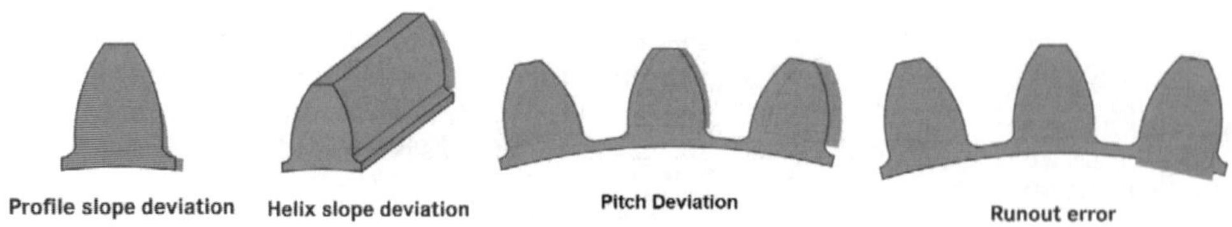

Figure 7: distortion criteria [3]

Profile slope deviation

Profile slope deviation, also called profile angle deviation, is the deviation from the actual slope of the tooth flank and the nominal slope before form deviation. Figure 8 shows the profile slope deviation measured on the left and right tooth flanks for all teeth of the five variants.

The reference steel has a distribution slightly shifted into the negative range with average values below zero. On both flanks, the profile slope deviation is distributed in an overall value range of approx. 20 µm.

In the case of the induction hardened steels, the differences between the average distortion behavior of the right and left flanks are significantly greater than for the reference. While the profile slope deviation on the left flank varies on average around 10 µm, the distribution of values on the right flank is shifted towards negative values (approx. -30 µm). It is noticeable that deviating values are measured on at least one tooth for each variant, i.e. 50LH on tooth 5, and these deviating values are marked with arrows. On the left flank, these are shifted towards negative values. For the corresponding tooth on the right flank, the values are above the mean.

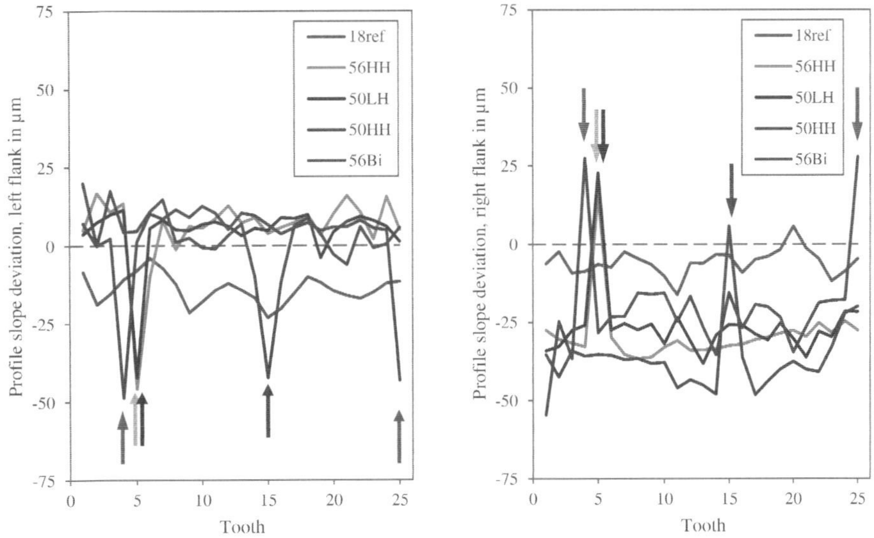

Figure 8: Average profile slope deviation $f_{H\alpha}$ of the left and right flanks

Helix slope deviation

Helix slope deviation, also known as lead angle deviation, is the deviation of the actual slope of a lead trace to the nominal slope deviation without influence of form. Figure 9 shows the helix slope deviation measured on both flanks.

For the reference steel, all the values of the helix slope deviation at the right and left flanks lie within a range of -5 µm to 20 µm. There is a continuous change in the values over the circumference. The distribution has one maximum and one minimum.

The deviations of the induction hardened variants are more or less at the same level as the reference. Variant 56HH is an exception here. Again, there is a continuous change in the values over the circumference. The difference between maximum and minimum is 40 µm. In contrast to the distribution of the profile slope deviation, no anomalies can be identified in the distribution of the helix slope deviation. In any case, no clearly deviating values can be recognized on the corresponding teeth.

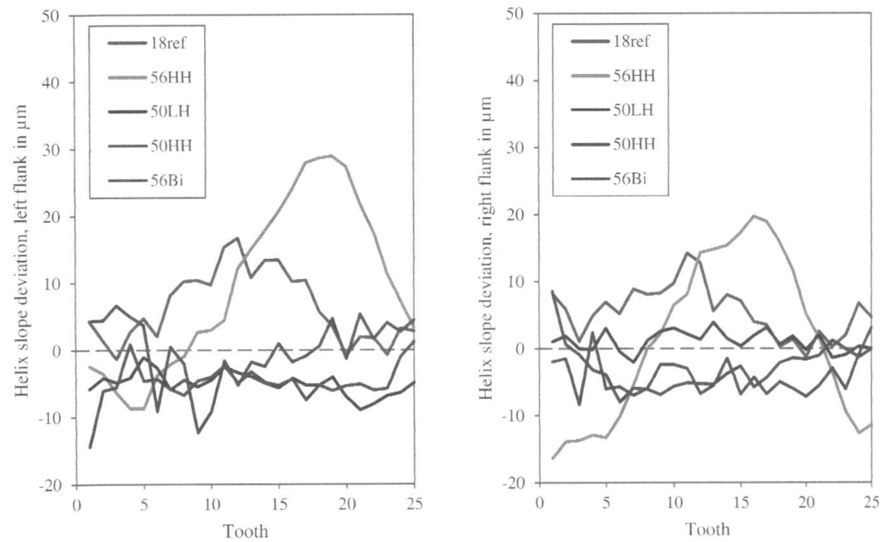

Figure 9: Average helix slope deviation $f_{H\beta}$ of the left and right flanks

Pitch deviations and runout error

Pitch deviation is the difference between the theoretical pitch and the actual measured pitch between teeth. Here, the pitch deviation is shown as both the total cumulative pitch deviation (sum of the single pitch deviations), as well as the adjacent pitch deviation (maximum absolute value of the difference between the currently measured pitch and the previously measured pitch). Runout error is the radial position deviation.

With regard to the total cumulative pitch deviation (Figure 10), no conclusion can be extracted about any difference between the reference and the surface hardened variants based on the data available. The reason for this is the large scatter between the two measurement positions (right and left flanks), particularly for 56HH. The measurements were taken at an axial height z of 3 mm and 22 mm.

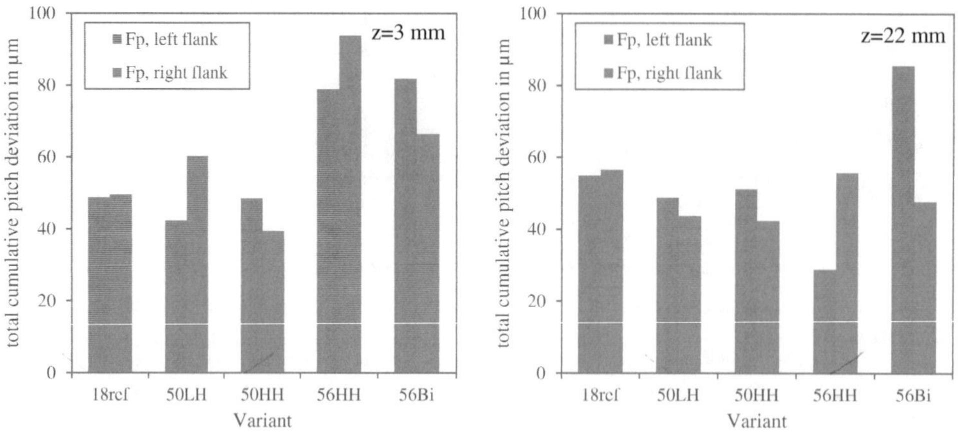

Figure 10: Total cumulative pitch deviation F_p in µm; z=3 mm (right); z=22 mm (left)

In contrast to total cumulative pitch deviation, the parameters maximum adjacent pitch deviation (Figure 11) and runout error (Figure 12) show much more significant distortion behavior. In this case, the values of the surface hardened gears are significantly higher than the reference. This result can be attributed to a greatly increased single pitch deviation on one tooth, which also exhibits unusual values for the profile slope deviation. This local deviation dominates the value of the two parameters maximum adjacent pitch deviation and runout error. This distortion behavior is present in all the surface hardened gears and indicates the start or end gap of the inductive heat treatment.

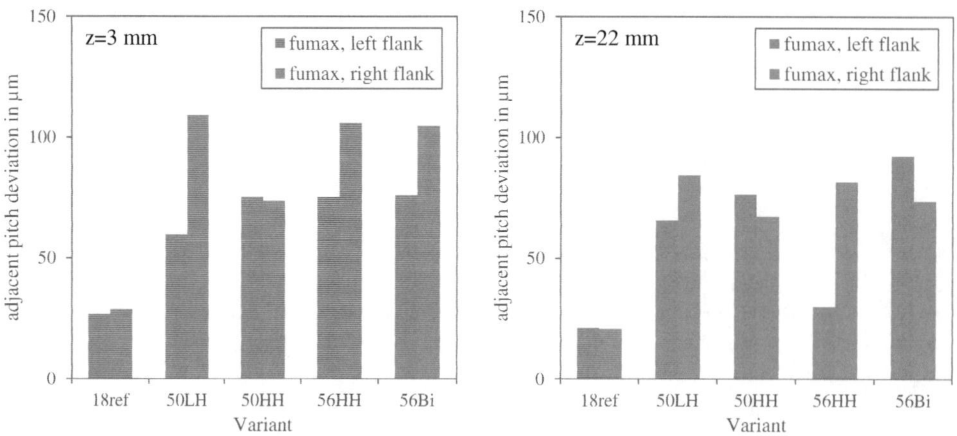

Figure 11: Maximum adjacent pitch deviation f_{umax} in µm; z=3 mm (right); z=22 mm (left)

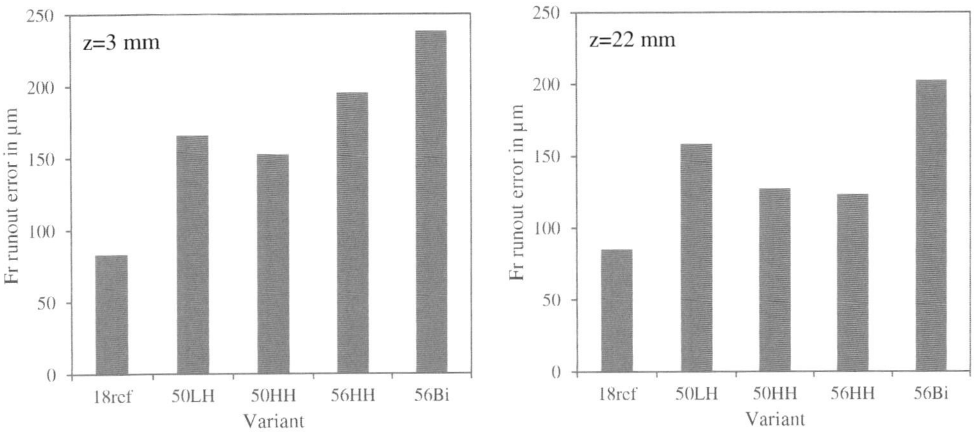

Figure 12: Runout error F_r in µm; z=3 mm (right); z=22 mm (left)

Conclusions

For all five variants, induction hardening using the gap-to-gap hardening technique resulted in the desired hardening pattern without overlapping in the area of the tooth tip. However, some points were identified during the metallographic investigations that should be optimized in the further course of the project. Although the opposing tooth flanks were actively cooled during induction hardening, the undesirable effect of back tempering [4][5] is found in all variants. This effect describes the undesired tempering of previously hardened areas. Heat is transferred by thermal conduction from a high temperature region to the opposite flank of the gear. The metallographic investigations showed that a decrease in SHD can be expected in these areas. The literature [2] mentions that this effect may be unavoidable depending on the tooth shape and process parameters. In these cases, the softening is insignificant and acceptable. Further investigations will show whether this local softening leads to premature failure of the gears.

There are differences in the hardness profiles at different positions along a single tooth (Figure 4 - Figure 6). This may be due to insufficiently precise positioning of the inductor in the tooth space. Asymmetrical positioning of an inductor results in a non-uniform hardness pattern due to electromagnetic proximity effect [5]. Differences can be observed both between the two flanks and between the tooth tip and tooth root. The differences in the area of the tooth root can be considered as insignificant. There are otherwise typical hardness curves after induction hardening. In some cases, a slight premature decrease in hardness occurred. In this case, the local austenitizing conditions were not sufficient to dissolve enough carbon in the austenite for optimum hardening. Below the transition zone, a minimum in the hardness curve due to additional tempering was not found in any case.

The comparison of the distortion behavior between the carburized reference and the induction hardened variants revealed certain differences. This primarily concerns the profile slope deviation and the adjacent pitch deviation. Increased concentricity was also determined because of these deviations on individual teeth. Overall, there are only a few references in the literature about distortion behavior due to induction hardening using the gap-to-gap technique. The reference [2] describes the change in the tooth profile due to the induction hardening, which leads to a corresponding change in the profile shape deviation. The study described in [4] mentions changes in shape at the last heating position without assigning them to a specific distortion parameter: "Shape distortion is particularly noticeable in the last heating position. The last tooth can be pushed out 0.1 mm – 0.3 mm." It suggests a modification to the procedure (avoiding the induction hardening in continuous circumferential order, the procedure followed in this initial phase of the project) to minimize distortion. Due to the measured deviations and even with a different induction hardening approach, the teeth grinding seems to be unavoidable to guarantee the required standards.

Some sources [1] suggest a method of hardening every second tooth, which would reduce the time between the heating of each flank of the first tooth by one half, but would also increase this time for every other tooth, and require the increased energy and cost associated with two revelations of the gear rather than one. Further research is needed on this topic to verify if the benefits to the first tooth outweigh the increased time and cost, as well as the increased probability of distortions on the other teeth.

Acknowledgements

This paper is part of a project that has received funding from the European research program of the research fund for coal and steel (RFCS 2020) under grant agreement No. 101034069.

The following parties are involved in the "WINDUCTION" project:

- SIDENOR I+D SA
- OVAKO SWEDEN AB
- ZF WIND POWER ANTWERPEN NV SA
- LEIBNIZ-INSTITUT FUR WERKSTOFFORIENTIERTE TECHNOLOGIEN-IWT
- INDUCTOHEAT EUROPE GMBH
- RHEINISCH-WESTFAELISCHE TECHNISCHE HOCHSCHULE AACHEN (RWTH AACHEN)

Many thanks to all partners for the work and research carried out in the context of this paper.

References

[1] Rudnev, V., Loveless, D., Cook, R., *Handbook of Induction Heating*, Marcel Dekker, Inc. (New York, 2003), p 309-317.
[2] Ingham, D. and G. Parrish, G., *The submerged induction hardening of gears,* Heat Treatment of Metals (UK 1998), p 1-8.
[3] *Quality Parameters and Tooth Flank Modification*, WENZEL Metromec AG, https://wenzel-metromec.ch/images/pdf/WENZEL_Poster_Gear_Quality_Parameters.pdf, accessed Dec 2024.
[4] Rudnev, V., "Induction hardening of gears and critical components," *Gear Technology* (2008), p 58-63.
[5] Rudnev, V., *Induction Heat Treatment of Gears and Gear-Like Components*. Inductoheat, Inc. (2019)

QDE 2025: Proceedings of the 3rd International Conference on
Quenching and Distortion Engineering
May 6–7, 2025, Vancouver, Canada
https://doi.org/10.31399/asm.cp.qde2025p0173

Cooling Requirements for Soft Magnetic Composite Impeders in Inductive Tube Welding Installations

Robert C. Goldstein[1], D. Scott Mackenzie[2], Sean Muyskens[1], Sahadev Khatri[1]
1 - Fluxtrol, Inc. Auburn Hills, USA
2-Quaker Houghton, Conshohocken, PA
rcgoldstein@fluxtrol.com

Abstract

Induction tube welding is commonly used for in-line welding of small diameter steel tubing. To improve the efficiency and quality of the tube welding process, impeders made of ferrite cores are used on the inside of the open strip in the region leading up to the welding point. Recently, several production trials with SMC cores replacing ferrite core have energy savings of 20-50% with the potential to increase production rate, reduce equipment downtime and improve metallurgical properties on small diameter steel tubing.

The challenge for implementation of the SMC materials is to properly cool them as the magnetic losses for these materials are much higher than for ferrites. To verify performance capabilities, a combination of electromagnetic modeling and physical simulation has been presented at previous conferences. In this paper, the importance of proper design of the coolant flow circuit will be discussed. A case story will be presented for an impeder design and the calculation results will be compared to measurements made on the physical test stand.

Introduction

Inductive welding is a popular method for making tubes used in a variety of industries for over 60 years. Over the past 20-30 years, there have been significant advancements in material handling, process monitoring and control, material cutting and solid-state electronics which improved the productivity of inductive tube welding systems and allowed for the production of higher quality tubing with lower processing costs. At the same time, one of the key components of the system, the impeder, has not changed significantly, which is limiting the performance of many of these systems.

Impeders are typically manufactured using a combination of constructive elements and soft magnetic materials. The most common soft magnetic materials used are ferrites. Ferrites have very high relative magnetic permeability and low magnetic losses, but a relatively low saturation flux density which is strongly temperature sensitive, are limited in size and availability, and are very susceptible to mechanical and thermal shock. The saturation flux density of the impeder core material is often the limiting factor in the system productivity for very small diameter tubes, smaller tubes with internal constructive elements where there is a limited amount of space for magnetic material or medium sized tubes with larger wall thicknesses. When the ferrite material is in magnetic saturation, there is an opportunity to significantly improve the welding system performance using SMCs [1-5].

Computer simulation conducted over the past several years has demonstrated that the appropriate use of SMC's impeders had the potential to dramatically improve welding system performance. The models were focused on the simulation of the tube heating process, incorporated magnetic flux controllers and quantified their role in improving the welding process. Comparisons were made between impeders using SMCs and those made with traditional ferrites for multiple tube sizes [2, 6].

In parallel with the computational efforts, tests were conducted in real tube mills using impeders containing SMC cores compared to ferrite cores. These tests were run on tubes with outer diameters between 12.7 and 27 mm and wall thicknesses between 1.2 and 3 mm. In each case, the required power was 20 – 50% lower for the SMC core for the same line speed. As the line speed increased for a given wall thickness, the relative energy savings increased for all cases. In the cases where longer duration testing was performed, the SMC impeder cores were reported to last several times longer than the comparable ferrite cores [1, 3-5, 7, 8].

Based on the promising theoretical and field-testing results, efforts were made to validate SMC impeder core designs outside of the tube mill production environment. Researchers at Technical University of Chemnitz in Germany led by Prof. Martin Kroll developed a physical welding test rig. This rig was developed for short duration tube welding testing and was capable of testing different impeder core designs in addition to varying mechanical forming conditions for smaller diameter tubing at speeds up to 60 m/min. Tests using this rig have verified the performance benefits of SMC impeders when ferrites are in saturation [9]. However, this test rig does have inherent limitations for line speed, available power and the ability to test for longer durations.

The main failure mode of impeders in the field are either due to thermal and mechanical stresses degrading the properties of the ferrites, the casing burning through causing a loss of cooling and mechanical failure due to physical contact between the moving metal tube and the impeder. The technical drawbacks of SMC's compared to the ferrites used for inductive tube welding are lower magnetic permeability at low fields and higher magnetic losses at the same magnetic flux density. The magnetic losses are proportional to both the frequency and magnetic field strength. With the potential to operate the machine effectively at higher magnetic flux densities, this means there is the potential for significantly higher total losses that will need to be removed from an impeder by cooling. This becomes the main differentiating risk factor in changing from a ferrite impeder core to one made using an SMC. To understand this relationship, magnetic losses were measured for some candidate SMC's at typical inductive tube welding frequencies up to around 1 T [10].

With the magnetic loss data and computer models for impeder magnetic loading in tube welding installations, a physical test stand was created to verify that SMC impeders could survive the electromagnetic conditions of a tube welding system under these higher loads for satisfactory durations. Fig. 1 shows the current test stand that is being used to carry out these tests. The major components of the test stand are an induction heating power supply (3, 25 or 40 kW depending upon impeder core design) capable of 150-440kHz, a single or three-turn induction coil, an impeder positioned within the coil, recirculating water to cool the coil, and various measurement devices to capture the coil current, voltage, and inlet and outlet temperature of the cooling water from the impeder which is cooled by a separate closed loop chiller. Numerous tests have been conducted on this rig and have verified the ability of properly designed SMC's impeder cores to survive under conditions exceeding the capability of ferrite impeders for durations of up to several hours. The rig has also demonstrated good correlation between computer models for electromagnetic parameters, magnetic losses and calorimetric measurements [11].

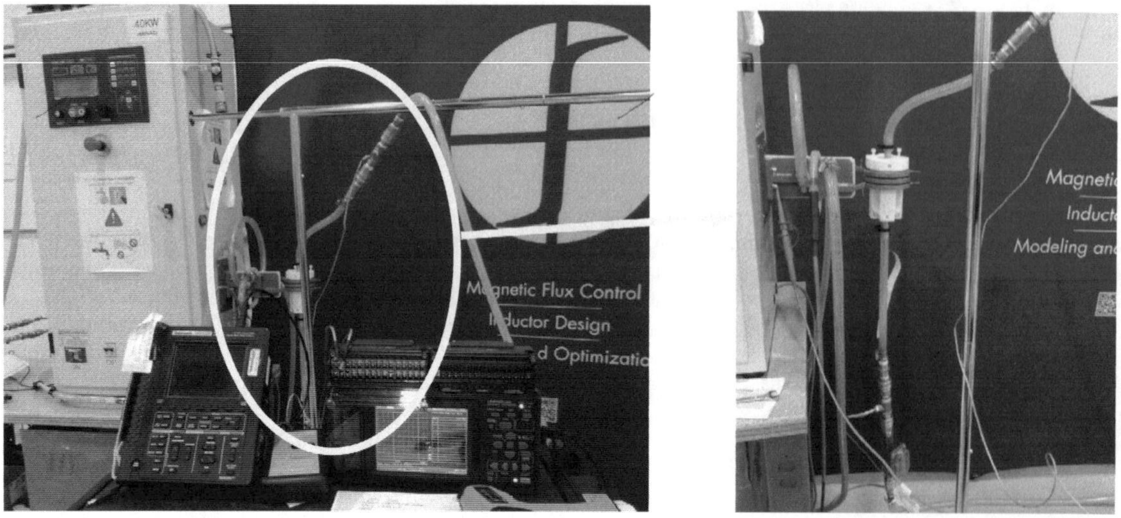

Figure 1. Impeder test rig.

Thermal initiated failures can occur due to either insufficient overall water flow leading to an unacceptable temperature rise in the bulk fluid in the impeder or due to a local temperature which exceeds the rated temperature of the component, which leads to physical deterioration of that region of the component, which can then lead to a catastrophic failure. Calculations and practical experience have shown that the greater the product of the cross-sectional area and the relative permeability of the magnetic core in the impeder casing, the higher the welding installation efficiency will be [8]. The current paper will focus on the cooling considerations to ensure reliable performance in the field for SMC impeder cores for smaller diameter tube welding installations.

Heat Sources in an Impeder

In an impeder for welding smaller diameter tubes, the heat sources that need to be removed include both electromagnetic losses in the impeder components along with thermal radiation from the hot weld seam. In many small tube welding installations, there is a large amount of water present in the welding region from cooling of the welding rolls along with in some cases water applied to the external surface of the impeder casing and the external surface of the welded tube. Thermal radiation from the welded tube to the impeder is governed by Equation 1:

Equation 1 - $Q = \sigma A_1 F_{12} \varepsilon_1 (T_1{}^4 - T_2{}^4)$

Where:

- **Q**: is the rate of heat transfer (W).
- **σ**: is the Stefan-Boltzmann constant (5.67 x 10^-8 W/m^2K^4).
- **A_1**: is the surface area of the hot component (m^2).
- **F_{12}**: is the view factor from component 1 to component 2 (dimensionless).
- **ε_1**: is the emissivity of the hot component (dimensionless, between 0 and 1).
- **T_1**: is the absolute temperature of the hot component (K).
- **T_2**: is the absolute temperature of the cool component (K).

Due to the relatively small region of high temperature localized to weld seam edges near the welding point, the integral value for radiant heat transfer will likely be relatively small and of a level that is relatively easily managed with moderate water cooling and in many cases this comes from the outside of the impeder. In addition, this value will tend to be smaller with SMC impeders due to the reduced electromagnetic end effect on the inner surface of the tube from the core not being in saturation which creates a reduction in the hourglass shape of the weld vee [X – Padua/HTS paper]. For the purposes of this study, we will not consider this factor in our analysis as it is likely to be the same or lower value for SMC impeders than for ferrite impeders, so the cooling requirements will not be negatively impacted by the material change.

Electromagnetic losses are viewed to be the larger source of heat in an impeder. The electromagnetic losses in the impeder include both magnetic losses in the soft magnetic material (either SMC or Ferrite) along with eddy current losses in any electrically conductive components in the assembly. The electrically conductive components in the assembly depending upon the impeder design may include:

- Copper or brass water fittings
- Copper or brass tubes in the assembly
- Ferrite or SMC impeder cores
- Steel rods

The value of the losses (both magnetic and eddy current) in all of these components are non-linear in the system due to the presence of the soft magnetic impeder cores with variable relative magnetic permeability which is dependent upon the machine set up and operating conditions. When the copper or brass water fittings are located well, the eddy current losses in these components is very small relative to other losses in the system and can be neglected in the analysis. This leaves the radial cross-sectional components under the induction coil as the main determinants of loss in the impeder system.

Heat Removal in an Impeder

In an impeder, the heat generated inside the impeder casing is commonly removed by a flowing liquid cooling medium. The type of cooling medium depends upon the specific mill and the tube being produced. In some cases, the cooling fluid flows through the impeder and exits at the other end of the impeder and flows down along the length of the welded tube. This type of impeder is called a "through flow" impeder. The most common cooling medium for through flow impeders is called mill water. Mill water is a process water used in tube mills that is a combination of water and oils that are used for manufacturing of the welded tube. In applications where the cooling is critical, the water for the impeder is often filtered to remove metallic contaminants like scale or metal filings and then put through a dedicated pump to increase the pressure and flow rate through the impeder. Clean water is rarely used for through flow impeders due to cost and environmental factors.

Through flow impeders can be broken down into 2 classes, hollow through flow impeders and solid through flow impeders. Solid, through flow impeders are the most simple case. Figure 2 shows a picture of some typical, small diameter through flow impeders (https://www.enrx.com/en/Induction-Products/Induction-heating-equipment/Consumables). There is a fitting on the right side of the images where fresh coolant is supplied. The diameter of the inlet cooling fitting is typically significantly smaller than the OD of the casing as the water supplying tube must exit at some point through the opening in the weld vee. It is also not necessary to have the fitting be the same internal diameter as the impeder core takes up a significant portion of the cross-section of the impeder inside of the casing. The impeder core is typically contained within the tube using a pin or other mechanical constraint. The water exits through the open end of the casing.

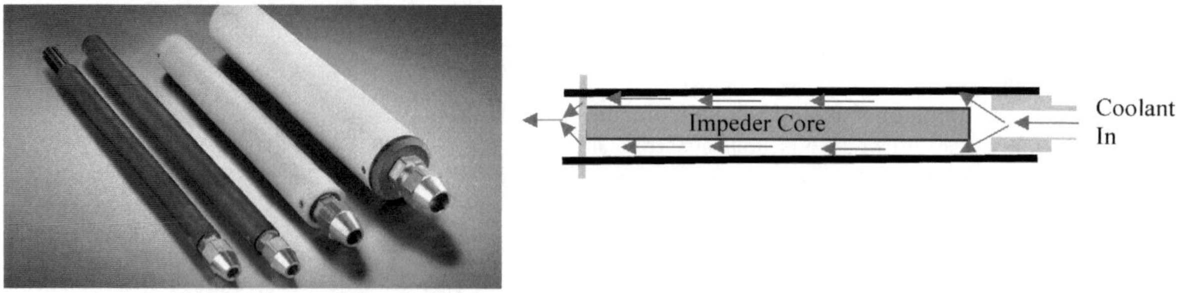

Figure 2. Typical small diameter through flow impeders (https://www.enrx.com/en/Induction-Products/Induction-heating-equipment/Consumables) and conceptual view of the water circuit.

In hollow through flow impeders, there is typically an internal brass rod that separates the inside of the impeder from components that run through the center of the impeder, such as a scarfing rod for removing the internal weld bead. In these designs, the water is generally supplied by a smaller tube(s) to the fitting, then the water splits and flows down the length of the impeder core (Figure 3). It is desirable to have the water flow on both the ID and the OD of the impeder core to improve the heat removal from the core. It is important in the design of these types of impeders, that the size of the coolant channels is closely considered to make sure that the total heat generated in the core can be removed effectively. The best designs will have the total sum of the cross-sections of the inlet cooling line(s) and exit hole(s) equal to or greater than the cross-sectional area of the space surrounding the impeder core, to insure good water pressure on the impeder core and the resultant higher flow velocities and resultant heat transfer coefficients for the coolant.

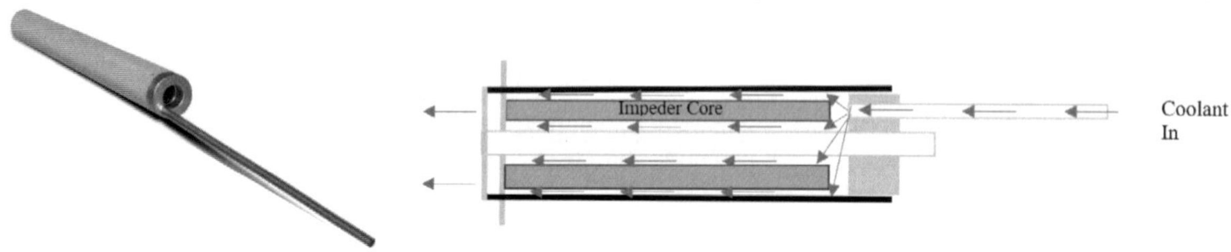

Figure 3. Typical small diameter hollow through flow impeders (Return Hollow Thru-Flow Impeders - Kent Corp) and conceptual view of the water circuit.

Return flow impeders can also be broken down into the same two classes, return flow impeders and hollow return flow impeders, depending upon if there is an internal component which also passes through the inside of the impeder, such as a scarfing tool. In both of these cases, the cooling circuit design is really critical in ensuring the reliable performance of the impeder and there can be multiple options for how the water should through flow the impeder and provide cooling for the internal components (magnetic core and central rod). Limiting the hydraulic resistance of the different regions of the flow circuit, especially the water supply lines, fittings and regions where the water changes direction are critical to ensuring the best cooling and impeder performance are achieved.

Case Study 13 mm OD SMC Solid Through flow Impeder

Impeder Design Considerations

The impeder OD is typically defined by the OD of the casing material, in this case 13 mm. The casing material typically has wall thickness between 0.75 and 1 mm depending upon manufacturer and material. In this case, a 1 mm thick-walled casing was selected, leaving 11 mm of space for the core and the cooling medium. The industry standard practice would be to use a 10 mm, fluted ferrite core. A commercially available 200 mm long TDK ferrite impeder core was selected for comparison that had 10 mm OD, 6 radial flutes, 1 mm wide and 2.5 mm deep. This led to a core cross-sectional area of 63.5 mm^2. The cross-section available for fluid cooling is 16.5 mm^2. Saturation flux density at typical tube welding operating conditions for this material is estimated to be 0.4 T.

The impeder core design target was to have nominally the same cross-section (+/-5%) for the core for the SMC impeder as the standard ferrite inside of the same casing, but to be able to operate for over 4 hours at a maximum local flux density of 0.8 T at a frequency of nominally 300 kHz on our impeder test rig. This level is approximately 2 times higher than for the TDK impeder core, which should lead to an approximately 4 times higher region of high process efficiency for the SMC impeder. Fluxtrol A was selected for this optimization as it has been used successfully in tube welding installations and has the highest maximum permeability and saturation flux density of the candidate materials considered [1, 7, 8].

Preliminary Design Process – Core Cross-Section Determination and Test Design

2D plane parallel thermal modeling using Altair Flux (altair.com) is first performed to determine the required impeder core cross-section [11]. The magnetic and thermal properties used in the model were taken from the Fluxtrol website [https://fluxtrol.com/fluxtrol-A]. A surface heat transfer coefficient of 10,000 W/m^2K as a design estimate and the core cross-section was modified to find achieve the design where the core temperature throughout the material remains below the short-term thermal limit for the material (300 C) at the targeted operating conditions with 30 C bulk fluid temperature of the cooling medium. The 10,000 W/m^2K is generally a conservative number for high velocity water flowing over a surface and is used to create an estimate for what the value may be with mill water that will likely be used in the field. The range of heat transfer coefficients for flowing mill water is not known by the authors at the time of this writing.
A user driven optimization converged upon 2 flute design on a 10 mm diameter core. Each of these flutes are 2 mm wide and 3.75 mm deep and located directly across from each other and cut through the unfavorable direction for thermal conductivity in the material [https://fluxtrol.com/magnetic-flux-controllers-in-induction-heating-and-melting]. Cross-sectional area of the core was 63.7 mm^2, or 0.3% higher than the ferrite. Predicted maximum internal steady state temperature was 287 C (Figure 4a).

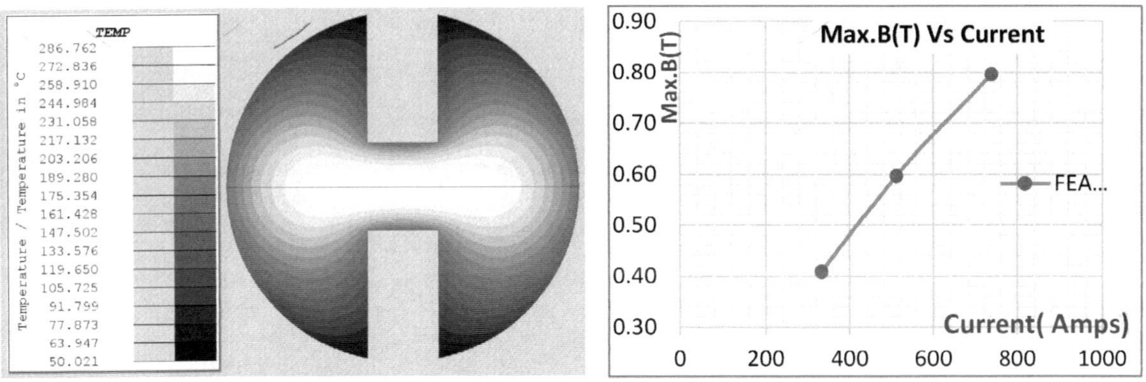

Figure 4. (a) Predicted steady state temperature in the SMC impeder core and (b) relationship between coil current and maximum flux density in the core

2D axisymmetric models using Altair Flux are used to determine the relationship between coil current and maximum flux density in the core as this value is frequency independent. The core diameter used in the model is adjusted to an equivalent diameter with an equal cross-section to the fluted core, in this case 9 mm. In the test design protocols, –5%, + 10% is allowed for frequency variation from the nominal target of 300 kHz. Calculations were made for 0.4, 0.6 and 0.8 T and are plotted in Figure 4b using the single turn coil on the test rig. The 0.4 and 0.6 T test points are additional, short duration testing (≈5 minutes with steady state delta T) prior to the longer duration testing at the highest target flux density.

Test Data and Comparison to Predictions

The flow rate through the impeder core was 12.1 lpm and the induction machine was tuned to between 304 and 309 kHz for the trials. The test currents were 333, 512 and 737 A, which correspond to flux densities of 0.41, 0.60 and 0.80 T respectively. During the initial portions of the trial, the impeder was run for 6 minutes 57 seconds at 0.41 T and 5 minutes, 46 seconds at 0.6 T. At the 0.8 T target level, the machine was run for 2 minutes, 45 seconds at the level, but the power supply was operating near a limit of the power supply, so it was decided to make a machine tuning adjustment before doing the long duration trial. After the adjustment, the trial was run for 5 hours, 12 minutes 59 seconds at 0.8 T. The temperature inlet and outlet temperatures for the trial are shown in Figure 5a. During the initial stages of the trial, the inlet and outlet water temperatures continue to rise as the temperature of the reservoir was below the regulation setpoints of the chiller. At approximately 6,500 seconds, the chiller began to regulate and control the inlet water temperature to within the target operating window.

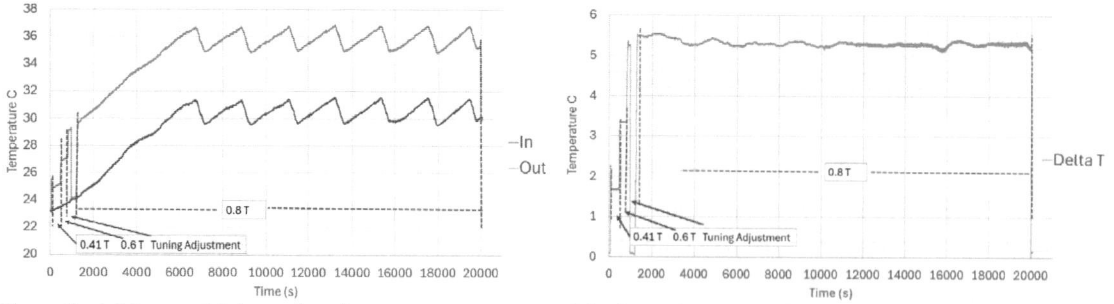

Figure 5. a) Measured inlet and outlet water temperature of the impeder versus time for the trial and b) respective Delta T's

The resulting temperature differentials (Delta T) are plotted in Figure 5b. The nominal values for the 3 levels are 1.7 C, 3.3 C and 5.3 C respectively. With the 12.1 lpm flow rate, this leads to calorimetric losses of 1.4, 2.8 and 4.5 kW respectively. Magnetic losses were calculated using the 2D models used to determine the relationship between flux density and current. The impeder core was separated into 20, 10 mm long segments. The average value of flux density in each 10 mm long segment was used to calculate the magnetic losses according to the formula for Fluxtrol A provided on the Fluxtrol website (https://fluxtrol.com/fluxtrol-A) and multiplied by the volume of the segment (0.636 cm^3). The losses in each section were then added together to determine the total losses for the impeder core at each loading level. The calculated values were 1.0, 2.1 and 3.6 kW respectively, which are approximately 20-30% below the measured values. Additional efforts are underway to attempt to better determine the source of the deviations between calculations and experiments.

Local Cooling Considerations

To better understand the internal temperatures in the core during the trial, the flow rates measured in the trials can be used as an input to CFD models to determine more realistic heat transfer coefficients in the system. The measured flow rate of the impeder was 12.1 lpm. A 3D, CFD model was created using SolidWorks with 1.5 million nodes. ¼ of the system was modeled due to symmetry (Figure 6a) to understand the axial flow conditions and the region of normalized flow. From the region of normalized flow where the maximum power densities in the impeder will be, a higher mesh density was used to get more accurate values for water velocities and heat transfer coefficients to compare to the earlier approximations used in Figure 4a.

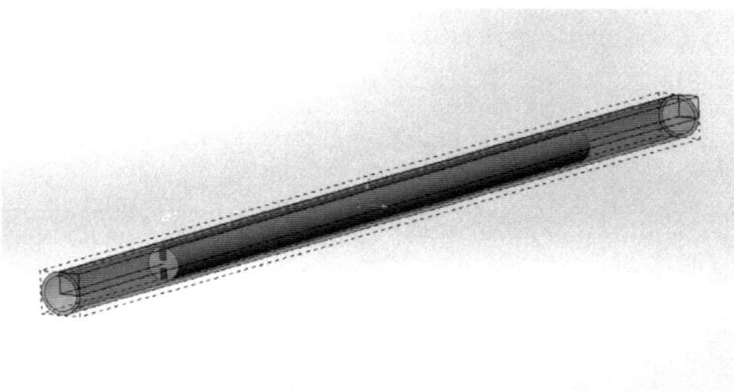

Figure 6. a)Model used for CFD analysis in SolidWorks

In the 3D model, water flows from the inlet on the right side to the outlet at the left side of the image in Figure 6. Figure 7 shows the axial cross-section cut through the center of the flute. We can see that the flow starts to distort approximately 10 mm prior to entering the impeder core and water velocity begins to increase. This relatively short transitory period can be explained as the cross-section for water flow in the region with the core is much smaller than in the region without the core. Once water begins to pass over the core, there is a transitory region to normalize the flow over the first 50 – 100 mm of the core. Flow is normalized over the second half of the impeder with flow velocities near 40 m/s in the center of the flute all the way to the exit of the core. At the end of the core, the flow is still not truly normalized at the exit, which is not surprising as the cross-section is much larger than in the core region.

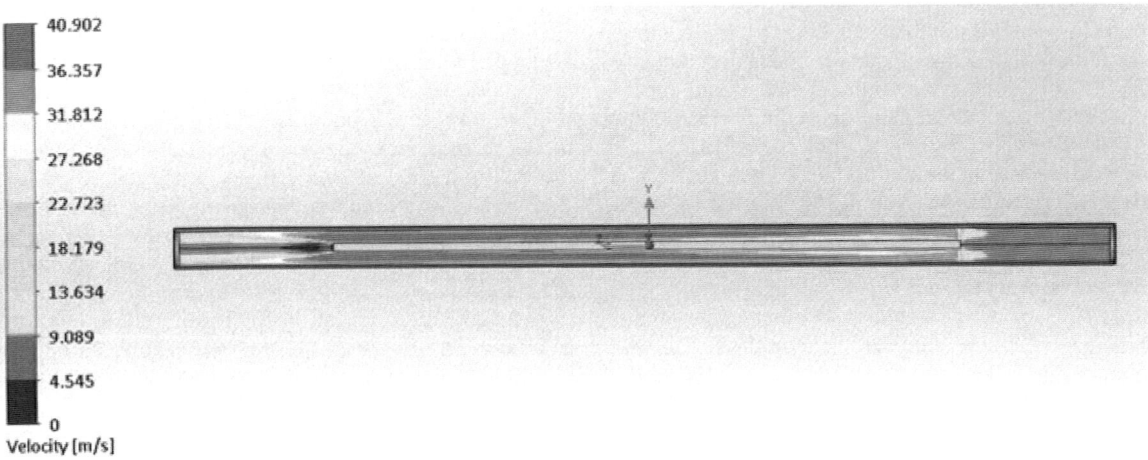

Figure 7. Water velocities in the cross-section of 3D CFD model through the center of the water cooling flute with the inlet on the right side of the image and the outlet on the left side of the image

The flow rate in the central, high mesh density section is shown in Figure 8a. The velocity in the regions without the slots is approximately half of the velocity in the regions with the slot due to the increased drag from the reduced distance between the two walls. The authors decided to evaluate the axial distribution of heat transfer coefficient at selected locations in the cross section identified at points A – F in Figure 8b and those values are plotted in Figure 8c. In general, the values for heat transfer coefficient remain fairly constant along the length of the impeder core at the various locations, despite some differences in the flow velocities predicted. The minimum in heat transfer coefficient appears to be in the region farthest from the fluted region (F), with a values of around 2000 W/m^{2}K. As we approach the fluted region, the values increase progressively from approximately 3000 (E) to just over 10,000 (C) W/m^{2}K. In the slot, the values at points A and B are on the order of 100,000 W/m^{2}K.

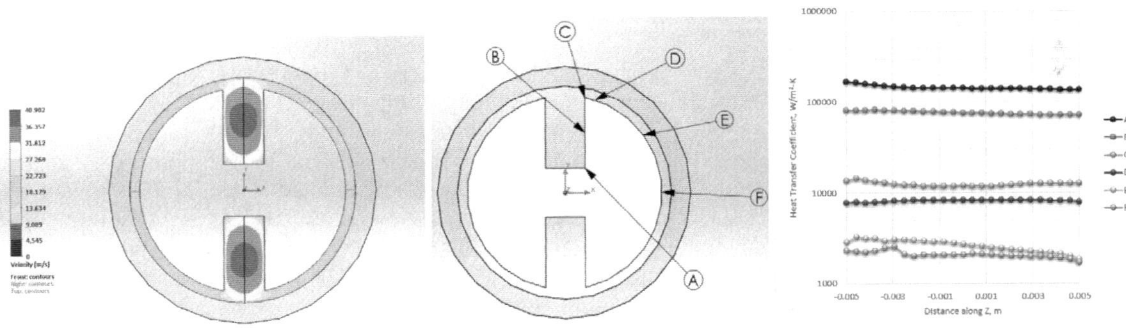

Figure 8. a) Water velocities in the cross-section of 3D CFD model through the center of the impeder in the high mesh density region, b) locations for heat transfer coefficient evaluation and c) curves for heat transfer coefficient along the length of the impeder core at the selected locations

The values in the model range from approximately 5 times lower to approximately 10 times higher than the uniform heat transfer coefficient approximations used to make Figure 4a. This would lead to significantly higher surface and internal temperatures in water regions far from the flute and significantly lower surface and internal temperatures in the fluted region. However, it is important to keep in mind that the evaluations in this section are made using a room temperature impeder core. As temperature of the impeder core rises due to the magnetic losses, the heat transfer coefficient will rise significantly as compared to the room temperature values.

Conclusions

SMC's have been shown to have significant benefits in small tube welding installations. The main technological hurdle involves insuring that there is sufficient cooling for the impeder core. The cooling circuit of the impeders can range from fairly simple to quite complex depending upon the type of tubular product that is being produced. Due to this a method combining computer models and experiments has been developed to test the ability of SMC cores to survive for extended durations in conditions representative of those present in a real tube welding installation.
A case story was presented using this method for a 13 mm OD through flow impeder core. The core was tested successfully at a maximum magnetic flux density of 0.80 Tesla for a period of over 5 hours continuously with no noticeable change in behavior over the duration of the trial. This was the expected result as the predicted internal core temperatures using an estimated uniform heat transfer coefficient on the surface were below the rated temperature value of the Fluxtrol A material used in the trials. Calorimetric measurements showed acceptable agreement with the computer models for magnetic losses in the core.
A 3D CFD model was created to better understand the flow characteristics inside of the impeder along with the variation in heat transfer coefficient on the surface of the impeder core. The results of the study showed moderate variations in the heat transfer coefficients at evaluation points in the axial direction. However, there were significant variations of well over an order of magnitude in the heat transfer coefficient in the cross-section of the impeder core. The high and low values of the heat transfer coefficient were both above and below the estimated value that is used in the standard workflow which will lead to differences in the temperature distribution inside of the core compared to the one predicted with the uniform heat transfer coefficient. It is also important to understand that the heat transfer coefficients predicted in the 3D CFD model are for a core with a room temperature surface; whereas in reality the core surface temperature in the area of high power density will be elevated, leading to increased heat transfer coefficients. Additional study is required to more strongly couple the CFD and thermal models to better predict the resultant internal temperatures in the core. It will also be beneficial to add additional components to the flow circuit (such as inlet water fittings), dimensional movement (both due to thermal and magnetic factors) and make some measurements for the material properties of "mill water" in order to more fully understand the flow and cooling conditions present in real impeders in the field. This data can be used for the further shape optimization of impeder cores to maximize the performance of small tube welding production lines.

References

[1] M.S. Milicevic, V.M. Milicevic; "Impeder for HF Welding of Steel Tubes", *IEEE Proceedings – Science Measurement and Technology*. June, 2002. Vol. 149, pp. 113-116. DOI: 10.1049/ip-smt:20020342

[2] Robert Goldstein, Magnetic Flux Controllers in Induction Heating and Melting, *Induction Heating and Heat Treatment*, Vol 4C, *ASM Handbook*, Edited By Valery Rudnev, George E. Totten, ASM International, 2014, p 633–645, https://doi.org/10.31399/asm.hb.v04c.a0005846

[3] Sean M. Muyskens, Tareq I. Eddir, Robert C. Goldstein; October 15–17, 2019. "Improving Inductive Welding System Performance with Soft Magnetic Composites." Proceedings of the *HT2019. Heat Treat 2019: Proceedings from the 30th Heat Treating Society Conference and Exposition*. Detroit, Michigan, USA. (pp. 63-69). ASM. https://doi.org/10.31399/asm.cp.ht2019p0063

[4] Muyskens, S.M., Eddir, T.I. and Goldstein, R.C. (2020), "Improving induction tube welding system performance using soft magnetic composites", *COMPEL - The international journal for computation and mathematics in electrical and electronic engineering*, Vol. 39 No. 1, pp. 185-191. https://doi.org/10.1108/COMPEL-06-2019-0232

[5] D. Günther, M. Kroll, R. C. Goldstein, A. Kunke; "Geometry optimization and efficiency increase of HFI welding processes while using SMC-impeders" *Proceedings of the XXth International UIE-Congress on Electrification of Industrial Thermal and Manufacturing Processes*. Nice (France), October 8-11, 2024

[6] Zhou, Yi; Baake, Egbert; Niedzwiecki, Igor et al., "Design of induction heating systems with soft magnetic composites", Proceedings of the 11th International Conference on Electromagnetic Processing of Materials, Shanghai, China, October 20-24th, 2024

[7] M.S. Milicevic, V.M. Milicevic; "IMPLEMENTATION OF OPTIMAL HF WELDING PROCEDURE OF STEEL PIPES FOR HIGH-QUALITY AND ENERGY-EFFICIENT WELDS", *PROCEEDINGS OF THE ROMANIAN ACADEMY, Series A, Volume 25, Number 1/2024, pp. 55–64*

[8] R. C. Goldstein, "Maximizing Small Diameter Induction Tube Welding Performance", *ASM HTS Webinar*. December 19th, 2024.

[9] Egger, C.; Kroll, M.; Kern, K.; Steimer, Y.; Schreiner, M.; Tillmann, W. Experimental Investigation of Temperature and Contact Pressure Influence on HFI Welded Joint Properties. *Materials* **2022**, *15*, 3615. https://doi.org/10.3390/ma15103615

[10] Vaishnava, P.P., & Goldstein, R.C. (2020). Magnetic core loss behavior at high fields in magnetic materials. *International Journal of Applied Electromagnetics and Mechanics, 63.*

[11] Muyskens SM, Morris DR, Yakey CJ, et al. Physical simulation and computational modelling for validation of soft magnetic composite impeder performance. International Journal of Applied Electromagnetics and Mechanics. 2024;75(2):107-118. doi:10.3233/JAE-230218

QDE 2025: Proceedings of the 3rd International Conference on
Quenching and Distortion Engineering
May 6–7, 2025, Vancouver, Canada
https://doi.org/10.31399/asm.cp.qde2025p0182

Reducing Heat Treat Distortion with Repeating Precision by use of 4 Dimensional High Pressure Gas Quenching

Thomas Hart
SECO/VACUUM Technologies LLC, Meadville, PA USA
tom.hart@secovacusa.com, +1-814-882-1606

Abstract

Heat treatment with oil, polymer or gas quenching has been the standard quenching practices used since heat treatment was discovered. This is because those methods were the only way to take a component and cool it fast enough to generate the desired microstructure. While these quenching methods have proven to be effective to create the desired microstructure, the resulting distortion created by these methods have made it difficult to conduct post manufacturing operations to correct the quenching distortion and produce a finished component. A newer method of quenching has been developed, practiced and proven to not only reduce heat treatment distortion, it can nearly eliminate it while providing precise repeatability for each component processed. This method is called 4 Dimensional Quenching (4DQ) where it utilizes high pressure gas as the quenching media with precise control of gas flow across the component. 4DQ provides the ability to fully control and/or adjust the quenching profile during a quenching cycle to optimize the quench and control how the component reacts throughout the full quenching process. This process has been proven on aerospace materials such as 9310, 8620, 440C, Ferrium C64 and Pyrowear to name a few.

Introduction

When materials are required to be hardened by quenching, this is a crucial (some would argue most critical) step in the manufacturing chain of a component. It takes place after heating the material to its austenitic range followed by rapid quenching, inducing the transformation of austenite into martensite. Throughout this process, components undergo significant temperature gradients across its geometry. Non-uniform quenching leads to thermal stresses and non-uniform microstructure transformation, resulting in deformations relative to their pre-heat-treated conditions. Common quenching medias include oil, water, salt, air and gas (nitrogen, argon & helium) and in these medias, components can be quenched as part of a larger batch with large thermal mass, a small batch with low thermal mass, or individually with the lowest possible thermal mass. Batch quenching, which is one of the most common methods to heat treat materials, the quenching media disperses unevenly among parts, causing each component to cool at varying rates depending on its position in the batch.

Manufacturers have long sought methods to mitigate distortion while quenching, as correcting distortion proves both difficult and costly. Among these methods, press quenching is a popular technique which quenches individual (or several small) parts at a single time, offering promising distortion control. However, press quenching has drawbacks such as safety concerns related to handling of hot components, environmental impacts due to oil usage/neutralization, and the necessity for specialized equipment and procedures for quenching components post-heat treatment. 4 Dimensional Quench (4DQ), like press quenching, is a technique that quenches and individual (or in some cases several) part(s) at a time isolating the quenching process to a single component while utilizing precise control of high pressure gas quenching to rapidly quench the material converting austenite to martensite and control part distortion.

Difficulties in the Correction of Heat Treatment Distortion

With traditional quench methods, one must engineer, design and manufacture the components so that they can tolerate the geometric change in size after heat treatment. The manufacturer of the component must plan for heat treatment distortion and leave excessive amounts of material on distortion prone geometric features (also referred to as machining stock) so that there is enough remaining material on the component to finish machine the component after heat treatment to the final shape. A manufacturer desires to leave as little machining stock on a component as possible since after the heat treatment hardening process the component is much harder than before heat treatment. Pre heat treatment material is considered to be in a "soft" condition, while after hardening the components' surface can be as high as 60+ HRC when case hardened. Due to these hardened conditions, the machining process requires special machines and tooling (such as surface/gear grinders [Fig. 1], hard mills/lathes [Fig 2]) to properly remove the hardened stock and precisely finish the component to its final dimensions. Therefore, the removal of hardened machine stock is time consuming and costly operation that manufacturers wish to avoid by leaving as little machine stock on surfaces as possible.

Figure 1 – Gear Grinding Machine

Figure 2 – Hard Turning Machine

Traditional Quenching

As mentioned in the introduction, traditional quenching medias include oil, polymer or gas. These medias can be paired with batch or individual quenching methods. In traditional quenching of steels, each process scenario has its positive and negative implications. This section will highlight the challenges associated with the various combinations of these methods.

Liquid Quenching

Liquid or liquid polymer quenching is a quenching method that takes place in separate chamber or tank whereby it receives a heated part or batch and utilizes the fluid to transfer the heat from the component(s) to the fluid and the heated fluid is transferred out of the chamber/tank to a heat exchanger, cooled down and re-circulated back to the component(s). This is the fastest way to remove heat from a component when compared to salt and gas. The speed in which the liquid quenching is good when fast transformation from austenite to martensite is needed, however quenching too fast can lead to parts cracking during the quenching process. An additional factor to consider is that a part is exposed to three (3) separate conditions (or stages) during liquid quenching; Vapor Stage, Boiling Stage, and Convection Stage (Fig 3.).

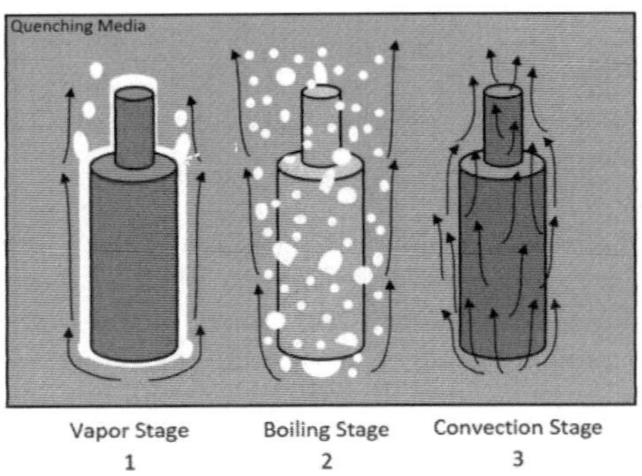

Fig. 3. Stages of Liquid Quenching [1]

The first stage (Vapor Stage), also known as the Leidenfrost phenomenon, generates a vapor blanket around the component thus insulating it from the quenching media. This vapor blanket is generated due to quenching media's inability to remove the high heat from the component, producing the vapor boundary which acts as an insulator during this stage. The first stage (when compared to the second) has a slower heat transfer, as the heat stored within the component must pass through the vapor barrier before it contacts the liquid and can be carried away.

The second stage (Boiling Stage) takes place when the vapor barrier collapses due to the temperature of the components surface lowering enough allowing the fluid to then begin boiling. During this stage, the components heat removal is at its highest rate and is experiencing its highest levels of stress. These internal stresses place great deal of force on the microstructure of the component which in turn subject it to its highest level distortion during the boiling stage.

The third and final stage (Convection Stage) takes place when boiling point is broken due to the temperature of the components surface lowering below the boiling point of the quenching media. Conduction and convection heat removal is now the mode heat transfer to the media and is the slowest of the previous two stages.

Two common liquid medias utilized in quenching are oil and water. Both provide the fastest quenching methods (water being the faster of the two) and in most all cases can fully transform a component when quenched. However, in addition to the distortion that occurs during the 3 stages of quenching, there are additional negative factors to consider when using these methods.

Oil quenching poses environmental concerns due to oil disposal and requires post-quenching cleaning procedures to remove residual oil. Post cleaning equipment is costly equipment and consume large (and valuable) manufacturing floor space. EPA regulations are also becoming more and more costly as government regulations require removal of the oil to be conducted in a safe manner. Use of oil quenching has safety risk to consider, if not properly maintained, a batch oil quenching furnace can catch fire. There are countless manufacturing facilities that have experienced an oil fire and either have had significant damages or in some cases had total losses. In addition to the risk to the operators and facility, insurance costs are greatly increased when oil is utilized for the quenching applications

Water quenching is commonly used for achieving high hardness in steel. However, as it is one of the fastest quenching methods, water quenching can induce significant thermal shock, leading to cracking and distortion. Using water for quenching requires careful control of cooling rates and designing of the part geometry.

Salt Quenching

Involves the heat treatment of immersing steel parts in molten salt during heating which rapidly heat components, providing one of the best heat transfer rates (conduction) compared to other heating types that rely on convection or radiation. During the quenching process, the parts are removed from the high heat molten salt bath and transferred to the lower temperature bath (which can range anywhere from 300°F [150°C] to 650°F [350°C], depending on desired metallurgical transformation desired). Quenching in salt only has 1 phase of quenching as the salt bath is set at the desired temperature before the components are submerged. The salt doesn't create a vapor barrier or boil, rather it relies on conduction as the heat transfer method and with proper agitation and natural buoyancy of the salt it can provide reduced distortion as compared to liquid.

Disadvantages to salt heat treatment do exist, regarding removal of the components, they require cleaning, and the salt can get become lodged in small holes within the components. Due to the molten salt exposure to the working environment, operators must take great care as to not overexpose themselves to the radiation they emit to the surrounding work environment. Additional care must be taken as the molten salt bath can explode if operated above 650°C and salt bath fires are difficult to subdue and extinguish. Salt can release cyanide gas so care must be considered in the surrounding working environment. Due to the reactions with salt, parts and fixturing must be free of contaminants as they will burn and can cause a fire. If/when a salt cools, it solidifies and when it reheated to be put back into service, it can be difficult to know if water has infiltrated the bath or tank, great care must be taken when heating the solidified salt to avoid a steam explosion.

Air Quenching

Some highly alloyed steels can be quenched with simple air as the quenching media. It is recommended to place the materials in a furnace to maintain improved control over the air cooling process. Advantages in air quenching include the ability to control the cooling speed for flexibility over the end hardness. When quenching in air, the pressure and exposure of the air flow, can be adjusted giving the operator control of the rate at which the components cools. This flexibility can be important in controlling toughness and durability. Air, like salt, provides one phase of quenching as there are no vapor or boiling phase reactions with the heated components. An additional benefits of air quenching is the cost, since air is free. Since air is a less aggressive quenching media, it promotes less distortion than free liquid quenching methods.

Disadvantages of air quenching can include cracking of certain materials and if a batch is removed from the heating chamber/source and left out in ambient air, this is the least controlled method of air quenching. Air also relative to oil, water and salt, has a low heat transfer coefficient and cannot through harden low alloy steel grades.

Intergranular Oxidation (IGO)

Both oil, water and salt quenching are widely used quenching methods heat treatment and are commonly paired with the atmospheric heat treatment furnaces. A disadvantage when processing components in an atmospheric constructed furnace, components will experience intergranular oxidation (IGO) due to the presence of oxygen within the furnace when at elevated temperatures. IGO is a negative phenomenon that occurs when oxygen atoms diffuse into the steel's surface reacting chemically with alloying elements. IGO when looked at under a microscope are tiny cracks (up to 0.007" deep) in a components surface. If IGO is not removed initially by vacuum heat treatment processing, it must be removed by post heat treatment machining. IGO negatively impacts the wear resistance, fatigue resistance and hardness of the treated surface [2]. Vacuum furnaces on the other hand, do not allow the presence of oxygen within its construction when operating, so it is worth noting that IGO is not introduced the components when heat treatment is conducted in a vacuum furnace.

Traditional High-Pressure Gas Quenching (HPGQ)

As opposed to air quenching which can be considered gas quenching, use of inert gasses such as nitrogen, argon, helium and in some special cases hydrogen can be used a quenching media. These gases are utilized in combination with batch vacuum furnaces. In a vacuum furnace, the chamber is manufactured to be vacuum tight where it does not allow oxygen into the chamber, so oxygen does not react with the components when heating. These chambers are also designed to handle backfilling of inert gasses when quenching is required. Backfilling of the chamber for quenching can be anywhere from 15 - 375 psi [2 - 25 bar abs.]. As the quenching pressure increases or gas type changes, the speed at which the heat transfer increases as well, reference [Fig. 4] that plots gas type and pressure vs. oil quenching.

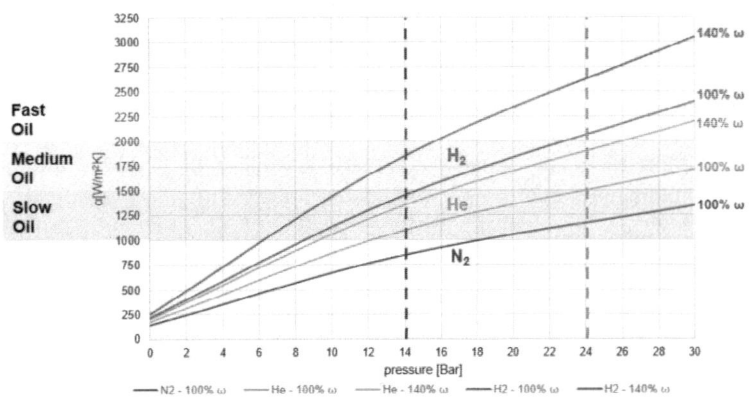

Figure 4 – Batch HPGQ vs. Batch Oil Quenching

Inert gas, like air, is a less aggressive quenching media, has one phase of quenching, and promotes less distortion than free liquid quenching methods. HPGQ can be controlled from the pressure, gas velocity and can be interrupted (isothermally) to allow core and surface to equalize before martensitic transformation.

Additional benefits to HPGQ in a vacuum furnace are:

1. High quality (clean and bright) parts after heat treatment
2. Very good process repeatability
3. No IGO
4. Full automation of the thermal processes
5. Easy process set up
6. On/Off furnace operation

7. Full SCADA system
8. Furnace flexibility with multiple processes
9. Compact Design
10. Safe and environmentally friendlily operation

Typical disadvantages to HPGQ are its limitations in quench speed as in most cases, it cannot quench as fast as oil and water which limits it ability to process certain materials and thick cross sections. Vacuum furnaces are more costly to produce than traditional atmospheric furnaces. Most vacuum furnaces must process parts in a batch and the disadvantages to batch processing are outlined in the batch quenching section.

Batch quenching

Batch quenching involves placing multiple components in a batch with a typical goal of optimizing the maximum number of components that can be processed in a furnace's allowable working volume. During heat ramp and soaking, each part within the batch experiences different characteristic based on its location. A part in the center of the batch will absorb heat slower than those that are on the outside of the batch. This requires batches to soak for extended period of time subjecting each piece to different temperature durations which introduce different microstructure conditions. When analyzing the gas flow during carburizing, the same phenonomnon occurs as the gas dispersion is not equal throughout the load leading to different absorption rates.

When analyzing both gas absorption with the variation in temperature, case depths can vary drastically as carbon diffusion rates increase when a component is at higher temperatures. Subsequently the opposite effect takes place during quenching as the parts in the center of the batch cool slower than those on the outside which can lead to a variation in hardness values. This inconsistency leads engineers to place wide ranging tolerance for metallurgical properties, including hardenss and case depths so that the variation coming from a full heat treatment batch will enough allowance to accommodate the deviations that take place. Since batch processing is the most accepted form of thermal processing, industry requirements have been forced into accepting case depth variations up to +/-30%.[#][4], [5].

Press Quenching

Individual quenching of a component is another common method quenching utilized for hardening after austenitizing. Press Quenching is one of the most widely used forms of individual quenching and is a specialized process designed to restrain/hold a part as quenching oil flows through/around the dies during the quench process and this restraining of the part during quenching is designed to the part as close to its pre-heat treated dimensions as possible. Common dimensions that require distortion control are runout, parallelism, concentricity, etc. and when press quenching is done in a proper manner, precise tolerances can be achieved, 0.001" – 0.002" [25 – 50 μm], in relation to the pre heat treatment dimensions. Description of a 4 stage Press Quench machine (Fig. 5)

Pos. 1 – Hot loading on lower die
Pos. 2 – Upper die compression and oil flow
Pos. 3 – Secondary (free) quench position
Pos. 4 – Discharge tank with conveyer for final quench

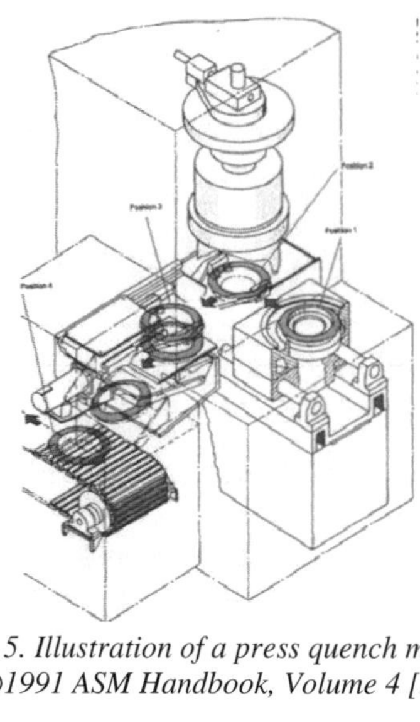

Figure 5. Illustration of a press quench machine, ©1991 ASM Handbook, Volume 4 [3]

Once the press quenching process is optimized, it proves to be an effective way to reduce distortion. A crucial detail to this method is the design, manufacturing and implementation of the special dies. These dies must be implements balancing the need for die contact and proper oil flow over with the part leading to proper microstructure, hardness all with minimal distortion. However, press quenching it not an easy process to perfect and can be quite costly. The first disadvantage to press quenching to note is that if the material is not being handled by specialized (and costly) automation equipment, the operators are subjected to the exposure to the austenitizing furnace which can be as high as 1750°F [950°C] when removing a component to be placed in the press quench. Additional safety concerns are the potential risk for fires as the surrounding areas (covered with oil) ignite causing the oil to combust. Regarding the retention of the component within the press, if it is not secured in the proper manner, it can be subject to severe distortion and even rupturing.

Each quenching method/combination discussed offers unique advantages and challenges, requiring careful consideration of factors such as material properties, distortion control, equipment requirements, and environmental impact in selecting the most suitable approach for specific applications.

4 Dimension Quenching (4DQ) with High-Pressure Gas

4DQ is a relatively new process when compared to oil, water, salt, air and gas as outlined the previous sections. However, this new method of quenching focuses on reducing and/or eliminating quench based distortion by use of special designed high pressure gas quenching chamber [Fig. 6] while quenching one (1) part at a time. The 4DQ system utilizes a cooling manifold [Fig. 7]with a gas flow pattern through specially position cooling nozzles envelopes a component from all sides. In addition to this envelopment, the platform in which the component is placed can be rotated during the quenching process. The combination of the envelopment (3D) plus the rotation of the component (1D) is where 4 Dimensional Quenching derives its name.

Figure 6 – 4DQ/Unloading Chamber

Figure 7 – 4DQ cross section

The dedicated nozzle arrangement that ensures uniform gas quench around the part and when paired up with the rotational feature, this gives engineers the ability to completely "engineer" the quenching profile of a component. The 4DQ chamber is a stand-alone chamber that does not have any heat source inside of it, so once a component is placed in the chamber, the only heat removal is of the component itself. With the manifold having specifically located nozzles (that direct the gas flow to the desired areas of the component and with rotation) the quenching is uniform/homogeneous where thick and thin cross sections cool at the same rate. This uniformity reduces the internal stresses that take place inside the component when it is rapidly quenching.

4DQ Specification

Utilizing 4DQ, allows for the best possible quench uniformity while providing full control of the quenching process. 4DQ with its ability to isolate the quenching gas to strictly the heated component, only 9 Bar quenching pressure can achieve quenching rates equal to fast oil [Fig. 8]. Since 4DQ has the quenching power equal to oil quenching (without the use of helium), it can stand up to the most difficult alloys with thick cross sections while achieving the desired core hardness.

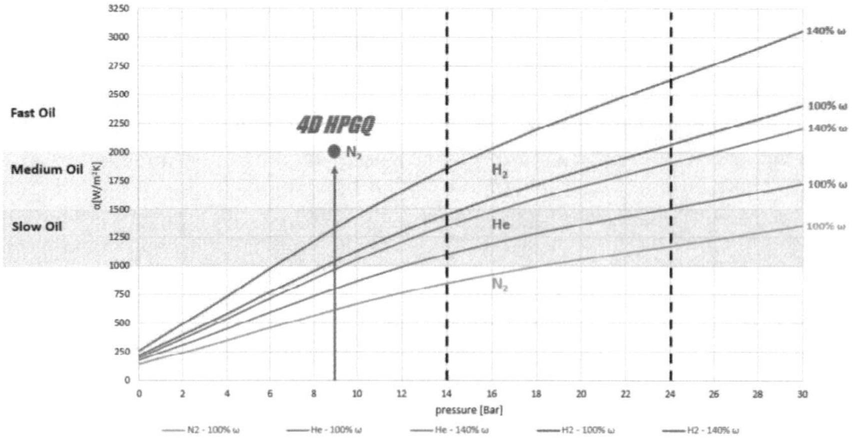

Figure 8 – 4DQ, Batch HPGQ, & Oil - Quench Speed Relationship

4D HPGQ Chamber – Adjustable Features

In addition to the aforementioned features, the following parameters are adjustable and can be varied in their combination:

Feature	**Description**
Quench Pressure	1 to 15 Bar abs. gas pressure
Gas Velocity	Cooling blower RPM control
Table Rotation	ON or OFF with RPM and directional adjustment
Table Oscillation	Angle or direction adjustable
Time Dependent Gas Flow	Controlled via time in seconds

An additional bonus to the quenching chamber's features, the process is completely observable and recordable for documentation of the quenching time. A sight port is placed on the chamber and has a camera mounted to record the 4DQ process. As shown in [Fig. 9], snapshots of a video pictured showing the transformation of a helix gear from its red hot color @ 1550°F [845°C] to its black color ~930°F [500°C] in 9 seconds.

Due to the control and adjustment of the 4DQ systems features, one can now use the quenching system as a tool to control the quenching process in ways that have never been available before. Engineers/manufactures can evaluate the quenching profile before a component is produced to create a quenching profile that is optimum for a particular part, and if needed it can further be optimized once that component has been manufactured with small (and very inexpensive) adjustments to the 4DQ system.

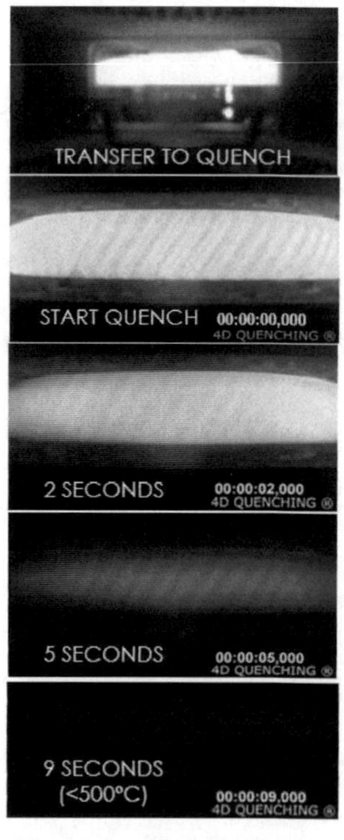

Figure 9 – 4DQ live Images

Engineering and modeling the 4DQ process

With 4DQ, one can engineer the quenching process, such that by Computational Fluid Dynamics (CFD), heat removal rates can be defined in specific sections of a components geometry so that the component quenches at the same rate for all geometric features. For example, cooling a thin cross section slower and a thicker cross section faster. When geometric features are cooled equally, they progress through their phase transformations at the same time. When phase transformations occur at different times, the part experiences different volumetric and strain values which causes a part to distort. In the CFD example below [Fig. 10] it shows a thin cross section component and the temperature gradient it can experience as it progresses through the quenching process. When the component is separated into sections, the 4DQ manifold can be engineered to produce more or less gas flow to the specific sections so they cool equally and therefore progress through their phase transformations at the same time. Therefore, by cooling the part uniformly during the quenching process with 4DQ, you can keep the part from experiencing significant amounts of volume expansion and strain wich keeps the part from experiencing distortion after the quenching process. Modeling of the quenching process can prove to be very beneficial when producing complex geometries that are prone to quenching distortion. These materials can include 9310, 8620, 440C, Ferrium C64 and Pyrowear which are expensive to produce and are very sensitive to heat treatment distortion. When these materials are called upon for service in the machines they operate in, it is beneficial to model their response to thermal distortion during heat treatment so that the necessary quenching method can be chosen so that the quenching method produces successful (and cost effective) post heat treatment machining activities.

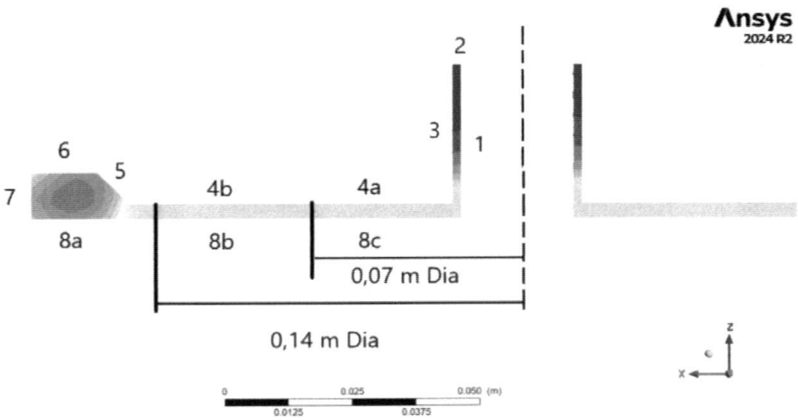

Figure 10 – CFD temperature gradient simulation

Case Study: Application of 4DQ in Hypoid Gears and Coupling Sleeves

A case study on hypoid hears and coupling sleeves will demonstrate the effectiveness of the 4DQ system in minimizing distortion and achieving dimensional consistency. Results illustrate the system's advantages over traditional quenching methods in terms of quality, repeatability, and cost-effectiveness.

Spiral Bevel Gear

In this section, we will review data from two (2) Case studies examples. In example #1 the 4D HPGQ process was completed on a series of spiral bevel gears (Fig. 11) with a 7.9" [200 mm] outside diameter. The requirements were to carburize and quenched the gears to obtain an effective case depth of 0.032" [0.8 mm] @ 550 HV and temper at 356°F [180°C].

Spiral Bevel Ring Gear Distortion Summary	
Feature	**deviation above mean**
Flatness	5 – 25 μm
Axial Runout	20 – 65 μm
Ovality	10 – 38 μm

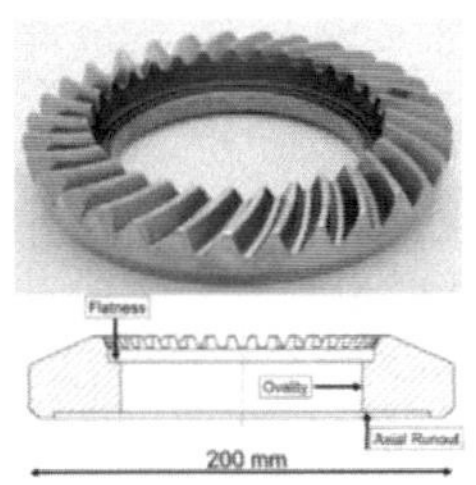

Figure 11 – Spiral Bevel Gear

Coupling Sleeve

In example #2 the 4D HPGQ process was completed on a coupling sleeve with internal teeth (Fig 12) to compare distortion between pre and post heat treatment conditions (Fig. 12). Ring Gear dimensions included a 6.3" [160 mm] outside diameter.

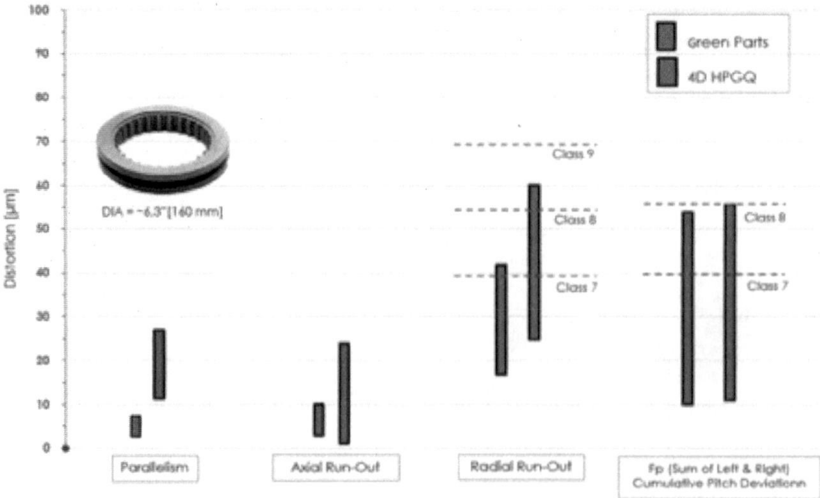

Figure 12 – Coupling Sleeve - Internal Teeth Distortion

In the data below the 4DQ process was optimized and tests were conducted on the coupling gear (fig. 13) made from 8620 with also with a tooth model greater than 3. The distortion was compared distortion between pre and post heat treatment conditions. The coupling gear's dimensions included a 6.0" [152 mm] outside diameter. Associated dimensions measured was radial runout, axial runout, total pitch deviation, total profile deviation and total helix deviation (fig. 14.a to 14.e).

Figure 13 – Coupling Gear

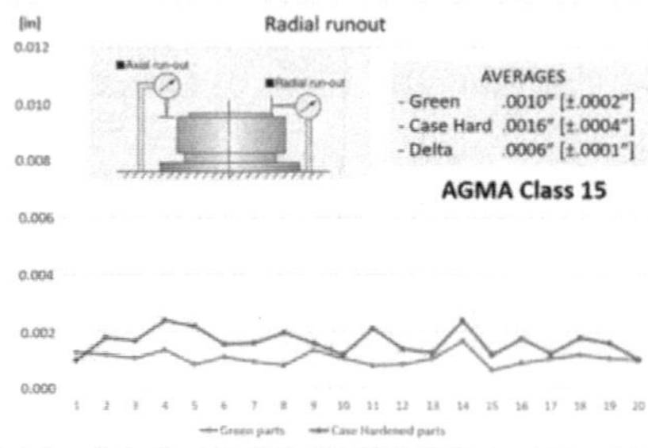

Figure 14.a – Radial Runout

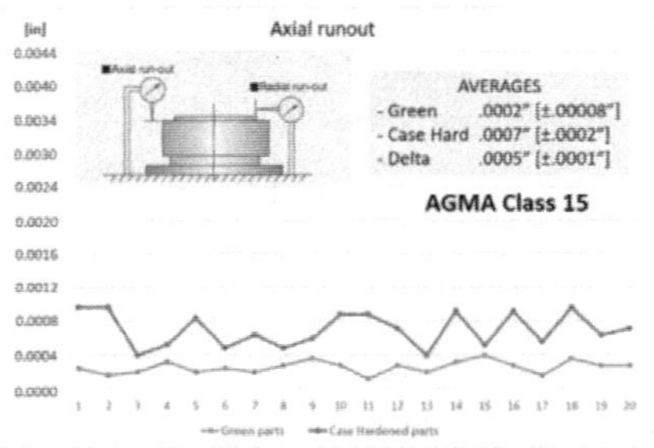

Figure 14.b – Axial Runout

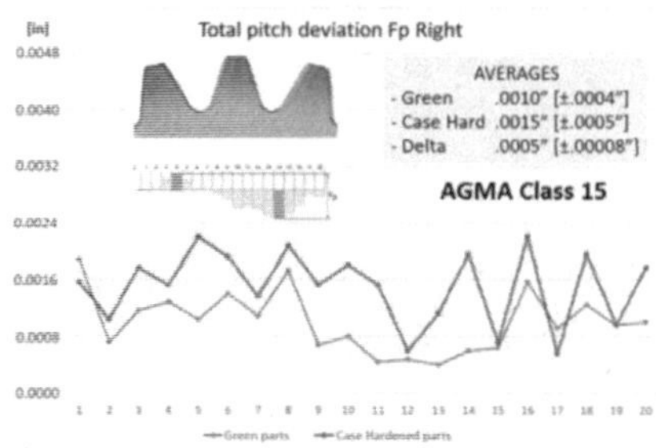

Figure 14.c – Total Pitch Deviation

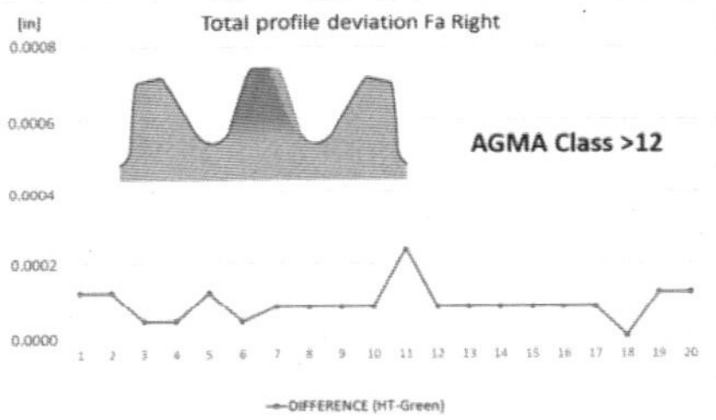

Figure 14.d – Total Profile Deviation

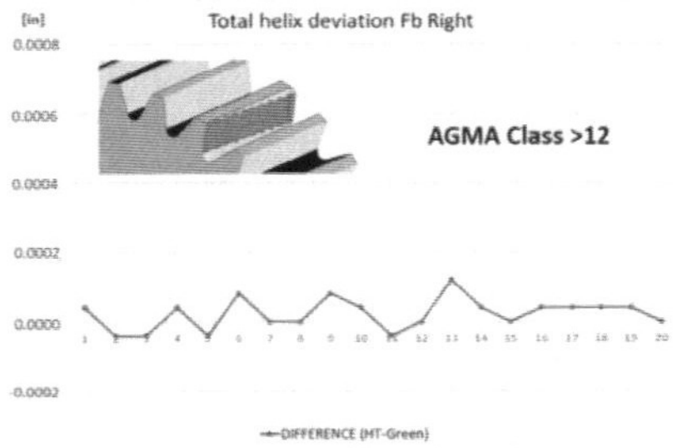

Figure 14.e – Radial Runout

In this trial, it shows that 4D HPGQ can maintain a 12 or higher AGMA Gear class quality rating which is on the same level as press quenching. In general, the 4D HPGQ process proved to that it can produce significantly improved distortion results from the hardening process in comparison to its pre heat treatment features. Comparatively speaking, the UCM when configured to its optimum quenching capacity, it can provide x5 the improvement in distortion to that of batch oil quenching. When distortion is reduced, it flows downstream to the subsequent machining operations. Whereas machining and grinding steps can be reduced and/or even eliminated altogether.

Conclusion

Costs of post heat treatment material removal are significant to the overall cost of a finished component, from costly machines, special tooling and skilled labor. It has been estimated in the mid 1990's that that in the automotive industry alone spends ~ 22.4 billion USD [4] on the correction of distortion to hardened components. 4DQ is a process that gives the end user a tool to not only reduce the distortion of quenched component, but they can also fully optimize the quench process due to the systems flexibility and customization. In addition to the distortion benefits, the system is environmentally friendly, safe to operate and can be implemented into a machining work cell

Acknowledgments

[1] Herring, D. H., Quenching Webinar, Industrial Heating, September 2010.
[2] Pye, D., What is Intergranular Oxidation?, Industrial Heating, October 2009.
[3] ASM Handbook, Volume 4, Heat Treating, ASM International, Quenching of Steel, Pg. 108.
[4] Bremen, 1995, "Economic consequences of distortion." IWT (Institut fur Werkstofftechnic), Germany

QDE 2025: Proceedings of the 3rd International Conference on
Quenching and Distortion Engineering
May 6–7, 2025, Vancouver, Canada
https://doi.org/10.31399/asm.cp.qde2025p0192

Comparison of Property Determination Methods

D. Scott MacKenzie, Sergio Gallegos
Quaker Houghton Inc., Conshohocken, PA, USA
scott.mackenzie@quakerhoughton.com

Abstract

The ability to accurately predict the properties of a heat treated part has been the Holy Grail of the industry. There are many methods and software packages that can predict hardness and properties of a heat treated part.

In this paper, we will compare several techniques of hardness prediction, from the very simple, to complex modeling, and compare the calculated results to those that have been published previously. Using "old-school" methods based on the Grossman H-Value, and Lamont charts, we predict the expected hardness for two different alloys (SAE 1045 and SAE 6140), in three different geometries - 1" (25mm), 3" (75mm), and 5" (125mm) round bars. Water is used as a quenchant. The resulting data is compared to that previously published by Bain and Grossman. Additional comparisons are made using different methods and software packages. Again, this data is compared to previously published data.

Introduction

In Grossman's Principles of Heat treatment [1] in Chapter 11, on "Variations of Hardening", he described a test, using different round bar diameters ranging from 0.5" (12.5mm) to 5" (125mm) of SAE 1045 and SAE 6140 steel. The composition of the bars is shown in Table 1. These bars were quenched in water. After austenitizing, and quenched in water, the bars were sectioned in half, perpendicular to the long axis of the bar. The sectioned surface was ground smooth, and Rockwell C hardnesses were taken at different radial distances. Several hardnesses were taken at each radial distance, and the data was plotted to create a "hardness penetration diagram."

Table 1. Steel compositions reported by Grossman [1] for the creation of the "Hardness Penetration Diagram."

Steel	C	Mn	P	S	Si	Cr	V
SAE 1045	0.48	0.6	0.022	0.016	0.17	...	...
SAE 6140	0.42	0.73	0.027	0.023	0.25	0.94	0.17

The results of the hardnesses at the surface, after quenching in water, are shown in Table 2. The complete hardness penetration diagram for each of the steels is shown in Figure 1.

Diameter		Surface Hardness (HRC)	
(in)	(mm)	SAE 1045	SAE 6140
0.5	12.5	59	59
1.0	25.0	58	59
2.0	50.0	41	55
3.0	75.0	35	53
4.0	100.0	30	50
5.0	125.0	24	48

Table 2. Measured [1] hardnesses at the surface of different diameter steel rounds of different composition.

The goal of the comparison was to show the effect of alloying elements on the depth of hardening. This concept of depth of hardening, and not the absolute hardness, is the basis for the modern concept of hardenability.

The progression of hardnesses from the exterior of the bar to the interior showed that the highest hardnesses were transformed to martensite. The intermediate hardnesses were a mixture of bainite, martensite and fine pearlite. The lowest hardnesses were pearlite of progressively coarser structures, with corresponding lower hardnesses.

This comparison between two different alloys was to demonstrate differences in hardenability between different steel grades. It wasn't until much later that the concept of quantifiable hardenability [2] [3] [4] was developed.

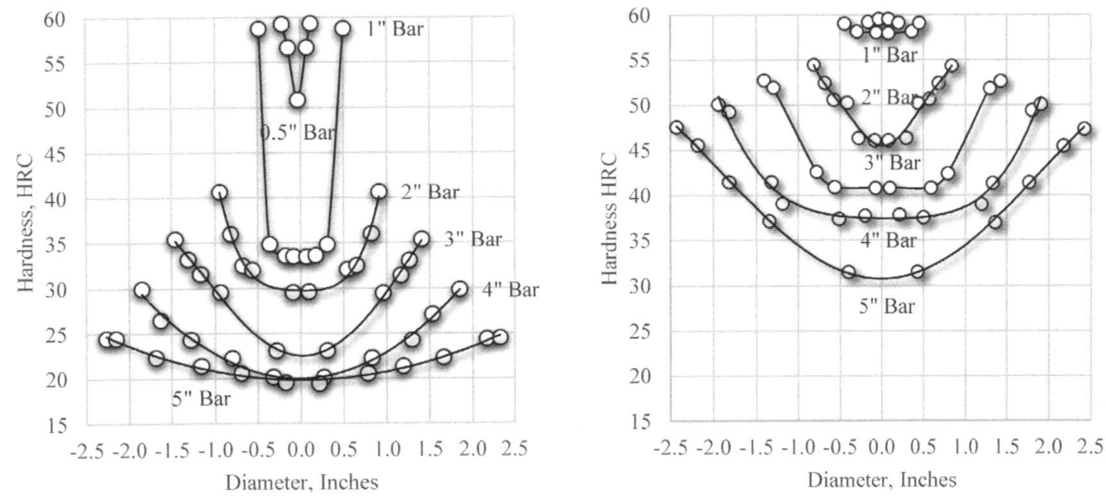

Figure 1. Hardness penetration curves for SAE 1045 (left) and SAE 6140 (right) [1].

The goal of this paper is to try and predict the hardness profile for these two alloys using several different methods, from the very basic "old-school" method to more modern modeling methods.

Quenchant Used

In Grossman's original paper, he described a water quench. However, the paper did not describe the agitation used, or the temperature of the water used. However, the inference is that the parts were quenched in room temperature water, with no agitation, other than light swirling by hand based on an earlier description. Based on this, the use of unagitated water was selected. The cooling curve of unagitated tap water at 25°C is shown in Figure 2. Using RISE SmartQuench Integra [5], the heat transfer coefficient as a function of temperature was calculated using an inverse method [6]. This is shown in Figure 3.

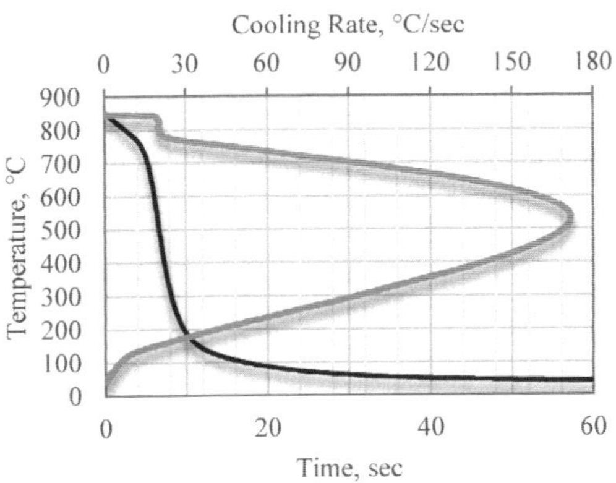

Figure 2. Cooling curve of tap water at 25°C, with no agitation. Curve measured using 12.5mm Inconel probe per ASTM D6200.

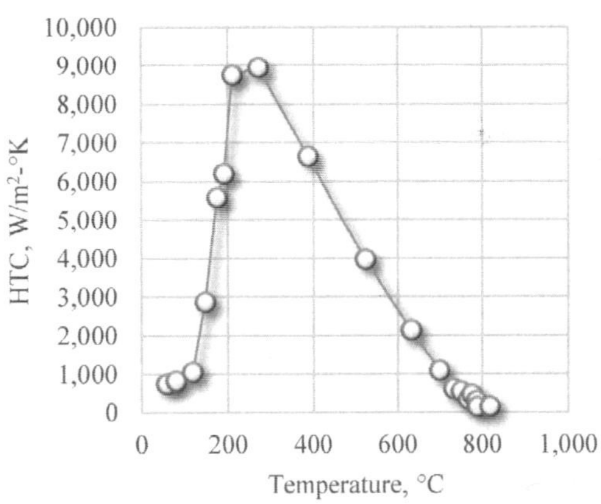

Figure 3. Heat transfer coefficient of unagitated water at 25°C, based on the cooling curve in Figure 2.

Alloy Physical Properties

For the different methods, different physical properties are required. Specifically, the Jominy end quench data is required, as well as the thermal conductivity and specific heat. For the compositions in Table 1, the physical properties were calculated using JMatPro [7]. The calculated Jominy data is shown in Figure 4, while the thermal properties are shown in Figure 5.

The Jominy curves calculated were verified against the published upper and lower bounds for each of the selected alloys [8], and found to be within the expected hardenability bands.

As expected, the hardenability of SAE 6140 is much greater than that of SAE 1045, confirming the depth of hardness penetration diagrams of Figure 1.

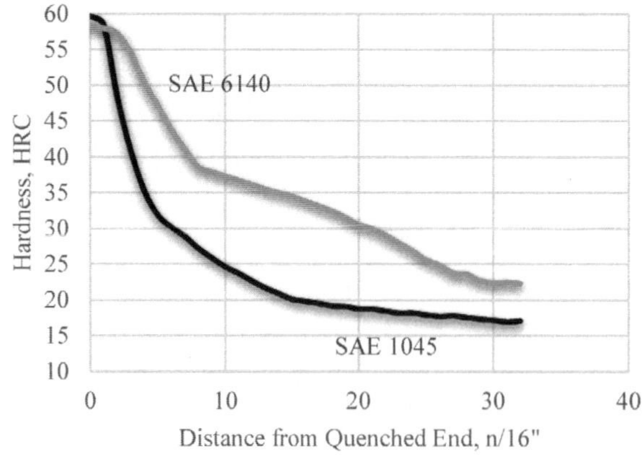

Figure 4. JMatPro calculated Jominy End Quench curves based on the compositions in Table 1.

The thermal conductivity (W/m-°K) of a material is a measure of the material's ability to conduct heat. If the thermal conductivity of a material is low, then the conduction of heat is slow. While the thermal conductivity of the two alloys is similar, the SAE 1045 shows a slightly higher thermal conductivity than the SAE 6140 due to the lack of alloying elements such as chromium.

The specific heat capacity (J/g-°K) of a material is the amount of heat needed to raise the temperature of a mass of material. In this case, it takes some number of Joules to raise one gram of material one degree Kelvin. The specific heat capacity varies with temperature. While the two alloys show similar heat capacities as a function of temperature, the heat of transformation from austenite to ferrite/pearlite is higher for SAE 1045 than SAE 6140.

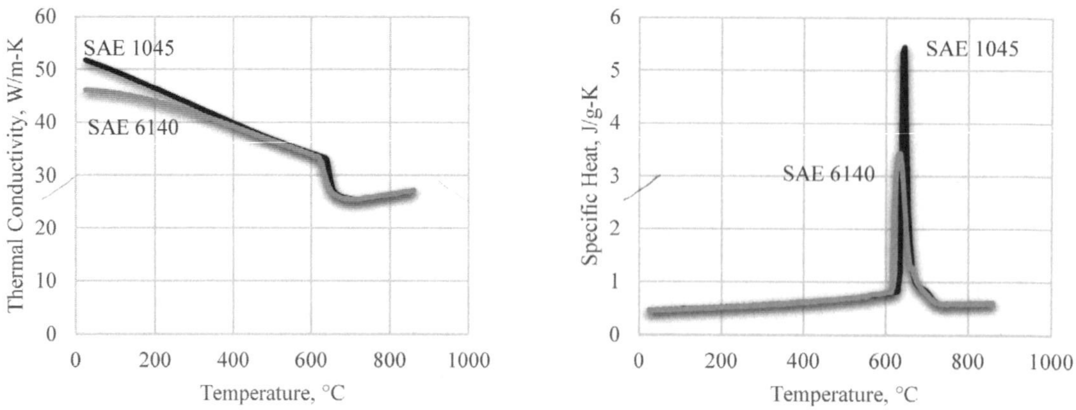

Figure 5. Thermal properties of SAE 1045 and SAE 6140, as calculated by JMatPro.

Parts Used

In this investigation, round bars of the different alloys were used that had diameters of 1" (25mm), 3" (75mm) and 5" (125mm). The length of the parts were seven times the diameter to eliminate end effects. This configuration provides heat extraction in the radial direction, with negligible heat transfer in the axial direction.

Investigation

As was stated earlier, the goal of this paper was to try and predict the hardness profile of different diameter bars and compare the results with the data set from Grossman (Figure 1). Three methods were used. The first method is based on the Grossman H-Value and Lamont charts [4] [9]. The second method uses SolidWorks [10] to perform a thermal simulation, and time-temperature curve. This time-temperature curve is then used to predict properties using JMatPro. The resulting properties are then compared to the reference data set.

Method I – Grossman H-Value

Historically, the Grossman H-Value has been used to describe the severity of a quenchant [9] [11]. It is a ratio of the heat transfer at the surface, divided by the conduction of heat on the interior of the part:

$$H = \frac{h}{2k}$$

Where h is the surface heat transfer coefficient (usually defined at 705°C or 1300°F) and k is the thermal conductivity. In English units, the H-value has units in^{-1}. The thermal conductivity of steel does not change appreciably over temperature, or from one alloy grade to another, the Grossman H-Value is approximately proportional to the heat transfer coefficient. Table 3. Severity of Quench (H) for various quenching media and agitation [12]. shows traditional values of the Grossman H-Value for different quenchants [12].

Table 3. Severity of Quench (H) for various quenching media and agitation [12].

Agitation of Fluid or Piece	Brine		Water		Oil	
	in.$^{-1}$	mm^{-1}	in.$^{-1}$	mm^{-1}	in.$^{-1}$	mm^{-1}
No Circulation	2.00	0.079	0.90	0.035	0.25	0.010
			1.00	0.039	0.30	0.012
Mild circulation or agitation	2.20	0.086	1.00	0.039	0.30	0.012
			1.10	0.043	0.35	0.014
Moderate circulation			1.20	0.047	0.35	0.014
			1.30	0.051	0.40	0.016
			1.40	0.055		
Good Agitation			1.50	0.059	0.40	0.016
			1.60	0.063	0.50	0.020
Strong Agitation			2.00	0.079	0.80	0.031
Violent Agitation	5.00	0.200	4.00	0.157	1.10	0.043

Although this method has been used in the industry for many years, it is not without problems. The biggest issue with the application of the Grossman H-value is the difficulty in quantifying agitation rates. The terms describing the agitation are not quantifiable and can result in errors. There is really no understanding of what is meant by "mild" or "violent" agitation.

Previously, there has not been a way of determining the severity of quench (H) from cooling curve data. Recently, a correlation of the Grossman H-Value to cooling curve data was presented [13]. In this method, the average heat transfer coefficient at 705°C was correlated to the Grossman H-Value. This method is based on the methodology of Kobasko [14] [15]. In this method, the dimensionless Biot number is used:

$$Bi = \frac{hL}{k}$$

Where h is the heat transfer coefficient; k is the thermal conductivity; and L is the characteristic length. In this instance, the measuring device is the ISO 9950 [16] or ASTM D6200 [17]. In each case, this is a 12.5 mm diameter probe composed on Inconel 600. For an infinite cylinder (L > 7D), the characteristic length, $L = 0.346r$ *(r is the radius of the probe or 12.5 mm for the ISO 9950 probe)*.

The thermal conductivity of Inconel 600 can be determined from the literature, so that the only remaining variable to determine is the heat transfer coefficient. The methodology to determine this is provided in [14] [15]. From [18], the H-Value as a function of cooling rate at 1300°F (705°C) has been determined (Figure 6). From the cooling curve in Figure 2, the cooling rate at 705°C, is 89°C/s, or 160°F/s. Examining Figure 6, the resulting H-Value is 0.85 in^{-1}. While the quench severity is slightly less than the traditional H = 1.0 for unagitated water, it was not thought to affect results significantly.

Once the H-Value has been determined, the hardness as a function of position can be estimated. This is accomplished using the Lamont charts [9] [19]. Lamont charts can be found in different locations, such as the Timken Practical Guide for Metallurgists [8]. These charts are used to estimate the approximate cross section hardness profile of quenched round bars based using Jominy End Quench data. An example of a Lamont chart is shown in Figure 7.

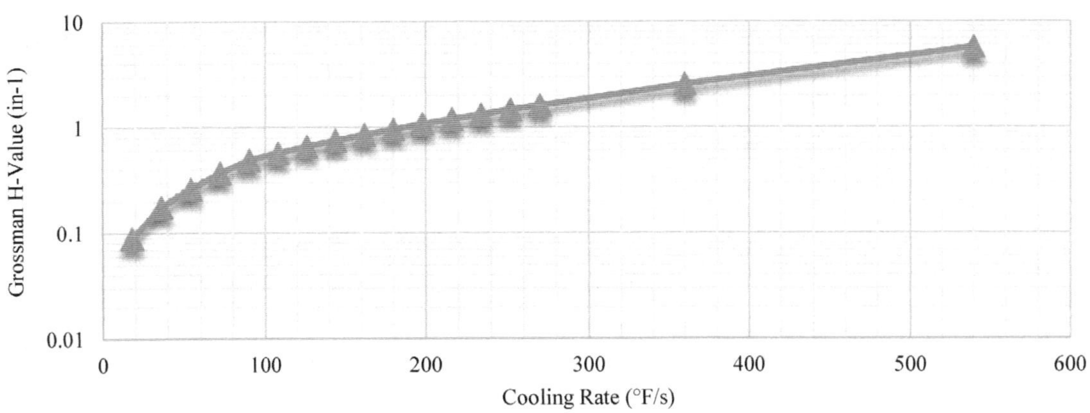

Figure 6. Correlation of the cooling rate at 1300°F and the Grossman H-Value [18].

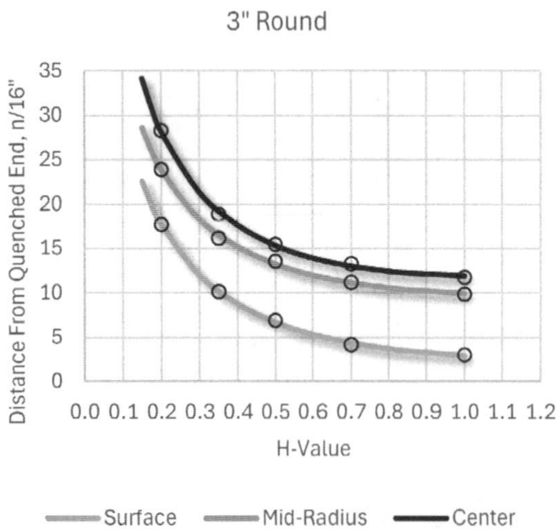

Figure 7. Example of Lamont [9] chart for a 3" (75mm) round bar. Data taken from [8].

Using the determined H-Value from the cooling curve and examining the Lamont charts for a 1" (25mm), 3" (75mm) and 5" (125mm) round bar, the Jominy values were extracted. These values are shown in TABLE. The hardness was determined from the Jominy curves in Figure 4 for both SAE 1045 and SAE 6140. The results of the analysis are tabulated in Table 5.

The results for SAE 1045 were mixed. The H-Value method tended to under predict the surface at a bar diameter of 1" (25mm), while over predict the surface hardness of the 5" (125mm) bar. The center (r = 0) values for all diameter bars showed a good correspondence to the measured values. The mid-radius predictions were within 3 HRC for the 3" (75mm) and 5" (125mm) bars, but showed a large difference for the smallest bar, with the H-Value hardness being 7 HRC higher than predicted.

Table 4. Jominy positions from Lamont Charts [9] for different Grossman H-Values.

Jominy Position from Lamont Charts									
H-Value	1.0" Round			3" Round			5.0" Round		
	Surface	Mid-Radius	Center	Surface	Mid-Radius	Center	Surface	Mid-Radius	Center
0.18	8.8	9.1	9.3	19.2	25.3	30.0	N/A	N/A	N/A
0.52	3.3	4.4	5.0	6.3	12.9	14.9	8.7	25.7	34.7
0.84	1.9	3.1	3.7	3.5	10.4	12.3	4.9	20.3	26.3
1.00	1.5	2.7	3.3	2.9	10.0	11.9	3.8	19.0	24.7

The SAE 6140 results showed a better correlation between predicted and measured. In general, the H-Value predictions were conservative, with the predictions being slightly lower than the measured curve. In all cases, the surface hardnesses predicted were within 2 HRC, while the center and mid-radius positions varied between 3-6 HRC softer than measured.

Table 5. Comparison of predicted hardness by H-Value compared to those measured by Grossman [1].

SAE 1045 H-Value Method									
r/R	1" (25mm)			3" (75mm)			5" (125mm)		
	H-Value	Grossman	Δ	H-Value	Grossman	Δ	H-Value	Grossman	Δ
1.0	48.3	58.7	10.4	37.5	35.5	-2.0	31.8	24.5	-7.28
0.5	40.6	33.5	-7.08	23.2	26.4	3.2	18.8	21.3	2.55
0.0	36.5	33.5	-3	22.5	22.5	0.0	17.7	19.5	1.8

SAE 6140 H-Value Method									
r/R	1" (25mm)			3" (75mm)			5" (125mm)		
	H-Value	Grossman	Δ	H-Value	Grossman	Δ	H-Value	Grossman	Δ
1.0	57.2	59.0	1.8	52.0	51.9	-0.1	47.0	47.3	0.3
0.5	54.5	58.0	3.5	37.0	42.6	5.6	30.6	36.5	5.9
0.0	51.2	58.0	6.8	35.7	40.8	5.1	24.8	31.2	6.4

Method II – SolidWorks™ [10] and JMatPro™ [7]

In this method, a 3D solid was created for each of the three round bars. The dimensions of the bars were the same as above, but with a length of seven times the diameter to minimize end effects. Physical properties were obtained from JMatPro™ (Figure 5) and incorporated into the solid. The heat transfer boundary conditions were the heat transfer coefficients from the unagitated water and calculated by SQ Integra [5] as shown in Figure 3. The solid was meshed and the thermal model solved.

The resulting time versus temperature curves, from an austenitizing temperature of 860°C, were then input into JMatPro, and hardness and microstructure predicted for each of the three bars, at the surface, mid-radius and center. The results are shown in Table 6.

For SAE 1045, the hardness predicted using SolidWorks and JmatPro for the 1" (25mm) bar were substantially harder at the mid-radius and center, than those measured by Grossman [1]. The surface hardness was under-predicted by a large margin. For the larger diameter bars, the predicted values were much closer to those measured by Grossman [1]. However, the predicted surface hardness was overly optimistic, with the predicted hardness much higher than those measured. The center and mid-radius values were very close to those measured. Since the predicted values were close to the measured values when the microstructure would be predominately pearlite and ferrite, it is possible that the thermal conductivity and specific heat is different for different microstructures or quench rates.

For SAE 6140, the hardnesses predicted matched those measured within 3 HRC for all positions and diameters. Overall, use of SolidWorks and JmatPro, at least for SAE 6140, is acceptable.

Table 6. Comparison of hardness values predicted by a thermal model utilizing SolidWorks™ and JMatPro™ (SW) compared to those measured by Grossman [1].

SAE 1045 SolidWorks and JMatPro									
r/R	1" (25mm)			3" (75mm)			5" (125mm)		
	SW	Grossman	Δ	SW	Grossman	Δ	SW	Grossman	Δ
1.0	51.6	58.7	7.1	39.9	35.5	-4.4	35.3	24.5	-10.8
0.5	50.7	33.5	-17.2	26.1	26.4	0.3	20.6	21.3	0.8
0.0	50.6	33.5	-17.1	25.4	22.5	-2.9	19.4	19.5	0.1

SAE 6140									
r/R	1" (25mm)			3" (75mm)			5" (125mm)		
	SW	Grossman	Δ	SW	Grossman	Δ	SW	Grossman	Δ
1.0	57.6	59.0	1.4	52.6	51.9	-0.7	49.1	47.3	-1.8
0.5	57.6	58.0	0.4	44.4	42.6	-1.8	34.9	36.5	1.6
0.0	57.6	58.0	0.5	44.2	40.8	-3.4	34.2	31.2	-3.0

Method III – Transvalor SIMHEAT™

In this method, the software package SIMHEAT was used to predict the hardness of the three bar diameters. The heat transfer boundary condition was the same as in Method II (Figure 3), and material properties from JMatPro™ were used for SAE 1045 and SAE 6140, of the chemistry originally reported by Grossman [1]. The Model II bar geometry was used in Method III. The meshing conditions are found in Table 1.

Table 7. Meshing details of Transvalor SIMHEAT™ simulation.

Mesh	1" (25mm) bar	3" (75mm) Bar	5" (125mm) Bar
Nodes	25465	50469	130866
Triangles	15576	18738	30626
Tetrahedrons	127348	268373	722318

The results of the simulation for the SAE 1045 and SAE 6140 are provided in Table 8.

The results of the simulation of SAE 1045 were mixed. For the 1" (25mm) bar, SIMHEAT correctly predicted the surface hardness of the bar. However, for the mid-radius and center, there was a difference of approximately 7.5 HRC, with the model predicting a harder center and mid-radius hardness than measured by Grossman. However, the software properly predicted the decrease in hardness at the center of the 1" (25mm) bar but was optimistic on the hardness.

For the 3" (75mm) bar, the results closely matched the center and mid-radius hardness of the reference data set. The predicted surface hardness was 4.5 HRC harder than the reference set. Overall, there was good correspondence with the measured and predicted hardness.

For the 5" (125mm) diameter bar, the surface hardness was over predicted, while the center and mid-radius hardnesses were under predicted, being softer than the reference data set.

Overall, the results for the predicted hardnesses for the SAE 6140 were significantly better than the results of the SAE 1045. The 1" (25mm) bar hardness was predicted to be within 2 HRC of the reference data set. This is to be expected with the high hardenability of SAE 6140.

The 3" (75mm) bar showed nearly identical results to the reference data set. The surface hardness was slightly higher at 4HRC than the data set, but the center and mid-radius predicted hardnesses were with 3HRC.

The 5" (125mm) bar predicted the same hardness at the center of the bar as the reference data set. However, the predicted mid-radius hardness was lower than the reference data set by 4HRC. The predicted surface hardness however was significantly higher than the reference data set at nearly 10HRC.

Table 8. Predicted hardness values by Transvalor SIMHEAT compared to reference data set.

SAE 1045 SIMHEAT									
r/R	1" (25mm)			3" (75mm)			5" (125mm)		
	SIMHEAT	Grossman	Δ	SIMHEAT	Grossman	Δ	SIMHEAT	Grossman	Δ
1.0	58.0	58.7	0.7	40.0	35.5	-4.5	41.0	24.5	-16.5
0.5	41.0	33.5	-7.5	25.0	26.4	1.4	16.0	21.3	5.3
0.0	41.0	33.5	-7.5	22.0	22.5	0.5	13.0	19.5	6.5

SAE 6140									
r/R	1" (25mm)			3" (75mm)			5" (125mm)		
	SIMHEAT	Grossman	Δ	SIMHEAT	Grossman	Δ	SIMHEAT	Grossman	Δ
1.0	57.0	59.0	2.0	57.0	52.6	-4.4	57.0	47.3	-9.7
0.5	57.0	58.0	1.0	40.0	42.6	2.6	32.0	36.0	4.0
0.0	57.0	57.0	0.0	38.0	40.8	2.8	32.0	31.2	-0.8

Discussion

Method I – Grossman Quench Severity H-Value

Historically, the Grossman H-Value describing quench severity is cited and used in virtually all texts that discuss hardenability and selection of quenchants. The use of the Grossman H-value has been used successfully for many years for predicting hardnesses of quenched parts. However, one of the main difficulties is inadequately defined agitation rates. There is no quantifiable agitation rate for "mild" or "violent" agitation. Further, the H-Value only refers to the quench severity at 705°C (1300°F). The H-value only considers one point in the entire cooling curve, instead of examining the entire quenchant path [20].

To predict hardness using the Grossman H-Value method and Lamont charts, several things must be known. Prediction of hardness depends on the accuracy of the Jominy End Quench data. This was not provided in the original Grossman reference [1]. Only the chemistry of the alloys was provided (Table 1). Further, the temperature of the quench, or amount of agitation was provided. It was inferred from the text that no agitation was used. Lastly, the grain size of the round bars was also not provided. An assumption that the grain size was an ASTM grain size of 7 was made. However, unless the variation of grain size is large, the grain size effect would be small between a grain size of 6-8.

The largest effect is the accuracy of the Jominy End Quench Data. For this study, the Jominy curve was calculated by JmatPro™, based on the chemistry and assumed grain size of 7. However, other Jominy End Quench calculators are available. One calculator is provided by Ovako [21]. This online calculator takes the chemistry and calculates the Jominy End Quench curve. A comparison of the JMatPro™ and Ovako calculated curves for SAE 1045 and SAE 6140 are provided in Figure 8.

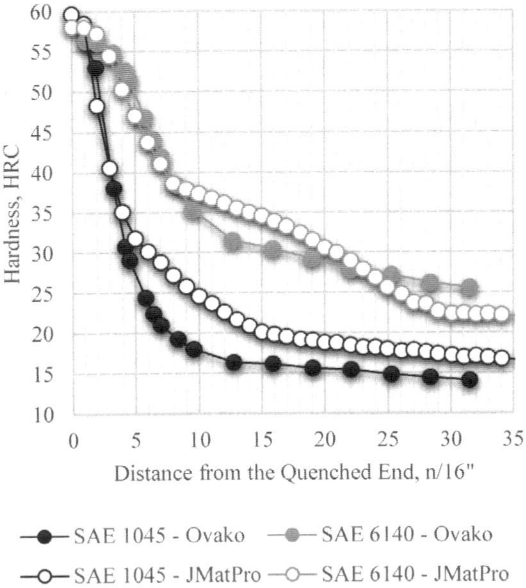

Figure 8. Comparison of the Jominy End Quench curves calculated by JMatPro™ or the Ovako website [21].

Both the Ovako and the JmatPro SAE 1045 curves are within the maximum/minimum range of Jominy data for SAE 1045.

Examining the Jominy curves for SAE 1045, the Ovako prediction shows a lower hardness response to a given Jominy n/16" position than that predicted by JMatPro™ in the range of J4 - J30. This would mean for the center of the 1" (25mm) bar, the center hardnesses would be lower than those predicted by JmatPro™.

The 3" (75mm) SAE 1045 bar would have a similar surface hardness as the reference data set. But the center and mid-radius hardnesses would be lower than the reference data set, and lower than the JmatPro™ predicted values.

The 5" (125mm) SAE 1045 bar would show lower hardnesses at the surface, mid-radius, and center of the bar when using the Ovako data, as opposed to the JmatPro™ data.

For the SAE 6140 results, the JMatPro™ data shows close results to those of the reference data. The Jominy data from Ovako shows a higher hardness in the J3 – J 7 than the JmatPro data, and a lower hardness in the J8 – J24 range. In this case, the JmatPro data better predicted values that compared favorably to the reference data set. Summary results comparing the Ovako and JmatPro results are shown in Table 9.

Table 9. Comparison results from different Jominy data set from Ovako [21] (O) and JMatPro™ (J), referenced against the Grossman [1] (G) reference data set.

SAE 1045												
Position	1" (25mm) Bar				3" (75mm) Bar				5" (125mm) Bar			
	n/16	O	J	G	n/16	O	J	G	n/16	O	J	G
Surface	1.9	56.0	48.3	58.7	3.5	35.0	37.5	35.5	4.9	26.0	31.8	24.5
Mid-Radius	3.1	38.0	40.6	33.5	10.4	17.0	23.2	26.4	20.3	15.5	18.8	21.3
Center	3.7	34.0	36.5	33.5	12.3	16.0	22.5	22.5	26.3	15.0	17.7	19.5
SAE 6140												
Position	1" (25mm) Bar				3" (75mm) Bar				5" (125mm) Bar			
	n/16	O	J	G	n/16	O	J	G	n/16	O	J	G
Surface	1.9	57.2	57.2	59.0	3.5	52.0	51.9	51.9	4.9	47.0	47.3	47.3
Mid-Radius	3.1	54.4	54.5	58.0	10.4	37.0	42.6	42.6	20.3	30.3	36.5	36.5
Center	3.7	52.0	51.2	58.0	12.3	36.0	40.8	40.8	26.3	24.4	31.2	31.2

Method II – SolidWorks™ and JMatPro™

Similar trends between Method I and Method II were observed. For SAE 1045, Method II over predicted the hardness profile of the 1" (25mm) bar. There was a considerable difference between the center hardness of the reference data set and the Method II results. Both Method I and II predicted the hardness profile of the 3" (75mm) bar, with Method II showing slightly overall higher hardnesses, compared to the reference data set. While the 5" (125mm) bar had good correspondence to the center and mid-radius values of the reference data set, the surface hardness predicted was considerably harder.

For the Method II data for SAE 6140, the method predicted the hardnesses at the center, mid-radius, and surface very well for the 1" (25mm) bar with 3HRC for the three positions. The 3" (75mm) bar, showed good correlation to all positions of the reference data within 3HRC, with the prediction at the center to be slightly harder. The profile of the 5" (125mm) bar predicted by Method II followed closely to the reference data set, with the center being predicted to be slightly harder.

Method III – Transvalor SIMHEAT™

In Method II, Transvalor SIMHEAT™ simulated the heat treatment and quenching process. JMatPro™ calculated the thermophysical data. The same thermophysical data set was used in both Method II and Method III.

For the SAE 1045, as the bar diameter increased, there was a greater discrepancy between the reference data set and the surface hardness. In the smallest bar, the center hardness was over predicted, while for the largest bar, the hardness was under predicted for the center. For the 3" (75mm), the center and mid-radius hardnesses were predicted to be close to the reference data set.

In general, for the SAE 6140, Method III tended to over predict the surface hardnesses for the 3" (75mm) and 5" (125mm) bars, while accurately predicting the hardness within 3HRC of the reference data set from Grossman for the mid-radius and center of the bars.

General Comments

The validity of any prediction method is dependent on accurate input data. Often, accurate data is unavailable. In this case, the method of agitation was missing. The grain size of the parts was not provided in the original reference. This results in errors in prediction of quenching response. The specifics of the Jominy End Quench data for the specific alloys were also unavailable. This missing data results in errors in the prediction of the reference set.

Thermophysical data was predicted from chemistry for each of the specific alloys. This is only an estimate and cannot replace actual data. Inaccurate thermal conductivity and specific heat can result in predictions that are either higher or lower than the predicted data. Actual measurements, as a function of the quench rate would likely improve the results of the prediction as compared to the reference data set.

Conclusions

In this paper, three methods were used to predict the hardness variation as function of position in 1" (25mm), 3" (75mm), and 5" (125mm) diameter bars. In general, all three methods predicted the hardness of the high hardenability SAE 6140 satisfactorily. In the case of SAE 1045, much larger deviations from the predicted hardness compared to the reference data set were observed.

One of the most important conclusions that can be made is the importance of accurate thermophysical and metallurgical data. Inaccuracy of the data results in errors in prediction of microstructure, hardness, and in complex models, the distortion and residual stress fields.

Acknowledgments

We would like the thank the Management of Quaker Houghton for permission to present this paper. We would also like to acknowledge the assistance of Transvalor, in helping us to properly set up the models for simulation. We would also like to thank the organizers of this event for accepting this paper for presentation and publication.

References

[1] M. A. Grossman, Principles of Heat Treatment, Metals Park, OH: American Society of Metals, 1937.

[2] W. E. Jominy, Hardenability of Alloy Steels, Cleveland: American Society for Metals, 1939.

[3] J. E. Jominy, "Standardization of Hardenability Tests," *Metals Progress,* vol. 40, pp. 911-914, December 1941.

[4] M. Asimov, W. F. Craig and M. A. Grossman, "Correlation between Jominy Test and Quenched Rounds," *SAE Trans.,* vol. 49, no. 1, pp. 283-292, 1941.

[5] Swerea IVF AB, SQIntegra Product Brochure, Monldal, 2012.

[6] I. Felde and T. Reti, "Evaluation of Cooling Characteristics of Quenchants by Using Inverse Heat Conduction Methods and Property Prediction," *Materials Science Forum,* vol. 659, pp. 153-158, 2010.

[7] Sente Software, "JMatPro," 11 November 2018. [Online]. Available: https://www.sentesoftware.co.uk/.

[8] Timken Company, Timken Practical Guide for Metallurgists, 14th ed., Canton, OH: Timken Company, 2015.

[9] J. L. Lamont, "How to Estimate Hardening Depth in Bars," *Iron Age,* no. 10, pp. 64-70, 1943.

[10] Dassault Systems, 10, Rue Marcel Dassault, 78140 Vélizy-Villacoublay, FRANCE.

[11] M. Grossman, M. Asimow and S. F. Urban, "Hardenability, its Relation to Quenching and Some Qualitative Data in Hardenability of Alloy Steel.," 1938.

[12] B. Liscic, H. M. Tensi, L. C. Canale and G. E. Totten, Eds., Quenching Theory and Technology, Boca Raton, FL: CRC Press, 2010.

[13] D. S. MacKenzie, "Selection of Oil Quenchants for Heat Treating Processes," in *23rd IFHTSE Heat Treatment and Surface Engineering Congress, April 18-21*, Savannah, GA, 2016.

[14] N. Kobasko, "Optimized Quenched Layer for Receiving Optimal Residual Stress Distribution and Super-strengthened Material," in *3rd WSEAS International Conference on Applied and Theoretical Mechanics*, Barcelona, Spain, December 14-16, 2007.

[15] R. Otero, "Calculation of Kobasko's Simplified Heat Transfer Coefficients from Cooling Curve Data Obtained with Small Probes," *J. ASTM,* vol. 4, no. 2, pp. 30-36, 2012.

[16] International Organization for Standardization, "Industrial quenching oils -- Determination of cooling characteristics -- Nickel-alloy probe test method," Geneva, 1995.

[17] ASTM International, Standard Test Method for Determination of Cooling Characteristics of Quench Oils by Cooling Curve Analysis, ASTM D 6200, Conshocken: ASTM, 2012.

[18] D. S. MacKenzie, "Determining Grossman H-value from cooling curve data," *Thermal Processing,* pp. 21-23, February 2020.

[19] M. A. Grossman and M. Asimov, "Hardenability and Quenching," *Iron Age,* vol. 145, no. 18, pp. 39-47, 1940.

[20] G. E. Totten, G. M. Webster, C. E. Bates, S. W. Han and S. H. Kang, "Limitations of the Use of Grossman Quench Severity Factors," in *17th Heat Treating Society Conference Proceedings Including the 1st International Induction Heat Treating Symposium*, Indianapolis, IN, 1997.

[21] Ovako, "Steel Navigator - Heat Treatment Guide," [Online]. Available: https://steelnavigator.ovako.com/heat-treatment-guide/. [Accessed 05 January 2025].

QDE 2025: Proceedings of the 3rd International Conference on
Quenching and Distortion Engineering
May 6–7, 2025, Vancouver, Canada
https://doi.org/10.31399/asm.cp.qde2025p0202

CFD Modelling of Gas Cooling for Sustainable Heat Treatment for Drivetrain Components

Vincent Dinh[1], Angelica Lantto[2], Pavel Ramirez Lopez[2,3], Sofia Larsson[1], Anders Olofsson[4] and Andreas Jansson[5]

[1] Department of Engineering Sciences and Mathematics
Luleå University of Technology, SE 971 87, Luleå SWEDEN

[2] Metallurgy Department, SWERIM AB
Aronstorpsvägen 1, SE 97437, Luleå SWEDEN

[3] Materials Department, KTH Kungliga Tekniska Högskolan
SE 100 44 Stockholm, Sweden

[4] Advanced Engineering Transmission Production, DXTGR
Scania CV AB, 151 87 Södertälje, Sweden

[5] Rock Drill Product Line - Underground Division
Epiroc Rock Drills AB, Klerkgatan 21, SE 701 91, Örebro, Sweden

Abstract

Drivetrain electrification is essential to reduce environmental impact from road transportation. One critical step is the production of higher-performing drivetrain components with tighter tolerances to reduce energy consumption and waste. This study focuses on replacing traditional gas-fired furnaces with sustainable low-pressure carburizing (LPC) and gas cooling methods. The project leverages advanced computational tools for predicting quenching outcomes (e.g., cooling rates, material properties, and distortions) to enable sustainable and efficient production. The approach involves CFD modelling of the chamber, splitting it into a 2D/3D sub model for the heat exchanger and a 3D model for the entire chamber. Parametric studies are conducted to optimize turbine RPM settings for uniform cooling and minimal distortions, providing detailed velocity profiles and cross-sectional flow insights. Components with all their design details are added for running a series of parametric studies to predict velocity and heat transfer characteristics in the chamber. The results highlight optimal turbine RPM settings for a more even cooling which would reduce distortion, marking significant progress toward predictive accuracy in gas cooling processes.

Keywords: numerical modelling, gas cooling, low-pressure carburizing, quenching, flow field prediction, sustainable manufacturing

Introduction

The transition to sustainable drivetrain manufacturing is essential for reducing environmental impact and meeting the demands of electrified transportation systems. Traditional gas-fired atmosphere furnaces with mineral oil quenching contribute significantly to energy consumption, dimensional distortions, and CO2 emissions. To address these challenges, low-pressure carburizing (LPC) combined with gas cooling has emerged as an environmentally friendly alternative, offering up to 80% lower CO2 emissions compared to conventional methods [1].

While LPC with gas cooling reduces environmental impact, controlling the gas flow dynamics in the cooling chamber is crucial for minimizing dimensional distortions. These distortions arise due to uneven heat transfer, particularly for slender or complex components with tight tolerances. Computational Fluid Dynamics (CFD) is increasingly recognized as a powerful tool for analysing gas flow behaviour and optimizing cooling performance, thus reducing the dependence on resource-intensive experimental trials [2].

Several previous studies have demonstrated the effectiveness of CFD for predicting gas flow and heat transfer in LPC processes and quenching applications. J. Wang used ANSYS Fluent to study flow and heat transfer coefficients during the quenching of an H13 steel tool in a vacuum furnace. The study highlighted the significance of temperature-dependent properties of gas and steel, such as thermal conductivity, specific heat capacity, and viscosity, which had a strong impact on the simulation results [3]. In a related publication, Yada and Watanabe (2013) combined ANSYS Fluent with thermochemical calculations to predict carbon gradients in steel during carburizing. Their simulations showed good agreement with laboratory experiments [4]. A more recent study in 2020, part of the "CFD-kyl" project within Värmebehandlingscentrum, applied CFD to model gas flow in an industrial LPC cooling chamber. The results revealed that components in critical positions, such as chamber corners, are at higher risk of non-uniform gas flow and uneven heat transfer, leading to distortions [5]. In addition, Macchion's work on gas quenching furnaces further supports the use of CFD for improving gas flow uniformity and cooling performance. This numerical study analysed flow and heat transfer fields in both empty and loaded furnaces, identifying key design improvements for enhancing flow uniformity and optimizing gas mixtures for efficient quenching. This research highlighted the direct relationship between gas flow uniformity and the quality of heat treatment processes [6]. These studies provide a strong foundation for applying CFD to LPC gas cooling processes, demonstrating its potential to predict velocity profiles, temperature distributions, and heat transfer characteristics.

The main objective of this work is to use CFD to analyse and optimize the gas cooling process in LPC heat treatment. Specifically, this study aims to simulate and understand the behaviour of gas flow in the cooling chamber to identify areas of non-uniform cooling that can cause distortions in automotive components.

Governing Equations

The basic mathematical formulation for gas flow and heat transfer used in fluent [7] can be described as in the following governing equations.

Continuity

$$\frac{\partial \rho}{\partial t} + \nabla \cdot (\rho \boldsymbol{U}) = 0 \tag{1}$$

Where ρ is the fluid density, $\boldsymbol{U}$ is the velocity vector, and t is the time.

Momentum

$$\frac{\partial(\rho \boldsymbol{U})}{\partial t} + \nabla \cdot (\rho \boldsymbol{U}\boldsymbol{U}) = -\nabla p + \nabla \cdot \overline{\boldsymbol{T}} + \boldsymbol{B} \tag{2}$$

Where p is the pressure, **B** is the body force and $\overline{\boldsymbol{T}}$ is the stress tensor, defined by

$$\overline{\boldsymbol{T}} = \mu\left[(\nabla \boldsymbol{U} + (\nabla \boldsymbol{U})^{\mathrm{T}}) - \frac{2}{3}\nabla \cdot \boldsymbol{U}\boldsymbol{I}\right]. \tag{3}$$

Here, μ is the molecular viscosity and I is the identity tensor.

Energy

The energy transport for the fluid is expressed as:

$$\frac{\partial(\rho H)}{\partial t} + \nabla \cdot (\rho \boldsymbol{U} H) = \nabla \cdot (\lambda \nabla T) + S_{\mathrm{h}} \tag{4}$$

where H is the total enthalpy, T is the temperature, λ is the thermal conductivity and S_h represents the heat source term.

Turbulence modelling

The $k - \epsilon$ turbulence model was employed to account for the turbulent flow induced by the cooling turbine. This model is widely used due to its robustness and computational efficiency for simulating industrial-scale flows. The turbulence kinetic energy k and its dissipation rate ϵ are determined from the following transport equations:

$$\frac{\partial(\rho k)}{\partial t} + \nabla \cdot (\rho \boldsymbol{U} k) = \nabla \cdot \left[\left(\mu + \frac{\mu_t}{\sigma_k}\right)\nabla k\right] + G_k + G_b - \rho \tag{5}$$

$$\frac{\partial(\rho \varepsilon)}{\partial t} + \nabla \cdot (\rho \boldsymbol{U} \varepsilon) = \nabla \cdot \left[\left(\mu + \frac{\mu_t}{\sigma_\varepsilon}\right)\nabla \varepsilon\right] + C_1 \frac{\varepsilon}{k}(G_k + C_3 G_b) - C_2 \rho \frac{\varepsilon^2}{k} \tag{6}$$

Here, G_k represents the turbulence generation due to velocity gradients, G_b is the generation due to buoyancy, and μ_t Is the turbulent viscosity:

$$\mu_t = \rho C_\mu \frac{k^2}{\varepsilon} \tag{7}$$

The constant used in the $k - \epsilon$ model are summarized as follows:

C_μ	C_1	C_2	σ_k	σ_ϵ
0.09	1.44	1.92	1.0	1.3

Numerical Modelling

The numerical modelling of gas flow in the furnace (figure 1) was divided into two main parts: the heat exchanger and the chamber. This approach is used to reduce computational time and make possible a parametric study.

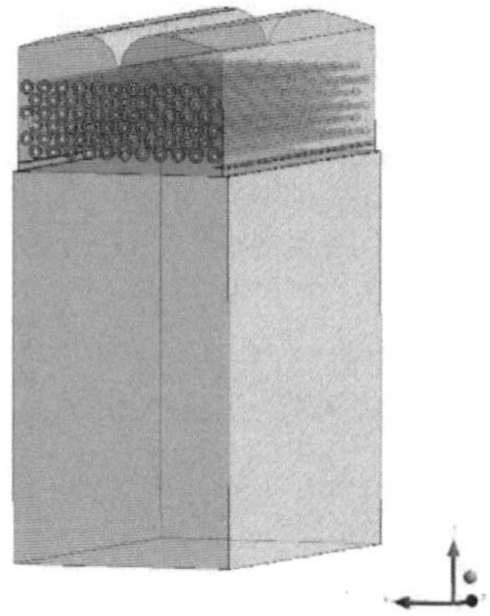

Figure 1: 3D model of the chamber and the heat exchanger

Heat Exchanger Modelling

Due to its geometric symmetry, the gas flow within the heat exchanger was initially simulated using both a 2D model (figure 2) and a 3D sectional slice (figure 3). As shown in figures 2 and 3, the difference in diameter of the circular holes is due to the presence of fins on the heat exchanger tubes, which are accounted for in the 3D case.

These two simulations were conducted for two different cases: one with a normal inlet velocity perpendicular to the inlet surface, and the other with an angled inlet velocity, designed to account for the curvature in the upper part of the heat exchanger. For both cases, constant velocity profiles were imposed at the inlet, while no-slip boundary conditions were applied on the heat exchanger walls. The outlet of the heat exchanger was treated as a pressure outlet.

The resulting outlet velocity profiles from these simulations were compared against experimental measurements for calibration and validation. Once validated, the accurate velocity distribution at the heat exchanger outlet was extracted and mapped as the inlet boundary conditions for the subsequent chamber simulations.

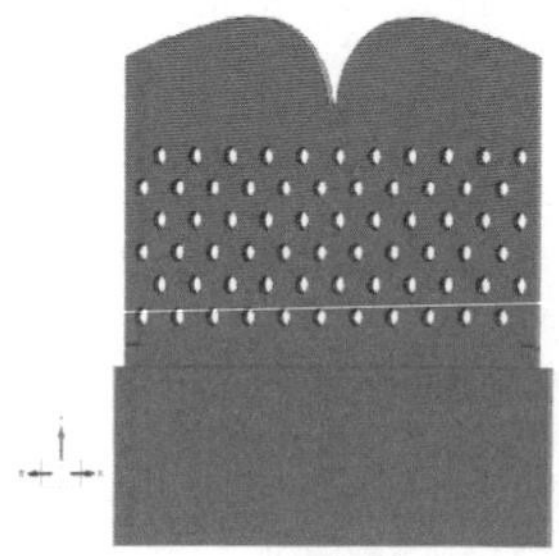

Figure 2: 3D section of the heat exchanger

Chamber Simulations

The gas flow within the chamber is simulated under various configurations to analyse its behaviour under different conditions, including an empty chamber, a chamber with a tray and components and a chamber with components only.

The mapped velocity profile from the heat exchanger served as the inlet boundary condition for these simulations. The chamber walls were treated as no-slip boundaries. At the chamber outlet, a pressure outlet condition was applied to allow the gas to exit the system. For simulations involving components, the surfaces of the components were also assigned no-slip boundary conditions.

The study starts with isothermal steady-state simulations at five turbine RPM levels to analyse gas flow patterns and flow uniformity. Next, isothermal transient simulations at three RPM levels capture the dynamic behaviour of the gas flow over time. Transient heat transfer simulations for two RPM levels focus on evaluating thermal performance and heat transfer distribution on the loaded components. If time allows; a more detailed tray simulation will be carried out for one transient case which could provide a more precise analysis, since the tray has been modelled as porous media to simplify the simulations.

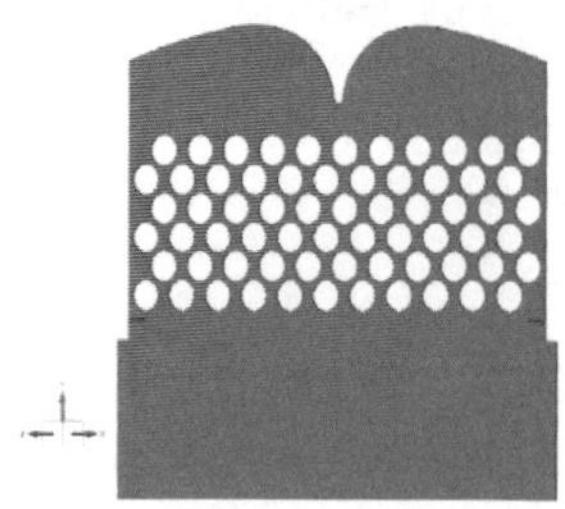

Figure 3: 2D section of the heat exchanger

Experimental Methods

The purpose of the experiments is to validate the outlet velocity profile of the heat exchanger to be calibrated and obtain measurements of gas flow parameters within the chamber, such as velocity and temperature distributions. The experimental setup consists of a cubical metal frame that holds a line of seven sensors. Each sensor is equipped with two small holes, through which a thin metal string is inserted to measure both the gas flow rate and temperature. This setup is designed to allow measurements in a 3D space by enabling controlled movement of the sensors in all three directions (x, y, z) within the chamber. The movement is automated and executed using Python code, providing precise positioning and systematic scanning of the measurement volume. In the measurement process, the sensors systematically capture data at multiple positions, recording both velocity magnitude and temperature across the chamber. This enables the generation of a 3D flow field and temperature map under various operating conditions, such as different turbine RPM levels. The setup allows measurements in an empty chamber and a chamber with only a tray, but not when the chamber contains both the tray and components due to lack of space in the chamber.

Results

As described in the Numerical Modelling section, the simulations of the heat exchanger were carried out using a 2D model and a 3D sectional model to extract the outlet velocity profiles, which were then applied as inlet boundary conditions for the 3D chamber simulation. The results presented in figures 5 and 6 show the velocity magnitude contours at the top inlet of the chamber and on a cross-section from the side for both cases.

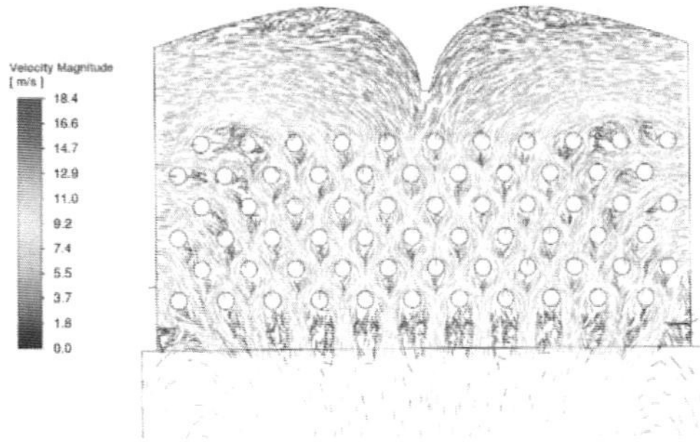

Figure 4: Vector plot of the 3D heat exchanger simulation with 0-degree inlet angle

Figure 4 presents a vector plot of the 3D heat exchanger simulation with no inlet angle condition. The plot illustrates the flow behaviour through the finned tubes, a detailed representation of the flow paths and the resulting velocity profile at the chamber inlet.

The results presented in figures 5 and 6 show the velocity magnitude contours at the top inlet of the chamber and on cross-section from the side for both the 2D and 3D heat exchanger case. In the first case (figure 5), where the inlet velocity profile is obtained from the 2D heat exchanger simulation, the velocity magnitude contours show a fairly uniform but suboptimal distribution of gas flow at the chamber inlet. The slice through the chamber highlights how the flow develops downstream, with the velocity gradually decreasing as the gas spreads across the chamber. In the second case (figure 6), where the inlet velocity profile is obtained from the 3D sectional heat exchanger simulation, the results show a significantly improved and more uniform distribution of velocity magnitude at the chamber inlet. The corresponding slice through the chamber further reflects a smoother and more consistent flow development. While the 3D model accounts for the finned tubes of the heat exchanger, the contours confirm that there is still no significant velocity variation in the z-axis at the chamber inlet, likely due to the flow mixing as it exits the heat exchanger. Figure 7 shows the chamber inlet velocity profile extracted from the 3D simulation with no inlet angle, captured from above. The consistent velocity profile along the z-axis highlights the even distribution of flow at the chamber inlet.

Figure 5: Velocity contours at the chamber inlet and mid cross-section using 2D heat exchanger profile

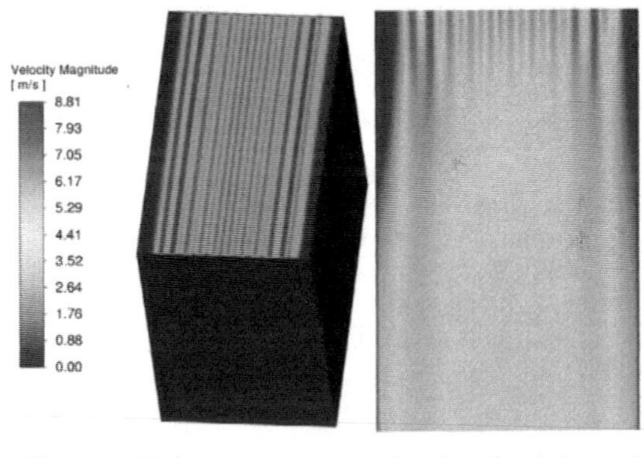

Figure 6: Velocity contours at the chamber inlet and mid-slice using the 3D heat exchanger profile.

Figure 7: Chamber inlet velocity profile

Preliminary conclusions

This study focuses on the application of CFD to analyse and optimize the gas cooling process in LPC heat treatment, with the aim of achieving uniform cooling and minimizing dimensional distortions in drivetrain components. The numerical modelling was divided into two parts: the heat exchanger and the chamber. By employing a 2D and 3D sectional model for the heat exchanger, the gas flow distribution and outlet velocity profiles were analysed, calibrated, and mapped as boundary conditions for chamber simulations. The chamber simulations are performed for various configurations and operating conditions to evaluate flow behaviour, cooling performance, and the influence of turbine RPM on flow uniformity. In addition, an experimental method was developed to validate the numerical results and obtain gas flow parameters such as velocity and temperature distributions within the chamber.

The experimental setup, consisting of a movable array of sensors capable of capturing 3D flow data, providing inputs for validating the numerical models. However, due to spatial limitations, measurements can only be performed in an empty chamber or a chamber containing a tray without components.

While significant progress has been made in modelling the gas cooling process and designing the experimental setup, the project remains ongoing with several tasks currently being addressed. These tasks include finalizing the planned simulations, such as isothermal steady-state, isothermal transient, and transient heat transfer cases, to comprehensively evaluate the gas flow and thermal performance. Experimental measurements of the heat exchanger outlet flow and chamber gas flow parameters are being conducted to validate the numerical results, ensuring accuracy and reliability. Additionally, efforts are underway to extend the simulations by including detailed tray models and components to improve the prediction of flow behaviour and heat transfer performance. Finally, post-processing of the results is ongoing to analyse key parameters such as velocity distributions, heat transfer coefficients, and temperature profiles, which are critical for further optimization of the gas cooling process.

Acknowledgements

The authors would like to thank the support from the Värmebehandlingscentrum (VBC) at SWERIM and the participant automotive producers for support throughout the project. VD and PRL also would like to thank Stefan Jonsson at KTH for the experimental setup. PRL would like to thank Sven Haglund, Johan Wendell and Eva Troell for support and coordination of the main project. This project has received funding from VINNOVA under Diarienummer: 2022-01656.

References

[1] Troell, E. (2022). *GASELL – Gaskylning och simulering för stålkomponenter med ökande krav för elektrifiering.* RISE, Sweden. Unpublished project application. Contact: eva.troell@ri.se

[2] Computational Fluid Dynamics (CFD) is increasingly recognized as a powerful tool for analyzing gas flow behavior and optimizing cooling performance, thus reducing the dependence on resource-intensive experimental trials. *MDPI Energies*. Available at: https://www.mdpi.com/1996-1073/16/4/1664, accessed June 2024.

[3] Wang, J., "Numerical Simulation of High Pressure Gas Quenching of H13 Steel," *Journal of Materials Processing Technology*, Vol. XX, (2008), p 188–194.

[4] K. Yada och O. Watanabe, "Reactive flow simulation of vacuum carburizing by acetylene gas," Computers & Fluids, p. 65–76, 2013

[5] H. Kristoffersen och P. Ramirez, "CFD-kyl," VbC Programforskningsprojekt, Sweden, 2020.

[6] Macchion, O., *CFD in the Design of Gas Quenching Furnace,* Technical Report, Royal Institute of Technology (KTH), Stockholm, Sweden, June 2005.

[7] ANSYS Inc., *ANSYS Fluent Theory Guid, Ansys Documentation, https://dl.cfdexperts.net/cfd_resources/Ansys_Documentation/Fluent/Ansys_Fluent_Theory_Guide.pdf, accessed June 2024.*

AUTHOR INDEX